METEOR SHOWERS

A Descriptive Catalog

METEOR SHOWERS

A Descriptive Catalog

by
Gary W. Kronk

Foreword by Dr. Fred L. Whipple
Smithsonian Astrophysical Observatory

ENSLOW PUBLISHERS, INC.

Bloy St. & Ramsey Ave.
Box 777
Hillside, N.J. 07205
U.S.A.

P.O. Box 38
Aldershot
Hants GU12 6BP
U.K.

Cover photo: This magnitude -5.0 Perseid flares as it ends near the Square of Pegasus on August 12, 1966. (Courtesy of Hal Povenmire, Florida Fireball Patrol)

Library of Congress Cataloging in Publication Data

Kronk, Gary W.
Meteor showers: a descriptive catalog.
p. cm. – (Enslow astronomy series)
Includes index.
1. Meteor showers–Catalogs. I. Title. II. Series
QB471.K76 1988 523.5'1 87-22159
ISBN 0-89490-072-2

Printed in the United States of America

10 9 8 7 6 5 4 3 2 1

This book is dedicated to
my wife Karen,
my newborn son David,
my mother Jacqueline,
my in-laws Richard and JoAnn Longhi,
and the memory of my father Richard.

Contents

Acknowledgements

I wish to express my gratitude and appreciation to those people around the world who helped me accumulate the data used in this book. Zdenek Sekanina (Jet Propulsion Laboratory, California) and Carl Murray (Queen Mary College, England) helped in my acquisition of the 39,145 radio meteor orbits used frequently throughout my investigations. Jack Drummond (Stewart Observatory, Arizona) supplied technical assistance in the methods of analysing the large amount of data at my disposal. I would also like to thank Jeff Wood (Western Australia), Paul Roggemans (Belgium), Hans Betlem (The Netherlands), and Robert Mackenzie (England), for observations obtained by their respective organizations, Robert Lunsford (California) for sending his detailed visual observations, Harold Povenmire (Florida) for his assistance in obtaining several hard-to-find documents, Sybil Csigi (University of Pennsylvania), for sending most of the American Meteor Society publications published between 1930 and 1968, Betty Eickhoff (Washington University, Missouri) for allowing me special privileges in the use of the university's Physics library, and Cathy Schaewe (Washington University, Missouri), for translations of critical French documents. Special thanks are given to Ruth Armes and Pete Simpson (Southern Illinois University-Edwardsville) for their initial encouragement in getting this book published, and to Cori Hartje (Washington University, Missouri) for *extensive* use of the Campus Computer Store's Apple LaserWriter. Very special thanks go to my wife, who showed excessive patience while I spent time at libraries or at my computers.

Preface

This book contains historical data on what the Author believes are the 112 most active radiants in the sky. Data from the majority of the visual, photographic, and radar studies have been utilized. The Author began his research with in-depth investigations into the orbits and day-by-day radiant movements of several major meteor streams, such as the Perseids, Orionids, and Geminids. After noting how the daily radiant movement was associated with steady changes in a stream's longitude of its ascending node (and, subsequently, its longitude of perihelion), a preliminary list of over 600 potential radiants were investigated.

Using photographic and radar data, the Author first determined the orbit of each potential meteor stream. The next stage was to establish the duration of activity, as well as the probable daily movement of its radiant. Finally, the "D-criterion" was applied to the tens of thousands of visual radiants to determine which stream they actually belonged to and, thus, how often each stream had been visually detected in the past. Following these investigations, a complete history was available for each stream—allowing this final selection of the strongest and most consistent meteor showers.

For each meteor stream, the Author has tried to determine what may be the first observations, probable dates of maximum activity, average radiant, daily radiant motion, and duration of activity. However, since observers are now frequently noting such characteristics as average meteor brightness, color, and train duration, the Author has tried to include this data whenever possible.

Each chapter represents a month and the showers are discussed in the chapter representing the month they experience maximum activity. There are some showers which have maximums falling on either the last day of a month or the first day of the next month (i.e., the Delta Aquarids), but the Author has placed these showers in the chapter containing the earliest date of maximum.

The showers are listed alphabetically by their constellation of radiation. This was done because most minor showers do not have well-established names. Visual estimations of a radiant are always accurate enough to isolate a shower's constellation of origin, but usually not the star—especially when several bright stars are nearby. The Author has tried to adopt either the most commonly used name or, on occasion, the most appropriate name for each shower.

Abbreviations & Symbols

To conserve space and prevent the numerous footnotes from overwhelming every page, the Author has utilized numerous abbreviations. The 54 major sources for this book are given in Appendix 4. For the numerous other sources quoted throughout, another set of space-saving abbreviations, involving the name of each journal, has been utilized and are as follows:

AJ=The Astronomical Journal.
AJP=Australian Journal of Physics.
AJS=American Journal of Science.
AN=Astronomische Nachrichten.
BAC=Bulletin of the Astronomical Institute of Czechoslovakia.
CAPS=Center for Astrophysics Preprint Series.
FOR=Flower Observatory Reprint.
JBAA=Journal of the British Astronomical Association.
MET=Meteoros.
MN=Meteor News.
MNRAS=Monthly Notices of the Royal Astronomical Society.
MRAS=Memoirs of the Royal Astronomical Society.
PA=Popular Astronomy.
PASP=Publication of the Astronomical Society of the Pacific.
SA=Soviet Astronomy.
SCA=Smithsonian Contributions to Astrophysics.
SEAN=Scientific Event Alert Network.
SSR= Solar System Research.
WGN=Werkgroepnieuws.

The following nine symbols are frequently used. The first three represent values used to indicate the location of a meteor shower radiant. The other six represent common values used to illustrate the orbit of a meteor stream, comet, or asteroid.

λ: Solar longitude (degrees). This is measured in degrees and represents the location of an object along the plane of the solar system as viewed from the sun. It is used to predict when Earth will encounter a meteor stream.

α: Right ascension (degrees). The astronomical equivalent of Earth's longitude.

δ: Declination (degrees). The astronomical equivalent of Earth's latitude.

q: Perihelion (astronomical units). The closest distance to the sun.

a: Semimajor axis (astronomical units). This represents one-half of the distance between an orbit's closest and farthest points from the sun.

Ω: Longitude of the ascending node (degrees). The longitude, as seen from the sun, when the orbit crosses the plane of the solar system.

ω: Argument of perihelion (degrees). The angle between the longitude of the perihelion point and the longitude of the ascending node.

e: Eccenticity. A proportional relationship between the perihelion distance and the semimajor axis.

i: Inclination (degrees). This represents how much an orbit is inclined or tilted with respect to the plane of the solar system.

Foreword

A popular idea early this century was that a considerable fraction of visual meteors comes from space on hyperbolic orbits. This idea was dispelled in the 1950s by the double-station observations of the Harvard Photographic Meteor Programs. We now know that almost all meteors are non-icy solids that have been ejected from comets. (A meteoroid that strikes the earth's atmosphere is a meteor and a group of meteors is a shower.) Some critical questions remain about the origins of meteors: Do any of the meteors arise from defunct comets that have lost or completely covered their ices, so that they look like asteroids? Or do any come directly from stony asteroids?

The observation of meteors and meteor streams will aid in clarifying these and other problems of comets and of the interplanetary complex of meteoroids, dust, gas, and ionized plasma. Precisely how do meteoroids spread out around the orbit of a comet that is ejecting them slowly? How long have the streams existed, particularly those associated with live comets? How much mass have these comets lost to the interplanetary medium? Are the particles that scatter sunlight to make the Zodiacal Light and the Gegenschein provided by comets alone, or do asteroids also contribute? Are the Zodiacal Particles continuously supplied to keep the system quasi-stable, or have they come largely from an exceptional comet or asteroid?

Observations of meteor streams, both by professionals and amateurs, are valuable to such research. Mr. Kronk's painstakingly thorough compilation and clear-cut presentation of detailed information about meteor streams is a unique guide for potential observers. Moreover, this book is an unequalled sourcebook for anyone who wishes to pursue the subject for pleasure, education, or research.

Dr. Fred L. Whipple

Director Emeritus, Smithsonian Astrophysical Observatory
Professor Emeritus of Astronomy, Harvard University
Author of *The Mystery of Comets* (1986 winner of the Phi Beta Kappa Science Award)

Introduction

Despite meteors having attracted man's curiosity for centuries, our knowledge of their proper origins only recently reached its 120th anniversary. The earliest recorded account seems to have been in 1809 B.C., during the era of Hsia Hoa Shih Ti Kuei, when "meteors fell like a shower at midnight" in China.[1] In Europe, the earliest recorded account seems to have been 652 B.C., when, according to Titus Livy in his *History of Rome*, there was a shower of stones on the Alban Mount.

The earliest theories of the origin of meteors varied widely. In China, for instance, meteors were thought to be messengers from heaven—their brightness and speed determining the importance of the message.[2] Indian tribes of California developed numerous ideas of meteor origins. The Wintu tribe considered them "to be the spirits of shamans who had died and were traveling to the afterlife." The Serrano, Cahuilla, Cupeño, Luiseño and Ipai tribes believed that a single, bright meteor was the cannibal spirit *Takwich* which "would travel about at night looking for wandering souls to eat." Some Pomo tribes thought meteors were fire dropping from heaven and some eastern California tribelets thought meteors were "stars which were changing places in the sky." One of the less glamorous theories developed by some southern California tribes was that meteors were feces from stars.[3]

Although an extraterrestrial origin was first suggested by Edmond Halley after seeing a great fireball in 1686, it was not until nearly the end of the 18th century that this idea was seriously pursued. Ernst Florens Friedrich Chladni of Czechoslovakia investigated the Pallas meteorite (found in Siberia in 1749) and the Tucuman meteorite (found in Argentina in 1783) and concluded in a book published in 1794 that neither object could have formed on Earth as a result of

---

[1]Imoto, S., and I. Hasegawa, *SCA*, **2**, No. 6 (1958), p. 137.

[2]Ho Peng Yoke, *The Astronomical Chapters of the Chin Shu*. Paris: Mouton & Co., 1966, pp. 136-138.

[3]Hudson, Travis, *Archaeoastronomy and the Roots of Science*. ed. E. C. Krupp, Colorado: Westview Press, Inc., 1984, pp. 39-41.

natural geological processes.[4] His theory of an extraterrestrial origin immediately drew heavy criticism, but in 1798, Heinrich W. Brandes and Johann F. Benzenberg (University of Gottingen), decided to try an experiment of observing meteors from two different locations. This simple method of triangulation proved to be quite a success for Chladni as the students found that meteors became visible at an average height of 97 kilometers.[5] They added that the distinct planetary velocities of several kilometers per second indicated an origin outside Earth's atmosphere.

The great Leonid meteor storm of November 13, 1833, provided a vital catalyst in the recognition of the shape of a meteor stream orbit. In early 1834, Denison Olmsted, a professor of mathematics at Yale College, stated that the hourly rate of several thousand meteors clearly revealed an apparent radiation from a fixed point. He concluded that the meteors had approached Earth in nearly parallel lines and that the apparent scattering across the sky was due entirely to the effect of perspective. Remembering reports of a fairly strong shower of meteors in November of 1832, Olmsted conjectured that a cloud of particles was responsible.[6] Subsequently, in November 1834, another shower was observed, admittedly more comparable with the weaker 1832 display than that of 1833, but it did confirm the existence of an annual meteor shower. The change in intensity was correctly theorized as due to a periodicity of some kind.

The 1836 discovery of the Perseid meteor shower of August proved fatal to Olmsted's "cosmical cloud" hypothesis and the struggle to arrive at a new workable theory was won over by Adolf Erman in 1839. Erman proposed that meteor showers were caused by closed rings of meteoric matter revolving around the sun. This theory explained that the showers occurred when Earth and the stream intersected, but it did not allow a clue as to the shape of the meteor stream orbits.

During 1861 Daniel Kirkwood proposed that meteor showers were the debris of old comets,[7] thus indicating the shape of Erman's rings was elliptical. This theory was confirmed in 1866, when Giovanni Virginio Schiaparelli published his discovery that the Perseid meteors were produced by the periodic comet Swift-Tuttle.[8] This comet had been discovered in 1862 and, although its period of nearly 120 years was much greater than that accepted for the Perseid orbit, the similarity in the orbital elements was too close for mere coincidence.

---

[4] Chladni, E. F. F., *Ueber den Ursprung der von Pallas gefundenen und anderer ihr aehnlicher Eisenmassen*, Riga, 1794.

[5] Hawkins, Gerald S., *Meteors, Comets and Meteorites*. New York: McGraw-Hill, Inc., 1964, pp. 5 & 17.

[6] Olmsted, Denison, *American Journal of Science Arts*, 26 (July 1834), p. 165.

[7] Olivier, p. 50.

[8] Schiaparelli, G. V., *Sternschnuppen*.

With the origins of meteors generally established, observers began to search the sky in the hopes of isolating other active meteor showers. Visual observations have, of course, been the most common during the last 120 years. Professional and amateur astronomers such as William F. Denning, Ronald A. McIntosh, Cuno Hoffmeister and Charles P. Olivier were not only prolific observers, but they also encouraged others to observe. Their combined efforts produced over 20,000 visual radiant determinations. Today's visual observers continue to be quite active.

Attempts have been made in the past to isolate the most noticeably active radiants and the most impressive work of the 19th century was that produced by Denning which utilized over 4000 visual radiants to isolate 278 active meteor showers.[9] Although Denning accumulated an exhaustive collection of 19th century radiants, the primary problem with this particular work was the theory applied to establish the active showers. Denning was a strong believer in the existence of stationary radiants—radiants which remained at the same position for months at a time. Although no one can deny that Denning was an excellent observer, his belief in the impossible stationary radiant theory plagued his statistical studies of active meteor showers well into the third decade of the 20th century, despite mathematical proof being put forth by several astronomers around the turn of the century. Subsequently many of his "established" annual showers were simply collections of sporadic, one-time-only radiants which occurred sometime within any particular five to nine month period.

The largest of the 20th century attempts to isolate active radiants was a list of Southern Hemisphere showers produced by McIntosh in 1935.[10] Containing 320 active radiants, the list's main fault was that it was not selective, and many of the radiants were based on only one or two nights of observations—making the probability of the inclusion of sporadic radiants quite high.

The more precise photographic and radio-echo methods, which had their primitive beginnings in 1891 and 1931, respectively, have been utilized by some astronomers since about the middle of this century to obtain precise data on radiants, velocities and orbits. Photographic orbits were used throughout the 1960's and into the 1970's in computer analyses that helped to better define the orbits of known meteor streams, as well as isolate several hither-to unknown streams. What made the comparison of these orbital elements dependable was the development of a mathematical formula by Richard B. Southworth and Gerald S. Hawkins in 1963.[11] This formula compared all variables in the orbital elements and produced the "D-criterion"—a factor indicating the level of probability that

---

[9]D1899, pp. 203-294.
[10]M1935, pp. 709-718.
[11]Southworth, R. B., and Hawkins, G. S., *SCA*, 7 (1963), pp. 261-285.

two or more orbits were related to one another (see Appendix 3).

Although the data tended to be more reliable than the straight visual data used by earlier researchers, these photographic surveys rarely extended over a period of two or three years. Since moonlight frequently hampered photographic observations it was not always possible to adequately cover any particular series of days in more than one year. Subsequently, several "new" minor meteor streams were based on data obtained in only one year. Although some attempts were made to find visual confirmations in past records, several major lists were frequently ignored, so that the visual evidence supporting some of these photographic streams was not statistically acceptable. In addition, although several photographic surveys were conducted by different countries and different groups of astronomers, the surveys were never looked at as a whole to confirm or deny the existence of these weakly supported radiants.

The newest method utilized in the study of meteor streams has been the use of radar equipment. Radio-echo surveys used in the Soviet Union, Australia, and the United States during the 1960's obtained very accurate data which could produce excellent orbital details. The problem here was that these surveys typically detected meteors below naked-eye visibility and few attempts were made to look at the extensive lists of visual radiants and photographic meteor orbits for an indication of naked-eye detectibility. Also, these surveys were typically only a year in duration, so that some "established" minor streams were not confirmed as potential annual showers.

Radar studies of meteor activity have finally provided the catalyst for confirming whether many minor meteor streams actually exist. Right at the dawn of radar meteor studies Charles P. Olivier wrote the following:

> "I confess that the more I work on meteors the less confidence I have in tables of minor radiants. Perhaps the true solution of the question will be to accept as real only those which can be detected with certainty every year, or at recurrent returns after the same number of years have elapsed. Meantime, however, I believe that the publication of lists, carefully derived, lays the foundation for future selection of those radiants which will stand the test and are thus justified."[12]

With the birth of radar studies, researchers now have a foundation with which they can build upon. This data has greatly increased our present knowledge of meteor astronomy. Numerous comet-meteor associations have been revealed, as well as the first asteroid-meteor associations. Several showers are now known to possess two or more simultaneously active radiants, while

---

[12]Olivier, Charles P., *PA*, **49** (1941), p. 550.

other streams intersect Earth's orbit on two occasions each year and, thus, produce two distinct showers.

The acquisition of data on average meteor magnitudes shows great improvement over methods used as recently as a decade ago. The key here is not only the careful estimate of the brightness of each meteor from each active radiant, but it is also important for each observer to provide an accurate estimate of his/her limiting sky magnitude. When combined, these estimates can provide a useful method of determining the size population within each meteor stream. Although tables of average meteor magnitude have been included throughout this book, the Author has not included a category for limiting magnitude since this value is, unfortunately, frequently left out when this data is published. Most observers, however, are known to observe under skies with limiting magnitudes of 5.5 to 6.5, and such values would fit about 90% of the data presented in this book. Whenever unusually high or low values are noted in tables containing overall consistent estimates of average magnitude, it can only be assumed that the observers were under skies with limiting magnitudes above or below the 5.5-6.5 range.

The duration of meteor trains can be of significant importance in the establishment of what a particular stream's meteors are made of. For instance, a meteor primarily composed of stone is much more likely to leave a dust train than one of iron. Here again the limiting magnitude of an observer's sky is important to note when estimating the percentage of a particular stream's meteors that are leaving trains. What may be of even greater importance is the estimation of the duration of a train. Currently groups around the world are divided three ways when it comes to a method of meteor train study. Some groups request observers to count all of a stream's meteors leaving trains, some groups request a count of the meteors with trains of one-second duration or longer (these are called persistent trains), while a few groups request a train duration estimate for each meteor seen. The first two methods are utilized to determine the percentage of a stream's meteors which possess trains, but the former can be considerably higher than the second and neither value can be compared to one another. The latter method can be broken down many different ways and can ultimately reveal the percentage of trained meteors to nontrained meteors, as well as a stream's average train duration. Obviously the latter method would be preferred, but due to the requirement of a fairly accurate estimate of train duration, the second method of counting only persistent trains could be utilized by more observers. The first method of counting all trained meteors is of little use since values could vary greatly depending on the experience of the observer.

Meteor colors are always of interest when studying various meteor showers, though it is unclear how significant the colors actually are. Some theories indicate colors are due to meteor speed or composition, while others indicate colors are dependent on dust and moisture levels in the air. There are even some professional and amateur astronomers who believe a meteor's color depends on the color sensitivities of observers' eyes. The latter seems less likely, as numerous estimations of the meteor colors within the major streams have frequently revealed very consistent percentages. Unfortunately, this particular statistic is frequently ignored by observers and more evidence is needed before the proper color-producing catalyst can be identified.

Chapter 1: January Meteor Showers

Zeta Aurigids

Observer's Synopsis

This minor meteor shower is visible during the period of December 11 to January 21, and reaches its maximum activity during December 31 (λ=279°), from an average radiant of α=77°, δ=+35°. Though the greatest activity seems to be detectable only by means of radar or telescopes, numerous visual radiants and photographic meteors indicate some activity is visible with the naked eye. A northern branch is also present during December 11 to January 15. Its maximum comes on January 2 (λ=281°), from α=65°, δ=+57°. The meteors of both showers are generally slow moving.

History

First detection of this shower was made by William F. Denning during December 24-31, 1885-1886, when a radiant at α=77°, δ=+32° was noted producing slow meteors. On December 27, 1863, Alexander S. Herschel was one of several people who observed a fireball from α=75°, δ=+30°. Calculations revealed Herschel's fireball possessed a velocity of 32 km/s, which is identical to results obtained by Denning after analyzing observations of a bright meteor observed from α=77°, δ=+30° on December 28, 1886.[1]

The true magnitude of this stream seems hard to determine, but some sense has been made out of the data after studying Zdenek Sekanina's radar data secured during the two phases of the Radio Meteor Project. The results seem to indicate a definite stream producing a shower during the period of December 14 to January 16; however, as can be viewed in the "Orbit" section below, the 1961-1965 survey possessed a smaller orbital inclination of nearly 7°, while the 1968-1969 survey indicated an inclination of about 11°. A possible explanation might involve the fact that the 1968-1969 session possessed a gap in the

---

[1]D1899, p. 244.

observational records that extended from December 21, 1968 to January 12, 1969. As a result, the period of maximum activity was missed and while the 1961-1965 survey revealed the stream to have crossed the ecliptic on December 30.9 (almost exactly at the time of the shower's expected maximum), the 1968-1969 survey showed the stream to have crossed the ecliptic on January 13.9.[2] Thus, it seems this gap not only reduced the number of meteors being observed, but also influenced the shower's apparent distribution of meteor activity.

A notable feature among records of meteor activity for the first half of January is that radiants are frequently present in Auriga, but, when the positions are plotted, it seems activity is likely to occur in a region whose dimensions are about 10°x30°. It should be kept in mind that this is based on radiants observed during the last 100 years. For any particular year the radiant seems to be less than 10° across. Examples of some of these observed radiants and fireballs are as follows:

Zeta Aurigid Radiants and Fireballs

Date (UT)	RA(°)	DEC(°)	Meteors	Source
1863, Dec. 27	75	+30	1	D1899
1885-1886, Dec. 24-31	77	+32	5	D1899
1886, Dec. 28	77	+30	1	D1899
1888, Dec. 24-29	88	+61	5	D1899
1891, Jan. 6-9	85	+61	5	D1899
1913, Jan. 2.82	75	+30	1	D1916
1921, Jan. 14.5	86	+42	?	H1948
1932, Jan. 1	82	+36	?	O1934

Interestingly, although most of these visual radiants originated from what might be considered as the "southern branch" of this stream, consulted lists of photographic meteors reveal a fairly well-defined "northern branch" consisting of 11 meteors detected during the Harvard Meteor Project conducted during 1952 to 1954. Eight meteors gathered from American and Russian sources form a fairly diffuse "southern branch."

The photographic northern branch seems to persist from December 11 to January 15, but may be active as early as December 1 and as late as January 21. Maximum activity seems to occur between December 28 and January 2, from an average radiant of α=65°, δ=+57°.

From the accumulated data, the Author sees the Zeta Aurigids as a split stream, with most photographic and visual data indicating the northern branch possesses a greater population of large particles. Most of the information

---

[2]S1973, pp. 257 & 260; S1976, pp. 274 & 291.

gathered on the southern branch reveals it contains primarily small particles, which are easily detectable using radar equipment; however, this branch has produced several fireballs in the past, which is a feature not present in the northern branch.

Orbit

The radar orbits which were determined from the two phases of the Radio Meteor Project conducted by Sekanina at Havana, Illinois, during the 1960's are as follows:

	ω	Ω	i	q	e	a
S1973	235.9	278.9	6.7	0.816	0.602	2.053
S1976	221.0	293.2	11.1	0.901	0.513	1.851

January Boötids

Observer's Synopsis

This fairly short duration shower reaches maximum around January 15 (λ=294°), from an average radiant of α=226°, δ=+44°. Meteors from this stream are generally observable during January 9-18.

History

The discovery of this meteor shower seems to have occurred during the 1870's. According to an analysis by William F. Denning, the Italian Meteoric Association plotted 7 meteors from an average radiant of α=221°, δ=+43° during January 1-15, 1872. On January 9, 1877, Denning himself plotted 6 meteors from α=221°, δ=+42°.[3]

The next apparent sighting of this shower came on January 10, 1937 (λ=288.9°), when A. Teichgraeber noted a radiant at α=225°, δ=+45°. This observation marked the only radiant listed in Cuno Hoffmeister's *Meteorströme*, which listed 5406 visual radiants. This radiant was designated 5061.[4]

Photographic studies have not revealed many members of this stream, possibly due to a lack of cameras operating around mid-January. In fact, the only apparent representative of this stream was detected on January 13, 1953, by cameras of the Harvard Meteor Project. The meteor had a magnitude of –0.3 and possessed a radiant of α=223.0°, δ=+43.9° The geocentric velocity was 27.3 km/s, while the height at which the meteor first became visible was 95.0 km.[5]

---

[3]D1899, p. 264.
[4]H1948, p. 249.
[5]MP1961, p. 35 & JW1961, pp. 100-101.

Radar studies have also detected this stream on three occasions. During January 16 to 19, 1957, Jodrell Bank observers, Dr. C. D. Watkins and Doylerush, detected a shower from a radiant roughly given as α=225°, δ=+25°; however, the large radiant diameter of 10° to 15° caused Jodrell Bank researchers to reevaluate the data and they subsequently arrived at an average radiant of α=233°, δ=+37°. Maximum was stated to have occurred on January 17 (λ=297°), when the hourly rate reached 25. There seemed to be evidence that the daily motion was roughly 2° eastward.[6]

The stream was again detected at Jodrell Bank in 1958. On this occasion, however, radio meteors were only detected on January 18 (λ=297.6°). The hourly rate reached 9, while the radiant appeared only 5° across. The radiant was determined to be α=237°, δ=+34°.[7] While analyzing all observations obtained during 1957 and 1958, G. C. Evans concluded that the January Boötids possessed a duration extending from solar longitude 290° to 304°, or roughly from January 11 to 25. Interestingly, during January 15-16, 1958, eight radar stations in the U.S.S.R. detected an unusual increase in meteoric activity.[8] Although no radiant was determined, the activity seems to add some support to Jodrell Bank's 1958 observation—at least in terms of enhanced activity.

The third detection of this stream using radar came during January 14 to 15, 1969, when the Radio Meteor Project at Havana, Illinois, detected 15 meteors from this stream. No apparent members were seen during January 13 or during January 16-17 and the radar had been shut down during January 1-12 and January 18-26. The average radiant was determined as α=225.8°, δ=+44.2°.[9] The average geocentric velocity was determined as 29.4 km/s.

Orbit

The strongest evidence for the existence of this short-duration meteor shower was obtained by Zdenek Sekanina in 1969, during the second session of the Radio Meteor Project. The orbit, which is based on 15 meteors, was given as follows:

ω	Ω	i	q	e	a
346.4	294.2	59.9	0.836	0.090	0.919

One photographic meteor has been detected from this stream. On January 13, 1953, a meteor was detected by cameras of the Harvard Meteor Project. The precise orbit was calculated by Luigi G. Jacchia and Fred L. Whipple and was published in JW1961. The result is at the top of the next page.

[6]E1960, pp. 281 & 292-293.

[7]E1960, p. 281.

[8]Routledge, N. A., *Nature*, **183** (April 18, 1959), p. 1088.

[9]S1976, pp. 274 & 291.

	ω	Ω	i	q	e	a
6095	349.1	292.9	56.3	0.690	0.177	0.839

The orbit is remarkably similar to the Aten-class of asteroids, or bodies with semimajor axes less than 1 AU. At the same time the short duration of the shower might indicate the stream is fairly young, so that the parent object may still be moving in this or a similar orbit.

Delta Cancrids

Observer's Synopsis

This shower possesses a typically long duration that is a major characteristic of ecliptic streams—extending from December 14 to February 14. Maximum occurs during January 13 to 17 (λ=292° to 296°) from an average radiant of α=128°, δ=+20°. A secondary center may lie about 5° to the south with a very weak maximum occurring around January 19 (λ=300°) from an average radiant of α=133°, δ=+14°. The daily motion of the Delta Cancrid stream is +1.0° in α and −0.2° in δ.

History

The earliest detection of this shower seems to have occurred during the 1870's. During January 1 to 15, 1872, members of the Italian Meteoric Association observed 7 meteors from an average radiant of α=130°, δ=+24°, while on January 12, 1879, a fireball was observed from α=133°, δ=+19°.[10]

Indications are that this is a fairly weak shower, whose presence has not been consistently noted in the literature. Cuno Hoffmeister's *Meteorströme* (1948) lists three probable radiants: one occurring in 1915, while the other two were observed in 1937. Although a few traces of this shower are found in a few sources published during the 1920's and 1930's, it should be noted that the extensive records of the AMS reveal no indication of activity from this radiant.

Perhaps the first solid evidence supporting this stream's existence came in 1971, while Bertil-Anders Lindblad examined the photographic meteor orbits obtained during the Harvard Meteor Project of 1952-1954. Seven meteors were found which indicated a duration of January 13 to 21 and an average radiant of α=126°, δ=+20°.[11]

Strong support of Lindblad's findings came in 1973 and 1976, when Zdenek Sekanina published the results of the two sessions of the Radio Meteor Project conducted at Havana, Illinois, during the 1960's. The first session

---

[10] D1899, p. 252.

[11] L1971B, p. 21.

covered the period 1961-1965 and detected 27 meteors during December 28 to January 30. The date of the nodal passage was given as January 13 (λ=292.2°), at which time the radiant was at α=123.7°, δ=+20.9°.[12] The second session covered the period 1968-1969 and detected 37 meteors during December 14 to February 14. The nodal passage came on January 17 (λ=296.4°) when the average radiant was α=129.8°, δ=+19.8°.[13]

After the details of the photographic and radar surveys had been published, visual observers began to search for the Delta Cancrids. The first came on January 19, 1974, when Bill Gates (Albuquerque, New Mexico) detected two meteors in 3 hours 46 minutes of observing using 7x50 binoculars from a radiant of α=128.25°, δ=+18.6°.[14] During January 1976, Norman W. McLeod, III, indicated surprise that the shower was for real. He described the individual meteors as being "Geminid-like, fairly slow and bright." During seven hours on January 16/17 and 17/18, McLeod (Florida) detected 12 meteors from this radiant.[15]

During 1977, several observations of this shower were published in *Meteor News*. On January 15/16, John West and George Shearer observed in Bryan, Texas, and saw seven Delta Cancrids in 2 hours 17 minutes, while Paul Jones (St. Augustine, Florida) saw two in 2 hours. On January 21/22, McLeod reported observations of six meteors in 4 hours, while Felix Martinez (Florida) saw four during the same period. The same two observers also observed on January 22/23, with the former seeing one meteor in three hours, while the latter saw none in 3 hours and 38 minutes.[16] An analysis of the available observations of 1974, 1975 and 1977, by Gates reveals probable activity levels of 2 to 3 per hour on the night of maximum.[17]

Rates continued to be low into the 1980's. During January 15/16 to 24/25, 1980, McLeod observed nine Delta Cancrids in 12 hours 55 minutes.[18] During 3 hours on January 16, 1983, Rick Hill (North Carolina) detected two of these meteors.[19] During 1984, Mark Zalcik (Edmonton, Alberta, Canada) observed one meteor in one hour on January 13/14, while Robert Lunsford (San Diego, California) saw one meteor during five hours on January 4.[20] Finally, during

---

[12] S1973, pp. 255 & 258.

[13] S1976, pp. 274 & 291.

[14] *MN*, No. 20 (Mid-March 1974), p. 3.

[15] *MN*, No. 26 (June 1975), p. 7 & No. 27 (August 1975), p. 5.

[16] *MN*, No. 36 (June 1977), p. 13.

[17] *MN*, No. 39 (January 1978), p. 3.

[18] *MN*, No. 50 (July 1980), p. 8.

[19] *MN*, No. 61 (April 1983), p. 7.

[20] *MN*, No. 65 (April 1984), p. 10.

1986, David Swann (Dallas, Texas) detected four meteors during three hours on January 11/12 and two meteors during one hour on January 14/15.[21]

Orbit

This stream was detected during both radar surveys conducted by Zdenek Sekanina as well as during a photographic meteor survey conducted by Lindblad. The orbits are as follows: .

	ω	Ω	i	q	e	a
S1973	287.9	292.2	1.2	0.425	0.777	1.901
S1976	291.3	296.4	1.5	0.397	0.783	1.829
L1971B	282.6	296.4	0.3	0.448	0.800	2.273

During C. S. Nilsson's radar survey conducted at Adelaide during 1961, 6 meteors were detected from a possible southern branch of this stream, with the following orbit being determined:

	ω	Ω	i	q	e	a
N1964	116.7	119.5	4.9	0.371	0.77	1.613

There seems to be no trace of visual observations from this southern branch, although a handful of photographic meteors do appear in MP1961. In fact, the photographic orbit listed above may actually be a combination of both the northern and southern branches.

Eta Carinids

Observer's Synopsis

The duration of this meteor shower extends from January 14-27. Maximum seems to occur around January 21, from α=160°, δ=–59°, with ZHRs of 2-3.

History

This stream was discovered by C. S. Nilsson (University of Adelaide, South Australia) from radio-echo observations made at Adelaide Observatory in 1961. Three meteor orbits indicated a radiant of α=156°, δ=–65°, and a possible nodal passage on January 19.[22]

The Eta Carinids were again noted when radio observations were made at Adelaide Observatory during 1969, by G. Gartrell and W. G. Elford. On this occasion the equipment operated only during January 21-23, but 3 meteors were again detected, this time from an average radiant of α=160°, δ=–63°. The authors noted a similarity between this stream and other streams detected in

---

[21]*MN*, No. 73 (April 1986), p. 9.

[22]Nilsson, C. S., Ph.D. Thesis, University of Adelaide (1962).

December and February and suggested they made up a family of high-inclination, low-eccentricity streams.[23]

The first visual observations of this stream were made by Michael Buhagiar (Perth, Western Australia). From observations made during 1969-1980, he determined the duration as January 14-28, with a maximum of only 1 per hour coming on January 21. The average position of the radiant was given as α=163°, δ=–57° when at maximum.[24]

Members of the Western Australian Meteor Section (WAMS) obtained an excellent set of observations of this shower during 1979. Meteors were observed during January 14-27, from an average radiant of α=160°, δ=–58°. Meteors were most numerous on the night of January 25/26, when the ZHR reached 1.77±0.25. Under skies with visual limiting magnitudes ranging from 5.0 to 6.5, observers detected 16 meteors and determined the average meteor magnitude as 3.75 (the brightest meteors were of magnitude 2). None of the meteors left trains and all seemed white in color. The shower was again observed during January 19-27, 1980. On this occasion, a maximum ZHR of 3.03±0.23 meteors was noted on January 21, from an average radiant of α=160°, δ=–59°. Although observed during 1982, the shower was virtually nonexistent during the interval of 1983-1986.[25]

Orbit

Two orbits have been determined from radio-echo observations made at Adelaide Observatory: Nilsson (N1964) determined one orbit from observations made in 1961, while Gartrell and Elford determined another one in 1969 (GE1975):

	ω	Ω	i	q	e	a
1961	0	119	70	0.98	0.59	2.38
1969	7	122	74.3	0.98	0.64	2.94

Eta Craterids

Observer's Synopsis

This weak shower is visible during the period of January 11 to 22. Maximum activity occurs around January 16-17, when the average radiant is α=176°, δ=–17°. The shower consists of generally rapid meteors, with streaks. The average magnitude is probably not above 4. It can best be seen from the Southern Hemisphere.

---

[23]GE1975, pp. 609-610.

[24]Buhagiar, Michael, *WAMS Bulletin*, No. 160 (1980).

[25]Wood, Jeff, Personal Communication (October 15, 1986).

History

First evidence for the existence of this stream occurred during the 1890s, with Corder making the first detection of a radiant in 1892, and Niessl detecting a fireball from the same area on January 16, 1895.[26] To date, Corder has remained the only *Northern Hemisphere* observer to detect a radiant from this shower; however, observers south of the equator have had better luck.

The New Zealand Meteor Section, led by Ronald A. McIntosh, was very active in the plotting of meteors during the latter half of the 1920's and throughout the 1930's. McIntosh had detected two probable Eta Craterid radiants in 1929: one on January 17, from α=177.7°, δ=–18.6°, and the other on January 21, from α=177°, δ=–11°. His colleague, Murray Geddes, had found one radiant at α=180.0°, δ=–15.6° on January 13, 1932.

No visual work even remotely close to the same magnitude as the New Zealand observations was carried out until the 1960s. During the period of 1961 to 1967, a systematic visual radiant survey was conducted at Waltair, India, by M. Srirama Rao, P. V. S. Rama Rao and P. Ramesh. It covered the months of November, December and January. During January 19-22, 1966, a radiant was found at α=178°, δ=–12°, which they estimated to possess a maximum hourly rate of 3.3 meteors per hour.[27]

The only other evidence supporting this stream's existence is present in the records of the AMS, where a magnitude –5 fireball, was given the designation AMS Fireball 2232. This fireball was detected on January 12.10, 1939, by three ships located in the Caribbean Sea between Jamaica and Colombia. Charles P. Olivier evaluated the observations and concluded that the radiant was at α=170°, δ=–19°.[28]

Not including Denning's grouping of this meteor shower with his stationary radiant called the Zeta Craterids, there have been two formal mentions of a possible annual or periodic radiant in this area. The first was made in 1935, when McIntosh included a radiant at α=177°, δ=–17° among his 320 southern hemisphere radiants (this radiant included McIntosh's two radiants mentioned earlier). The second was in 1964, when Olivier listed 194 "fireball radiants". Radiant 5008 (which included fireball 2232 mentioned earlier) showed a maximum on January 14 (±3 days) from α=170°, δ=–19°.[29] It was based on a projection involving 4 meteors.

There has never been a trace of this stream in any of the radar studies conducted in the last four decades. The most ambitious Southern Hemisphere

---

[26]D1899, p. 258.

[27]RR1969, pp. 767-774.

[28]Olivier, Charles P., *FOR*, No. 49, p. 11.

[29]Olivier, Charles P., *FOR*, No. 146, p. 12.

radar studies were conducted by C. S. Nilsson, during 1961, and G. Gartrell and W. G. Elford, during 1968-1969. The former easily covered the dates of maximum activity, while the latter only operated during January 21-23, 1969. It should be pointed out that two sessions of the Radio Meteor Project, conducted by Zdenek Sekanina in the Northern Hemisphere during 1961 to 1965 and during 1968-1969, also showed no evidence supporting the existence of the Eta Craterids. Sekanina's first study did not reveal any members during the period of January 11-19, while his second study revealed no reasonably close matches during January 13-17.

Extensive photographic studies conducted during this century have revealed only one member. On January 21, 1953, a 1.3-magnitude meteor was detected by two Baker Super-Schmidt cameras located in New Mexico and operated by the Harvard Meteor Project.[30] The radiant was calculated to have been $\alpha=177°$, $\delta=-19°$.

The conclusion that might be considered from the above data is that this radiant possesses a low hourly rate (hardly above the rate for sporadic meteors) and may be erratic or periodic in nature, rather than annual. The stream is nearly devoid of bright meteors and probably possesses an average magnitude no greater than 4. It is definitely a shower that can best be observed in the Southern Hemisphere, since observers in the Northern Hemisphere will probably see less than half the meteors that can become visible from that radiant every hour. The radiant reaches the meridian at about 4 a.m. (local time) on those January mornings of activity, and the altitude for most Northern Hemisphere observers never exceeds 35°.

Orbit

With only one photographic meteor being detected thus far, the elements for the orbit should be considered questionable. The meteor comes from MP1961 and the orbit is as follows:

ω	Ω	i	q	e	a
113	121	136	0.41	0.73	1.50

This, of course, represents a possible indication of a very short-period, elliptical orbit. Parabolic orbits were computed for the seven visual radiants and fireballs, with the average orbit being as follows:

ω	Ω	i	q	e	a
75.5	117.0	145.2	0.61	1.0	∞

No matter which orbit is closest to the truth, it is apparent that the Eta Craterid stream is in a high inclination, retrograde orbit. There seems to be no known comet orbit which might be linked to this stream.

---

[30]MP1961, pp. 15 & 37.

Rho Geminids

Observer's Synopsis

The duration of this shower extends from December 28 to January 28, with the radiant moving about +1.1° in α and –0.2° in δ daily. At maximum on January 8, the shower's average position is at α=108°, δ=+32°. A secondary maximum seems to occur on January 21 from α=125°, δ=+25°.

History

First detection of this stream came during January 1 to 15, 1872, when members of the Italian Meteoric Association plotted 8 meteors from an average radiant of α=109°, δ=+34°.[31] Other observations are as follows:

Rho Geminid Radiants

Date (UT)	RA(°)	DEC(°)	Meteors	Source
1897, Jan. 23-24	126	+19	7	D1899
1924, Jan. 6	114	+30	?	*Obs.*, 47
1921, Jan. 14.5	109	+30	?	H1948
1931, Jan. 10	109	+29	?	H1948
1933, Jan. 15-16	107	+30	?	Ö1934

The first suggestion that a meteor stream might be producing a regular meteor shower from Gemini in mid-January, was made by Richard B. Southworth and Gerald S. Hawkins during 1963. They examined photographic meteor orbits obtained during the Harvard Meteor Program of 1952-1954, and identified 4 meteors as representing the Rho Geminids. They concluded that the shower was active during January 15 to 28 from an average radiant of α=109.4°, δ=+32.3°.[32]

Bertil-Anders Lindblad conducted computerized stream searches during 1971. Once again the meteors were from the Harvard Meteor Project of 1952-1954. The most significant search utilized 865 precise meteor orbits and revealed 6 meteors over the period January 15-27 from an average radiant of α=110°, δ=+29°.[33]

In addition to visual and photographic detections, the Rho Geminids are also present in two radar surveys. Zdenek Sekanina recognized this stream during the Radio Meteor Project of 1961-1965. A total of 13 meteors were noted during December 28-January 16. The apparent nodal passage came on January

---

[31] D1899, p. 249.

[32] SH1963, p. 269.

[33] L1971A, p. 8.

7.9 (λ=287.0°) the radiant was at α=108.8°, δ=+31.5°.[34] Despite the fact that photographic data plainly showed activity to January 27, the radar was not in operation during January 20 to 25, and did not fully cover the period of the shower's activity during January 17 to 19. However, during the 1968-1969 survey, Sekanina noted 25 meteors during the period January 13-28. The nodal passage came on January 20.8, when the radiant was at α=125.1°, δ=+24.9°.[35] On this occasion the radar did not operate during December 21 to January 12, and, although Sekanina named this stream the "January Cancrids," it seems identical to the Rho Geminids, as can be seen in the "Orbit" section below.

The Author's analysis of the raw orbital data obtained from both sessions of the Radio Meteor Project reveals the stream's daily motion to be +1.1° in α and −0.2° in δ. Although the radiants and orbits of the photographic and radar data are very similar—certainly indicating an association—there seems to be an indication that two distinct populations of meteors exist. Concerning the population of radar meteors, it is interesting that a trend seems to exist which involves a slow decrease in the semimajor axis during the shower's period of activity.

Orbit

This stream was detected during both sessions of the Radio Meteor Survey conducted at Havana, Illinois, by Zdenek Sekanina in the 1960's. The following orbits were determined:

	ω	Ω	i	q	e	a
S1973	266.3	287.0	6.4	0.594	0.734	2.229
S1976	271.4	300.2	3.7	0.576	0.684	1.823

Using precise meteor orbits obtained during the 1952-1954 Harvard Meteor Project, the following photographic orbit was determined by Bertil-Anders Lindblad during 1971.

	ω	Ω	i	q	e	a
L1971A	243.4	301.4	3.5	0.77	0.710	2.66

Alpha Hydrids

Observer's Synopsis

The duration of this shower extends from January 15 to 30. At maximum on January 20, meteors can be expected to reach an hourly rate of 2-5 from a radiant of α=140°, δ=−9°.

---

[34]S1973, pp. 255 & 258.

[35]S1976, pp. 274 & 291.

History

The first apparent observation of this shower was made by Ronald A. McIntosh on January 15, 1929. Four plotted meteors revealed a radiant of α=132.2°, δ=–7.0°.[36] The shower does not appear in any records prior to the 20th century, nor in the major radiant catalogs of Ernst Öpik (1934) and Cuno Hoffmeister (1948). McIntosh's observation remained the only detection of this shower by Southern Hemisphere observers up to 1935, so that it was not included in his "An Index to Southern Meteor Showers," which appeared in the *Monthly Notices of the Royal Astronomical Society* in that year.

The first determination of this shower's hourly rate was made during January 20/21 and 24/25, 1971, by three members of the American Meteor Society. Karl Simmons (Florida), John West (Texas) and T. Recascino (New Jersey) detected maximum rates of 2-3 during this period when the limiting magnitude was 5.5-6.0.[37]

Perhaps the greatest knowledge of this meteor shower has come from the observations of the Western Australia Meteor Section (WAMS) made during the period of 1979 to 1986. During that first year, observers detected Alpha Hydrids during January 20-30, and determined the date of maximum as January 25. The average radiant was α=140°, δ=–9°, while the maximum ZHR was given as 8.51±1.84. Other details uncovered included an average magnitude of 3.15 (based on 54 meteors) and the fact that 10.2% of the meteors left trains. The meteors also tended to display three colors: 56.2% were white, 37.5% were yellow and 6.3% were blue. During 1980, Alpha Hydrids were detected during January 12-27, with a maximum ZHR of 1.90±0.19 occurring on January 16 from α=136°, δ=–10°.[38] The hourly rates indicated possible irregular activity coming from this radiant. Observations in the following years revealed this to be true, with hourly rates barely attaining 2 per hour during 1982, 1984 and 1986, while rates reached 4-5 in 1983 and 3-4 in 1985.

Members of the WAMS succeeded in determining a series of radiant positions during 1980, which display a definite eastward movement. Using only those days for which 2 or more positions could be averaged, the Author determined four positions: α=133°, δ=–10° on January 16/17; α=138.8°, δ=–10.3° on January 18/19; α=141°, δ=–9° on January 22/23; and α=146°, δ=–10° on January 26/27.

The Author has sought to determine the orbit of this stream and has succeeded in finding 2 photographic members (one in MP1961 and one in CF1973) and 19 radio members (M1986). Two apparent streams were revealed.

---

[36] Olivier, Charles P., *FOR*, No. 5 (1930), p. 14.

[37] *MN*, No. 5 (March 1971), p. 9.

[38] Wood, Jeff, Personal Communication (October 15, 1986).

The first radiant is composed of 4 radio meteors and the 2 photographic meteors and is apparently responsible for the visual activity. The duration is January 10-28. The nodal passage comes on January 17, with the radiant then being at α=135.9°, δ=–10.0°. The radiant's daily motion is +0.98° in α and –0.57° in δ. The other stream is made up of 15 radio meteors and is active during January 11-30. The nodal passage occurs on January 21 from α=149.0°, δ=–9.1°. The daily motion is +0.88° in α and –0.51° in δ. It is possible that, since the radar data never adequately covered the period of January 20-25, the maximums of both streams may occur later than the indicated dates of the nodal passages.

Orbit

Two photographic meteors (one in MP1961 and one in CF1973) and 19 radio meteors (M1986) have been uncovered and two streams appear to be present. The following represents these two orbits, with "A" being composed of 4 radio meteors and 2 photographic meteors and "B" being composed of 15 radio meteors:

	ω	Ω	i	q	e	a
A	117.8	115.9	50.9	0.289	0.935	4.446
B	139.7	120.2	60.1	0.159	0.911	1.787

Quadrantids

Observer's Synopsis

This meteor shower is generally visible between December 28 and January 7, with a very sharp maximum of 45 to 200 meteors per hour occurring during January 3 and 4 (λ=282.8°). The radiant is normally located at α=229°, δ=+49°, but there seems to be an occasional variation—possibly due to Jupiter's influence. The meteors tend to be bluish and possess an average magnitude of about 2.8.

History

The first observation of the Quadrantids seems to have occurred on the morning of January 2, 1825, when Antonio Brucalassi (Italy) remarked that "the atmosphere was traversed by a multitude of the luminous bodies known by the name of falling stars."[39] Other observations were made on January 2, 1835, by Louis François Wartmann (Switzerland), and on January 2, 1838, by M. Reynier (Switzerland).[40]

---

[39]Herrick, Edward C., *AJS (2nd series)*, 39 (1840), p. 334.
[40]Herrick, p. 334.

First mention that early January activity might be annual came in 1839, when Adolphe Quetelet (Brussels Observatory)[41] and Edward C. Herrick (Connecticut)[42] independently made the suggestion. The meteor shower became known as the Quadrantids because of its emanation from a now obsolete constellation called Quadrans Muralis (the Mural Quadrant)—located on some 19th-century star atlases near the point of meeting between Hercules, Boötes and Draco.

Details of the characteristics of this shower are not available from the time of its discovery, although the lack of published observations during the 1840s and 1850s probably indicates minor activity. The first major useful observation of this shower came in January 1863, when Stillman Masterman, of the United States, determined the first accurate radiant point as α=238°, δ=+46° 26'.[43] The following year, Professor Alexander Stewart Herschel of England was met with the unusually high rate of 60 meteors per hour at a time when the radiant was at an average height of only 19 degrees! According to J. P. M. Prentice, the ZHR was 131.[44] Although Herschel's high hourly rate did not become an annual event, it did help to stimulate interest in the shower in the years that followed.

Observations of this meteor shower since 1864 have revealed the earliest activity to be present on December 28, while the stream last appears on January 7. However, the shower is particularly noteworthy for a very sharp rise to maximum activity during January 3 and 4. Keith B. Hindley (England), using observations of the British Astronomical Association (BAA) made during the period 1965 to 1971, noted rates were higher than half the maximum rate for only 16 hours.[45] A study of BAA, British Meteor Society (BMS) and American Meteor Society (AMS) observations by the Author reveals that rates tend to drop below ten per hour just one day before and after the peak of shower activity. Furthermore, the rise and decline of the shower activity before and after January 3 and 4 tends to be more gradual and this apparently confirms the frequently held suspicion that this stream is made up of both a diffuse and a compact component. Additional support of this theory was gathered by Hindley in 1971. Using an IBM 360/65 computer at the University of Liverpool, telescopic observations of the Quadrantids were examined. The computer revealed a normal radiant diameter of 8 degrees, which contracts to less than a degree at the time of maximum.[46]

---

[41]Quetelet, A., *Catalogue des principales apparitions d'étoiles filantes*. Brussels, 1839.

[42]Herrick, Edward C., *AJS (2nd series)*, **35** (1839), p. 366.

[43]Fisher, W. J., *Harvard College Observatory Circulars*, (1930), No. 346.

[44]Prentice, J. P. M., *JBAA*, **63** (April 1953), p. 182.

[45]Hindley, Keith B., *Sky and Telescope*, **43** (March 1972), pp. 162-164.

[46]Hindley, K. B., *JBAA*, **82** (1971), p. 63.

Investigating 122 observations of this shower made between 1864 and 1953, Prentice concluded that the normal ZHR of the Quadrantids was 45.[47] However, this is far from being a consistent figure, as he pointed out that very strong returns occurred in 1909 (ZHR=202) and 1922 (ZHR=79), while very weak returns were noted in 1901 (ZHR=17), 1927 (ZHR=20) and 1940 (ZHR=21).[48] One figure Prentice did find more consistent was that of Earth's heliocentric longitude at maximum. This was determined as 282.9°,[49] but it should be stressed that this represents only visual observations. During 1947 to 1951, radio-echo observations were conducted at Jodrell Bank and the average time of maximum occurred when Earth was at a heliocentric longitude of 282.5°.[50] This difference between the times of the maximum of visual and radio-echo meteors illustrates the dispersion due to the Poynting-Robertson effect. According to Hindley, the dispersion factor amounts to 68 minutes of time per magnitude,[51] thus allowing the radar maximum to occur 6.3 hours before the visual maximum.

Interestingly, the hourly rate has continued to fluctuate in more recent times, though the range has been less prominent than in the past. ZHRs from the last 20 years are given below.

Quadrantid Maximum ZHR

Date	ZHR	Observers	Issue
1965	190	BAA	*JBAA*, **82**
1966	75	BAA	*JBAA*, **82**
1967	65	BAA	*JBAA*, **82**
1968	110	BAA	*JBAA*, **82**
1970	153	BAA	*JBAA*, **82**
1971	108	BAA	*JBAA*, **82**
1975	101.2	NMS	*MN*, No. 27

Abbreviations in the above table are: BAA, British Astronomical Association; NMS, Nippon Meteor Society.

The Quadrantids have not been consistently studied by visual observers. Cold weather prevalent in northern latitudes has frequently been cited as the main reason behind this. Another factor has been the very sharp maximum the shower displays, which frequently causes even the most diligent of observers to miss the activity only because they are in the wrong longitude. A final factor is related to

[47]Prentice, p. 179.

[48]Prentice, pp. 180 & 182-184.

[49]Prentice, p. 179.

[50]Hawkins, G. S., and Mary Almond, *MNRAS*, **112** (1952), p. 225.

[51]Hindley, K. B., *JBAA*, **82** (1971), p. 63.

the general faintness of this shower's meteors, thus requiring exceptional observing conditions for the main activity of the shower to be noted.

Quadrantid Magnitudes and Trains

Year(s)	Ave. Mag.	# Meteors	% Trains	Observer(s)	Source
1933	3.15	118	5.0	Dole	*PA*, **41**
1960-1976	2.81	142	—	McLeod	*MN*, No. 32
1967-1982	2.75	32	—	West	*MN*, No. 61
1971	1.63	1445	5.7	BAA	*JBAA*, **82**
1984	2.76	135	7.4	Lunsford	Personal Comm.
1985	2.57	244	2.4	Lunsford	Personal Comm.

Observers quoted in the above table are the following: Robert M. Dole (Maine); Norman McLeod, III (Florida); John West (Texas); British Astronomical Association; Robert Lunsford (California).

Examination of the photographic and radio-echo data by the Author reveals the average Quadrantid radiant to be at α=229.5°, δ=49.4°, but it should be pointed out that this is strictly an *average*. As mentioned earlier the shower does not possess a sharply defined radiant. In 1953, George E. D. Alcock and J. P. M. Prentice pointed out that "it has always been difficult to determine the radiant of the Quadrantid shower".[52] To correct this, they carried out a program in 1952 to obtain duplicate observations of radiants. On January 3, they established the existence of 13 active radiants,[53] thus demonstrating the complexity of the region.

Other studies have shown that the region is even more complex, since the same radiants are not necessarily active from year to year. The earliest mention of this was in 1918, when W. F. Denning and Mrs. Fiammetta Wilson noted their surprise to find the main radiant in January 1918 to be about eight degrees north of the normal radiant. They stated that a more northern radiant had been suspected in January 1916 and 1917, "but the data at the time were regarded as insufficient." Independent confirmations of the 1918 radiant came from several observers in England and it was noted that a weak shower actually did occur from the normal radiant as well.[54] This apparent change in active radiants from year to year is probably a by-product of the perturbations experienced by the stream every 11.86 years from Jupiter.

The planet Jupiter frequently appears in literature concerning the Quadrantids. In addition to the radiant changes mentioned, it has been linked to the initial appearance of the shower in the early 19th century and to the

---

[52] Alcock, G. E. D., and Prentice, J. P. M., *JBAA*, **63** (April 1953), p. 186.

[53] Alcock and Prentice, p. 187.

[54] Denning, W. F., and Wilson, Fiammetta, *MNRAS*, **78** (January 1918), pp. 198-199.

occasional irregularity in hourly rates. Also considered a result of perturbations is the slow retrogression of the ascending node, an occurrence which attracted four studies between 1958 and 1972. The subsequent calculated rates of the nodal retrogression were 0.31 degrees/century,[55] 0.41degrees/century,[56] 0.54 degrees/century,[57] and 0.6 degrees per century.[58]

One of the first studies of the long-term gravitational effects of Jupiter on the stream was conducted by S. E. Hamid and M. N. Youssef in 1963.[59] They took six doubly photographed meteors from 1954 and applied the secular perturbations of Jupiter during the last 5000 years. They noted that both the present inclination of 72 degrees and the perihelion distance of about 1 AU were at their lowest values of 13 degrees and 0.1 AU, respectively, 1500 years ago. About 4000 years ago, these values were very similar to what they are today, with the inclination being 76 degrees and the perihelion distance being about 1 AU. As a study of why the meteor stream is composed of at least two branches, the authors examined the change in the stream's distance from Jupiter over the last 5000 years. Prior to today's distance of only 0.3 AU, the stream was found to have been farthest from Jupiter 1500 years ago and only 0.2 AU away about 4000 years ago. The authors speculated that the stream's parent comet was captured by Jupiter about 4000 years ago and shortly thereafter it developed meteors along its path. "Because an appreciable number of these meteors, which now form the Quadrantids, did not suffer another close approach to Jupiter, the shower is observed to be compact."

Later in 1963, as a by-product of this study, Hamid and Whipple suggested a possible common origin for the Quadrantids and the Delta Aquarids, as 1300-1400 years ago the orbital planes and perihelion distances were very similar.[60] "Also," they added, "the physical characteristics of the meteoroids belonging to the two streams appear to be similar, as judged by their light curves."

In 1979, Iwan P. Williams, Carl D. Murray and David W. Hughes essentially repeated the Hamid-Youssef study, but they used a stream model and ten "test" meteoroids scattered about the orbit. Their study basically confirmed the earlier study back to 1500 years ago, but found the inclination and perihelion distance to closely reflect today's values only 3000 years ago. The study also indicated that, "casual observations of the original meteoroids at any time in the

---

[55]Hindley, K. B., *JBAA*, **80** (1970), p. 479.

[56]Hughes, David W., *The Observatory*, **92** (April 1972), pp. 41-43.

[57]Hamid, S. F., and Mary N. Youssef, *SCA*, **7** (1963), pp. 309-311.

[58]Hawkins, Gerald S., and Richard B. Southworth, *SCA*, **3** (1958), pp. 1-5.

[59]Hamid and Youssef (1963), pp. 309-311.

[60]Hamid, S. E., and Whipple,F. L., *AJ*, **68** (October 1963), p. 537.

interval 200-1000 yr before the present would not have revealed them to be members of the same stream. 1690 and 1300 yr ago they started off with similar orbits; these then separated, only coming together again in the last 200-150 yr."[61] The authors added that the parent comet probably underwent two major disruptions—one 1300 years ago and the other 1690 years ago.

The future of the Quadrantid stream was also examined in the 1979 study. The authors noted that the inclination would remain near 72 degrees and that the perihelion distance eventually will exceed 1 AU. Therefore, the authors predict Earth will no longer encounter the stream by the year 2400.

Orbit

Based on 25 meteors obtained from MP1961, BK1967, W1954 and HT1977, the following orbit was revealed:

ω	Ω	i	q	e	a
170.7	282.6	71.4	0.975	0.614	2.526

During 1961 to 1965, Zdenek Sekanina conducted a radar study of meteor streams from Havana, Illinois. Primarily covering meteors with magnitudes below naked-eye visibility, the orbit was as follows:

	ω	Ω	i	q	e	a
S1970	168.1	282.3	70.3	0.974	0.682	3.064

Finally, an analysis of the Quadrantid orbit was conducted by Ken Fox during 1985 and was published in F1986. It revealed the stream to have been in the following orbit 1000 years ago:

	ω	Ω	i	q	e	a
950	157.4	301.3	29.7	0.09	0.97	3.07

As can be seen (and as discussed briefly in the "History" section) the stream was quite different in the past. The shower's maximum occurred at the beginning of August from a radiant of α=341.1°, δ=–12.8°. A further analysis by Fox revealed no contact between this stream and Earth's orbit 1000 years from now.

Additional January Showers

Canes Venaticids

The discovery of this meteor shower is primarily due to Zdenek Sekanina. During the two sessions of the Radio Meteor Project, the presence of this stream was apparent each time. The surveys indicated this stream was active during

---

[61] Williams, Iwan P., Murray, Carl D., and Hughes, David W., *MNRAS*, **189** (1979), pp. 483-492.

January 13-30. The 1961-1965 survey indicated a nodal passage on January 24.2 (λ=303.7°), from α=111.8°, δ=+9.7°. The 1968-1969 survey revealed a nodal passage on January 15.5, however, the radio equipment did not operate during January 18-26, so that the date might be somewhat biased.

The two orbits determined by Sekanina are listed below. The orbit from S1973 is based on 9 meteors, while the orbit from S1976 is based on 11 meteors. As noted, the second session is biased toward an earlier date of maximum, thus, the first orbit (S1973) is probably closer to the truth.

	ω	Ω	i	q	e	a
S1973	66.0	123.7	6.1	0.751	0.666	2.249
S1976	70.2	114.8	1.4	0.700	0.770	3.045

By examining the raw meteor orbits used by Sekanina, the Author noted that the orbit in S1973 apparently included several meteors with inclinations above 10° (subsequently lowering the average declination). Only one such meteor orbit was present in the S1976 orbit sample. By combining all of the meteors from both radar sessions, the author determined a new average orbit and located several new members that had been previously missed. The orbit that follows is based on 18 radar meteor orbits.

ω	Ω	i	q	e	a
70.6	122.5	4.5	0.727	0.656	2.113

The average radiant is α=113.4°, δ=+12.6°, while the daily motion is +0.97° in α and −0.35° in δ.

January Draconids

The evidence supporting this stream's existence is scant, but what makes it most interesting is that the available observations seem to point to a fairly short-duration shower. The greatest support for this stream appeared during Zdenek Sekanina's 1969 session of the Radio Meteor Project. A total of 32 meteors were detected during January 13-17 from an average radiant of α=245.9°, δ=+62.4°.[62] The established orbit was as follows:

	ω	Ω	i	q	e	a
S1976	185.8	295.3	44.9	0.979	0.449	1.775

Possible visual observations are rare, possibly due to the short duration. During the 19th century, Giuseppe Zezioli (Bergamo) plotted several meteors during January 16, 1867-1869, which Schiaparelli derived a radiant of α=244°, δ=+64°, while William F. Denning's investigation of the records of the Italian Meteoric Association revealed six meteors plotted from α=241°, δ=+63°, during January 1-15, 1872.[63] During this century observations have also been scarce.

[62] S1976, pp. 274 & 291.

[63] D1899, p. 266.

R. Kingman (Bristol, England) plotted 6 meteors from α=245°, δ=+64°, during January 16-24, 1928. In Hoffmeister's *Meteorströme*, a radiant designated 2877 was observed on January 13, 1937, from a position of α=236°, δ=+59°.

Among all of the photographic lists, only two meteors appear in MP1961. Designated 6112 and 10081, they were detected on January 13, 1953 and January 13, 1954, respectively, and indicate an average radiant of α=236°, δ=59°. Although the orbit is very similar to that determined by radar, the orbital inclination is about 15° greater.

ω	Ω	i	q	e	a
187.5	293.0	60.5	0.98	0.73	3.63

Alpha Leonids

The strongest evidence for this stream's existence comes from the two sessions of the Radio Meteor Project conducted by Zdenek Sekanina during the 1960's. The duration of the stream definitely covers the period of January 13 to February 13, but it may actually begin as early as December 28. Maximum occurs sometime during the final week of January, with the average radiant then being near α=156°, δ=+9°.

The stream probably is a good example of a telescopic shower; however, traces of visual activity have been found in the records of the AMS and in H1948. In fact, the British Meteor Society's radiant catalog even gives the maximum ZHR as 10.[64] In addition, a handful of photographic meteors have been found in W1954 and MP1961.

The stream's low orbital inclination has apparently made it difficult to firmly establish the stream's ascending node, thus, although the shape of the orbital plane is firmly established, the ω and Ω are differing by nearly 180°. The available orbits are as follows:

	ω	Ω	i	q	e	a
S1973	325.6	302.9	1.8	0.142	0.898	1.395
S1976	143.2	130.7	3.7	0.198	0.801	0.993

Gamma Velids

Although Cuno Hoffmeister seems to have determined the first radiant for this shower on January 12, 1938 (α=132°, δ=–47°), this stream was basically ignored until 1979, when members of the Western Australia Meteor Section (WAMS) began systematic observations of it.

During 1978-1979, the WAMS continuously observed the skies during December 19/20 to January 6/7. The first Gamma Velids were noted on January

---

[64]Mackenzie, Robert A., *BMS Radiant Catalogue*. Dover: The British Meteor Society (1981), p. 11.

1/2, and their numbers reached a ZHR of 8.24±0.81 on January 6/7. The mean radiant position was given as α=125°, δ=–49°. Based on 27 observed meteors, it was concluded that the average meteor magnitude was 2.89, while 3.7% of the meteors left trains. Concerning colors, it was estimated that 10% of the meteors were orange, 10% were yellow, 20% were blue and 60% were white. More extensive observations during 1979-1980, revealed the Gamma Velid shower was active during January 1-17, with the mean radiant being α=125°, δ=–48°. A maximum of 7.06±1.36 was reached on January 3.[65]

The Author has analyzed observations of the WAMS obtained during 1982-1986, and has concluded that a relatively flat maximum ranging from 5 to 9 meteors per hour occurs during January 5-8. It was also apparent that the rise to maximum was fairly rapid, while the later decline was much slower. The following parabolic orbit was computed by the Author using the average radiant of α=125°, δ=–48° and an average maximum activity date of January 7.

ω	Ω	i	q	e	a
34.1	106.5	64.1	0.898	1.0	∞

The Southern Hemisphere radar surveys which were conducted at Adelaide Observatory during 1961 and 1969, did not operate during the first half of January.

[65]Wood, Jeff, Personal Communication (October 15, 1986).

Chapter 2: February Meteor Showers

Aurigids

Observer's Synopsis

This meteor shower is generally visible between January 31 and February 23, with a maximum of about 2 meteors per hour occurring during February 5-10. The radiant is normally located at α=74°, δ=+42°. The meteors are slow. Although the average magnitude is between 3 and 5, the shower is known for bright fireballs. The radiant seems to possess a daily motion of +0.7° in α and +0.3° in δ.

History

The discovery of the Aurigids should be credited to Robert P. Greg (Manchester, England), who, during 1845 to 1863, managed to plot 13 meteors from a radiant of α=76°, δ=+40° that seemed active between February 9 and 17.[1] On February 18, 1868, Giuseppe Zezioli (Bergamo, Italy) noticed 18 meteors from a radiant near Capella. Giovanni Virginio Schiaparelli derived a radiant of α=74°, δ=+48° from those observations.[2]

William F. Denning (Bristol, England) pointed out on several occasions that a meteor shower with a radiant just 5° south-southwest of Alpha Aurigae was active during February.[3] He considered the shower's duration to be from the 7th to the 23rd, and gave a radiant derived from his own observations as α=75°, δ=+41°. He said, "It often furnishes bright meteors in the evenings...." During 1912, he further elaborated on his Aurigid observations, saying that his radiant of α=75°, δ=+41°, was derived from 7 bright meteors plotted between 1876 and 1903. The meteors were described as slow and left trains.[4]

Some of the most notable Aurigid fireballs of this century follow:

---

[1]Newton, H. A., *AJS* (2nd Series), **43** (1867), p. 286.

[2]Denning, W. F., *MRAS*, **53** (1899), p. 243.

[3]Denning, W. F., *Observatory*, **26** (March 1903), pp. 137-138.

[4]Denning, W. F., *MNRAS*, **72** (May 1912), p. 633.

Aurigid Fireballs

Date (UT)	RA(°)	DEC(°)	Magnitude	Source
1901, Feb. 13.9	72	+41	"brilliant"	MNRAS, **72**, 426
1910, Feb. 17.8	72	+43	–5	MNRAS, **72**, 426
1920, Feb. 17.9	72	+43		Obs., **43**, 166
1935, Feb. 27.8	80	+46	–5	FOR, No. 60
1970, Jan. 31.4	62.1	+37.6	–9.1	CAPS, No. 665

The first visual study of this shower was conducted during 1962 and 1967, by a group led by V. Znojil (Public Observatory, Brno, Czechoslovakia). In 1962, observations were made from Mt. Klet on February 7/8 and 10/11. Two groups of observers applied different observing methods: one used no optical aid and observed with a limiting magnitude of 6.0, while the other group used 8-cm binoculars, which gave a limiting magnitude of 10.8. The naked-eye observers detected hourly rates of 1.5±0.5 on the 7/8 and 1.1±0.4 on the 10/11. The average meteor magnitude was noted as 2.9 on the first date and 4.9 on the second. The binocular observations revealed 11 members, with the average Aurigid magnitude being 9.0. During 1967, observations were made from Brno on February 4/5, 6/7, 8/9 and 10/11. Only binoculars were used. The results were that 7 Aurigids were detected and the average magnitude was 8.3.

Znojil's analysis of the Aurigid observations of 1962 and 1967, primarily discussed the apparent lack of small particles within the stream. He pointed out that the ratio of Aurigid meteors to sporadics changed markedly between visual and binocular observations, yet the change was always consistent. For instance, visual observations on February 7/8, 1962, revealed 21 Aurigids and 127 sporadics, and the February 10/11 visual observations revealed 24 Aurigids and 131 sporadics. On the other hand, binocular observations made on 2 days in 1962, revealed 11 Aurigids and 279 sporadics, while for the 4 days in 1967, they revealed 7 Aurigids and 234 sporadics. Znojil concluded that the small particles of this stream came about as a result of "fragmentation and cosmic erosion." He added that there was a possibility that the small particles were coming from a radiant that "is slightly displaced from the normally given position...."[5]

Recent observations have not been numerous. During February 8 to 14, 1970, Bill Gates (Albuquerque, New Mexico) acquired 11 counts which revealed hourly rates of only 0.6, while Martin Hale (New York) and Rick Hill (North Carolina) obtained 33 counts during February 5 to 17, 1972, which revealed an hourly rate of only 0.2. Gates described the meteors as being "very slow" and generally yellow. He also specifically pointed out the very low activity

---

[5]Znojil, V., *BAC*, **19** (1968), pp. 301-306.

in the early seventies.[6] In contrast to these low hourly rates, the Western Australia Meteor Section obtained a ZHR of 7.32±1.07 during February 2-7, 1980. The date of maximum was given as February 4, while the radiant was α=79°, δ=+39°.[7]

In examining the above observations, the Author has concluded that the daily motion of the Aurigid radiant is about +0.7° in α and +0.3° in δ.

Orbit

Based on one meteor obtained from M1976 the following orbit was revealed:

ω	Ω	i	q	e	a
193.5	311	3.3	0.976	0.52	2.00

Capricornids-Sagittariids

Observer's Synopsis

This daylight shower possesses a duration extending from January 13 to February 28. Reaching maximum between January 30 and February 3 (λ=309° to 313°), from an average radiant of α=299°, δ=–15°, this stream may be related to the Theta Ophiuchids of June.

History

The complete history of this daylight stream is contained in the details accumulated during the two sessions of the Radio Meteor Project during the 1960's. Zdenek Sekanina conducted the surveys and was able to isolate this stream in the two sets of data.

During the 1961-1965 survey, 26 meteors were detected in the period of January 13 to February 28. The indicated nodal passage was February 2.7 from an average radiant of α=299.0°, δ=–15.2°. Sekanina showed that a good probability existed that this stream was a twin branch of his Scorpiid-Sagittariid branch of June (see the Theta Ophiuchids of June), with the D-criterion being given as 0.149. He also suggested a possible relationship to the Apollo asteroid Adonis, with the D-criterion being 0.318.[8]

The 1968-1969 survey revealed 29 meteors during the period of January 15 to February 14. The nodal passage came on January 29.6 (λ=309.1°), at which time the average radiant was α=298.9°, δ=–14.2°. The resulting orbit made the identification with the June stream seem more plausible (D-criterion of

---

[6]Gates, Bill, *MN*, No. 39 (January 1978), p. 3.

[7]Wood, Jeff, Personal Communication (October 15, 1986).

[8]S1973, pp. 255 & 258.

0.119) and the suspected identification with Adonis was also strengthened (D-criterion of 0.199).[9]

The radiant catalog of the British Meteor Society gives the maximum hourly rate as 15.[10]

Orbit

Orbits for this stream were determined from data accumulated from each session of the Radio Meteor Project during the 1960's. The orbital elements are from S1973 and S1976.

	ω	Ω	i	q	e	a
S1973	60.0	313.3	6.8	0.314	0.842	1.991
S1976	69.8	309.1	6.2	0.415	0.758	1.712

Chi Capricornids

Observer's Synopsis

This daylight stream is detectable during the period of January 29 to February 28. At maximum on February 13, the average radiant is at α=315°, δ=–24°.

History

The discovery of this stream was made by Zdenek Sekanina during the 1961-1965 phase of the Radio Meteor Project conducted at Havana, Illinois. A total of 15 radio meteors were detected between January 29 and February 28. Sekanina found Earth to cross the stream's node on February 13.6, at which time the radiant was at α=314.3°, δ=–23.7°. The stream's geocentric velocity was determined to be 26.8 km/s.[11]

The Chi Capricornids were essentially confirmed during 1969, when G. Gartrell and W. G. Elford, using the Adelaide, Australia, radio meteor system, detected 3 meteors during February 10-17, which possessed an average radiant of α=316°, δ=–21°.[12]

Sekanina concluded that the stream possessed a close association with the Apollo asteroid Adonis (D-criterion=0.185) and suggested it might be the twin branch of the Sigma Capricornids of July (D-criterion=0.187).[13] Gartrell and

---

[9]S1976, pp. 274 & 291.

[10]Mackenzie, Robert A., *BMS Radiant Catalogue*. Dover: The British Meteor Society, 1981, p. 11.

[11]S1973, p. 258.

[12]GE1975, p. 596.

[13]S1973, pp. 264 & 279.

Elford suggested a possible association with the periodic comet Honda-Mrkos-Pajdusakova. The orbits of each object are listed in the "Orbits" section below for comparison. Gartrell and Elford suggested that Brian Marsden and Zdenek Sekanina's 1971 discovery of strong nongravitational forces influencing this comet's motion should not rule out "the possibility of related meteor streams with significantly different orbits...."[14]

Orbit

Radar equipment has been the sole means of studying this stream and the following orbits indicate surprisingly similar elements.

	ω	Ω	i	q	e	a
S1973	242.5	144.4	6.8	0.355	0.789	1.684
GE1975	246	144	4.5	0.36	0.82	2.083

The orbits of the Apollo asteroid Adonis (2101), periodic comet Honda-Mrkos-Pajdusakova (1954 III) and the July meteor shower called the Sigma Capricornids are as follows:

	ω	Ω	i	q	e	a
Adonis	41.06	351.21	1.37	0.442	0.764	1.873
HMP 1954	184.1	233.1	13.2	0.556	0.815	3.005
σ Cap	290.3	106.9	2.1	0.431	0.758	1.782

Alpha & Beta Centaurids

Observer's Synopsis

The duration of these meteor showers extends from February 2 to February 25. Maximum occurs around February 8. The Alpha Centaurids emanate from α=216°, δ=–60°, while the Beta Centaurids have a radiant of α=208°, δ=–58°. Despite the closeness of the radiants, they do have differences. The Alpha Centaurids possess maximum hourly rates of 3, while the Beta Centaurids can reach hourly rates as high as 14. The Alpha Centaurids have an average magnitude of 2.45.

History

This discovery of this shower should be attributed to Michael Buhagiar (Western Australia), who obtained observations of both Centaurid radiants during 1969-1980. In his "Southern Hemisphere Meteor Stream List" of 1980, Buhagiar listed two radiants which reached maximum on February 7. Radiant 290 was active during February 6-8, from α=206°, δ=–57°, while radiant 299 was active

---

[14]GE1975, p. 613.

during February 5-9, from $\alpha=214°$, $\delta=-64°$. Both radiants were referred to as "Beta Centaurids."[15]

Although both radiants have continued to be observed following the publication of Buhagiar's meteor stream list, observers have rarely distinguished between the two radiants during the same year. During 1979, members of the Western Australia Meteor Section (WAMS) managed to observe the "Alpha Centaurids" during February 2-18. At maximum on February 7, the radiant at $\alpha=216°$, $\delta=-59°$. During 1980, the same group observed members of the "Alpha Centaurids" during February 2-24. They noted that maximum came on February 8, from $\alpha=209°$, $\delta=-58°$.[16] The 1979 radiant obviously represents the true Alpha Centaurids, while the 1980 radiant is the Beta Centaurids.

Although it is possible that Cuno Hoffmeister observed this shower ($\alpha=210°$, $\delta=-57°$) on February 2, 1938, while in South Africa,[17] the observational history of this stream essentially began in 1969, so that the characteristics of each shower are hard to determine. Nevertheless, some interesting details should be noted. First of all, the Alpha Centaurids are apparently a consistent shower, with Buhagiar assigning an hourly rate of 3, and WAMS observers detecting high rates of 2 (ZHR calculated as 8.56±4.94) in 1979. The Beta Centaurids are apparently variable in activity, according to Buhagiar, with his 1969-1980 observations revealing high rates of 10 meteors per hour. WAMS observers obtained maximum rates of 11-14 per hour (ZHR calculated as 28.48±4.88) during a one-hour interval on February 8/9, 1980.

Characteristics of the Centaurid meteors have also been gathered in recent years. During 1979, 20 Alpha Centaurids revealed an average magnitude of 2.45, while, during 1980, 169 Beta Centaurids revealed an average magnitude of 1.64 (the latter number is an approximation by the Author based on a table published in the October 1980 issue of *Meteor News*).

The Alpha Centaurids may have been detected by radar at Adelaide Observatory during 1969. G. Gartrell and W. G. Elford operated the radar system during February 10-17. Two meteors were noted from a radiant of $\alpha=223°$, $\delta=-61°$, with the date of nodal passage being determined as February 15.[18] Assuming these meteors are members of the Alpha Centaurids, then this stream possesses an inclination near 105°, and a semimajor axis near 2.5 AU. This identification would also indicate that the radiant's daily motion is very close to +1° in α. The movement in δ can not be determined from the available observations.

---

[15]Buhagiar, Michael, *WAMS Bulletin*, No. 160 (1981).

[16]Wood, Jeff, Personal Communication (October 15, 1986).

[17]H1948, p. 246.

[18]GE1975, p. 619.

Orbit

Using radiants averaged from the above data, the following parabolic orbits were computed by the Author for the Alpha and Beta Centaurid streams:

	ω	Ω	i	q	e	a
Alpha	344.6	139.5	107.9	0.968	1.0	∞
Beta	357.3	139.5	108.3	0.986	1.0	∞

The two-meteor radar orbit listed in GE1975 most likely represents the Alpha Centaurids. It is as follows:

	ω	Ω	i	q	e	a
Alpha	344	146	105.0	0.97	0.61	2.5

Delta Leonids

Observer's Synopsis

The activity of this stream persists from February 5 to March 19. The shower reaches maximum on February 22, from an average radiant of α=156°, δ=+18°. The ZHR is 3, while the average magnitude of the meteors is near 2.86. A possible telescopic southern branch may have a duration extending from January 13 to February 24, with a maximum on February 3 and an average radiant of α=135°, δ=+8°.

History

This shower appears to have been first noted during the 20th century, since it is completely absent in the 19th century records of William F. Denning, Alexander S. Herschel, Robert P. Greg and Eduard Heis.

The first observation of a shower from this stream appears in the records of Denning for February 19-March 1, 1911. Seven meteors were plotted from an average radiant of α=155°, δ=+14°. The meteors were described as slow, with trains.[19] A meteor six times the brightness of Venus was observed by 16 observers on February 28.8, 1910. The radiant was determined as α=155°, δ=+16°.[20]

Additional visual observations belonging to this meteor shower were obtained in 1924 and 1930. In the former year, J. P. M. Prentice plotted several meteors from α=155°, δ=+13°, during February 25-28.[21] During 1930, two independent observations were made from opposite sides of the Atlantic. On February 19, Cuno Hoffmeister (Germany) detected a radiant at α=149°,

---

[19]D1912B, p. 634.

[20]D1912A, p. 428.

[21]Denning, W. F., *The Observatory*, **47** (March 1924), p. 98.

δ=+18°,[22] while observations by Balfour S. Whitney during February 20-21, revealed a radiant at α=154°, δ=+21°.[23]

The strongest arguments for this stream's existence have been its detection using both photography and radio-echo methods during the 1950's and 1960's. The former technique was best carried out during the Harvard Meteor Project of 1952-1954. The data was analyzed using a computer during 1971, by Bertil-Anders Lindblad (Lund Observatory, Sweden), with 24 photographic members of the Delta Leonids being detected. He gave the period of visibility as February 5 to March 19, and determined the average radiant position as α=159°, δ=+19°.[24] Although Lindblad suggested a relationship with either radiant 120 or 129, which appeared in Denning's 1899 "General Catalogue of the Radiant Points of Meteoric Showers and of Fireballs and Shooting Stars observed at more than one Station" as two long-duration stationary radiants, the Author has found no individual observations from these two showers which occur at a time and position that corresponds to the Delta Leonid radiant. It should also be pointed out that the Author believes several members of Lindblad's Delta Leonids are probably members of the Beta Leonids of March.

An ambitious radio-echo survey was carried out at Havana, Illinois, during 1961-1965. Called the Radio Meteor Project, it was headed by Zdenek Sekanina, who published the results during 1973. Sekanina isolated the Delta Leonid stream and said they existed during the period of February 9 to March 12. The date of nodal passage was given as February 19.9 (λ=330.7°), while the average radiant was α=154.3°, δ=+18.3°.[25] Combining the 8 radio meteors used to establish this orbit, as well as the photographic meteors, the Author determined the radiant's daily motion as +0.93° in α and −0.38° in δ.

It should be noted that the 1968-1969 session of the Radio Meteor Project detected a stream similar to the Delta Leonids, but the orbit's argument of perihelion and ascending node were about 180° off. In addition, the duration was given as January 13 to February 24, the date of the nodal passage was February 2.3, and the radiant position was α=135.2°, δ=+7.5°.[26] These figures all possess notable differences from that already established above for the Delta Leonids and, when compared to that stream's radiant ephemeris, this stream lies about 15° to the south. No trace of this southern branch of the Delta Leonids appears in visual and photographic records, which implies that it may be a possible telescopic shower.

---

[22] H1948, p. 227.

[23] Olivier, C. P., *FOR*, No. 8 (1931), p. 14.

[24] L1971B, pp. 20-21.

[25] S1973, p. 258.

[26] S1976, p. 275.

Observations by members of the Western Australia Meteor Section (WAMS) have supplied some of the most valuable information on the Delta Leonids in recent years. Jeff Wood, director of the WAMS, has analyzed several years of observations and has concluded that maximum occurs on February 26, from α=158°, δ=+17°. He gives the duration as February 1 to March 13.[27] In 1979, Delta Leonids were observed during February 22 to March 3. A maximum ZHR of 3.08±1.78 came on February 25. The radiant position was then given as α=159°, δ=+19°. The average magnitude of seven observed meteors was 2.86. During 1980, activity was observed during February 15 to March 9. A maximum ZHR of 2.59±0.57 came on February 22, at which time the radiant was at α=159°, δ=+18°.[28]

Compared to other currently active meteor streams, the Delta Leonids appear to possess a fairly short history. During 1985, Ken Fox (Queen Mary College, England) investigated the past and future orbits of 53 meteor streams. He found that the Delta Leonid orbit does not come in contact with Earth's orbit 1000 years in the past or future.[29] Thus, this stream is only a temporary feature as far as Earth is concerned.

Orbit

An orbit based on 24 photographic meteors detected during the Harvard Meteor Project (1952-1954) was published by Bertil-Anders Lindblad in L1971B and is as follows:

	ω	Ω	i	q	e	a
L1971B	259.0	338.1	6.2	0.643	0.747	2.618

As noted earlier, several members of the Beta Leonids of March may have been included in Lindblad's data and there is some possibility that a few other meteors belong to other less distinct streams. The Author has derived the following Delta Leonid orbit, based on 6 photographic meteors. A fairly strict D-criterion of less than 0.10 was utilized in order to determine the orbit of the stream's primary core.

ω	Ω	i	q	e	a
267.7	334.4	5.4	0.581	0.760	2.416

An orbit based on 8 radio meteors detected during the 1961-1965 session of the Radio Meteor Project was published by Zdenek Sekanina in S1973, and is as follows:

ω	Ω	i	q	e	a
266.4	330.7	4.8	0.612	0.687	1.954

---

[27] Wood, Jeff, Personal Communication, (October 24, 1985).

[28] Wood, Jeff, Personal Communication (October 15, 1986).

[29] F1986, pp. 522-524.

Additional February Showers

Sigma Leonids

The strongest evidence supporting this stream's existence comes from the 1961-1965 session of the Radio Meteor Project. Zdenek Sekanina isolated 16 meteors during the period of February 9-March 13. The date of the nodal passage was given as February 26.2 (λ=337.1°), at which time the radiant was α=169.4°, δ=+14.4°.[30] Sekanina gave the orbit as

	ω	Ω	i	q	e	a
S1973	283.8	337.1	8.1	0.468	0.748	1.856

Searches through past records have revealed a few possible observations of this shower: On March 3.5, 1886 (UT), William F. Denning observed a 2nd-magnitude stationary meteor at α=176°, δ=+9°;[31] Ernst Öpik plotted several meteors during February 29 and March 1, 1932, which indicated a radiant of α=168°, δ=+15°;[32] During February 27-28, 1947, Vincent Anyzeski plotted 4 meteors from α=163°, δ=+14°.[33] In addition, three photographic meteors were found in MP1961. These meteors were detected on February 5, 1953 (α=146°, δ=+22°), February 3, 1954 (α=147°, δ=+20°) and March 1, 1954 (α=167°, δ=+12°). Their combined orbits average out as

ω	Ω	i	q	e	a
279.7	323.3	6.0	0.473	0.803	2.405

Since the shower's official announcement in 1973, by Sekanina, only one notable observation of activity has been made. During February 23-March 11, 1979, members of the Western Australia Meteor Section observed this shower and noted that a maximum ZHR of 1.23±0.67 came on February 25, from α=170°, δ=+7°.[34]

The Author combined the radio and photographic meteors and noted a daily motion of +0.93° in α and −0.40° in δ.

---

[30] S1973, pp. 255 & 258.

[31] D1923A, p. 40.

[32] Ö1934, p. 35.

[33] *FOR*, No. 75, p. 12.

[34] Wood, Jeff, Personal Communication (October 15, 1986).

Chapter 3: March Meteor Showers

Eta Draconids

Observer's Synopsis

The duration of this shower persists from March 22 to April 8. Maximum seems to occur between March 29 and 31 (λ=7°-9°), from an average radiant of α=244°, δ=+62°.

History

Observations of the Eta Draconids seem confined to the 20th century. In Cuno Hoffmeister's book, *Meteorströme*, five German-observed radiants are listed which apparently belong to this stream. The first observation was made on April 2, 1910, when 20 meteors were detected from a 2°-diameter radiant at α=247°, δ=+63°. Another observation came on April 1, 1911, when 12 meteors came from a 2°-diameter radiant at α=246°, δ=+69°. The three additional radiants were at α=244°, δ=+61° (April 5, 1919), α=253°, δ=+54° (April 7, 1931), and α=251°, δ=+59° (March 23, 1936).[1]

Two members of the American Meteor Society (AMS) have also detected activity from this stream. On April 4.3, 1940, Donald Faulkner (Stetson University, Florida) coordinated two teams of students to observe the Eta Aquarids. The groups were located at Daytona Beach and Altoona, and all observed meteors were plotted. Faulkner's evaluation of the data revealed that both stations had detected meteors from a radiant at α=257°, δ=+56°.[2] On March 31.6, 1951, Philip Burt (Memphis, Tennessee) plotted 9 meteors from a radiant of α=247°, δ=+63°.[3] The records of the AMS are probably the most extensive collection of visual radiants in the world for the 20th century, yet it seems a puzzle that only two observations of the Eta Draconids appear in their records. On the other hand, the Author has noted that when meteor observations are made

---

[1]H1948, pp. 198-253.

[2]Olivier, Charles P., *FOR*, No. 63 (1943), p. 43.

[3]Olivier, Charles P., *FOR*, No. 85 (1952), p. 17.

during March (and they are rare), they are usually of the southern portion of the sky. The same is true for the early half of April.

The most impressive collection of data on this stream to date was described by T. L. Korovkina, V. V. Martynenko and V. V. Frolov in a paper published in 1971. The Eta Draconids were one of 23 meteor showers detected during observations in March 1969. Participating in the survey were members of the Yaroslavl Society of Amateur Astronomers and the Yaroslavl division of the Astronomical and Geodetic Society of the USSR. Observers were split into two groups, with one observing at Krasnye Tkachi during March 24-30, and the other observing at Rybinsk during March 6-16 and 25-29. The limiting magnitude of the sky during these observations was between 5 and 5.5. The observers at the former village had set their objective as searching for radiants of minor meteor showers and they were the successful observers of the Eta Draconids.

E. A. Malakhaev observed the first possible radiant on March 26.99, when 9 plotted meteors indicated a 1.0°-diameter radiant at α=231.0°, δ=+56.2°. On a scale of 1 to 5, the accuracy of this radiant was given as 3. A similar value was also assigned to a radiant detected by Malakhaev, N. V. Smirnov and T. A. Kopycheva during March 27.98. Based on 8 meteors, the position was given as α=237.0°, δ=+60.0°, while the radiant diameter was given as 2.0°. Two excellent radiants were determined during the following two nights: on March 28.91, Smirnov, Kopycheva and V. K. Karpov plotted 16 meteors (1 stationary) from a 2.0°-diameter radiant at α=241.0°, δ=+61.5°, while, on March 29.88, Smirnov, Kopycheva and L. M. Afanas'eva plotted 17 meteors (1 stationary) from a 1.5°-diameter radiant at α=245.5°, δ=+63.5°.[4]

The 1969 survey was repeated on a smaller scale during 1973. Smirnov and T. L. Korovkina published their evaluations of the visual data in 1975, and indicated that observations were primarily conducted during March 24-30 by members of the Yaroslavl Amateur Astronomers Society in Krasnye Tkachi. Overall, the 1973 observations revealed less activity from the Eta Draconids than was detected in 1969, but two radiants were nevertheless noted which might be related in some way. The first radiant was seen by N. A. Tsarev and Smirnov during March 25-29. Four meteors came from an area 1.0° across at α=237.0°, δ=+61.0°. The second radiant was seen by Tsarev and B. M. Belyakov during March 24-28. Six meteors came from an area 1.5° across centered at α=255.0°, δ=+64.5°.[5]

A search through the various records of photographic meteor orbits by the Author has revealed no possible members of this stream; however, among

---

[4]KM1971, p. 100.

[5]SK1975, p. 98.

Zdenek Sekanina's 39,145 radio meteor orbits, 15 probable members were found which seem to indicate two distinct streams. The first stream is based on 7 meteors. The indicated duration is March 22 to April 9, while the average radiant is α=247.0°, δ=+61.9°. The second stream orbit is based on 8 meteors, with an indicated duration of March 24 to April 8, and an average radiant of α=250.1°, δ=+54°. Neither of these streams were noted by Sekanina.

Orbit

Searching through the 39,145 radio meteor orbits obtained by Zdenek Sekanina, the Author has located 15 meteors which may be associated with the Eta Draconid stream. The meteors form two distinct groups, as is demonstrated by the two orbits below. Orbit "A" is based on 7 meteors, while orbit "B" is based on 8.

	ω	Ω	i	q	e	a
A	196.5	8.9	38.2	0.984	0.592	2.412
B	202.6	13.1	48.0	0.967	0.736	3.662

The orbits bare a striking resemblance to comet Abell (1954 X), the of which follows:

	ω	Ω	i	q	e	a
1954 X	194.4	2.3	53.2	0.970	1.001	-1720

During 1981, Jack D. Drummond (New Mexico State University) computed the theoretical meteor radiants for 178 long-period comets. Comet Abell was listed as producing a radiant at α=254°, δ=+57° on March 23 (λ=2.3°). The closest approach between the orbits of the comet and Earth was given as 0.01 AU.[6]

Beta Leonids

Observer's Synopsis

The duration of this meteor shower extends from February 14 to April 25. Maximum occurs around March 20 (λ=0°), at which time the radiant is located at α=177°, δ=+11°. The maximum ZHR is probably 3-4.

History

The Beta Leonids were apparently first observed by E. R. Blakeley (Dewsbury), when he plotted five slow meteors from an average radiant of α=175°, δ=+10° during March 16-26, 1895.[7]

---

[6]Drummond, Jack D., *Icarus*, 47 (1981), p. 505.
[7]D1899, p. 258.

Although the Beta Leonids were not widely reported on an annual basis, some sources reveal they were active. In Cuno Hoffmeister's 1948 book *Meteorströme*, five visual radiants are listed as having been observed during the 1930's. Radiants 1607 and 1623 were detected in 1931. The former radiant was detected on March 19 (λ=357.0°), from α=179°, δ=+8°, while the second radiant was observed on March 21 (λ=358.7°), from α=182°, δ=+15°. Radiants 2960 and 3108 were both observed On March 23, 1933: the former occurring at a solar longitude of 1.0°, from α=187°, δ=+12°, while the second occurred from a solar longitude of 0.7°, from α=183°, δ=+10°. Finally, radiant 5029 was observed on March 20, 1936 (λ=358.1°), from α=182°, δ=+15°.

The Beta Leonids were missed by radio-echo surveys until the 1968-1969 session of the Radio Meteor Project. Zdenek Sekanina determined the shower's duration as extending from February 14 to April 25. He said the nodal passage occurred on March 23.0 (λ=1.9°), at which time the radiant was at α=180.7°, δ=+11.5°.[8] Sekanina had referred to this stream as the March Virginids. It should be noted that the nodal passage may have actually occurred earlier than indicated, since the radar equipment was not in operation during March 20-22—the probable date of maximum activity.

The most recent observations of this shower come from members of the Western Australia Meteor Section (WAMS). In 1980, they detected shower meteors during March 14-23. Maximum came on March 19, when the ZHR reached 3.46±0.86. The radiant was determined as α=176°, δ=+15°.[9]

A search through the various lists of photographic meteors reveals 14 probable Beta Leonids (gathered from MP1961 and BK1967). The indicated duration is February 24 to April 11. Interestingly, two clusters occur in the data: the first is around March 5-6 (λ=344°-345°), while the second is around March 18-19 (λ=358°-359°). These two clusters cause the average orbit to possess a nodal passage date of March 12, at which time the radiant position is α=168.4°, δ=+12.6°. This strange grouping of the data distorts the average radiant position of this shower. The actual position for March 18-19, based on 3 photographic meteors, is α=173°, δ=+9.7°.

The existence of two dates of maximum in the photographic data is difficult to account for. Also difficult to explain is how so many photographic meteors could have been overlooked in the various computerized stream searches conducted during the 1960's and early 1970's. A possible explanation for the latter question is found in one of Bertil-Anders Lindblad's 1971 computerized searches.[10] He had identified 24 meteors as belonging to the Delta Leonids of

---

[8]S1976, pp. 276 & 292.

[9]Wood, Jeff, Personal Communication (October 15, 1986).

[10]L1971B, pp. 20-21.

February. The Author finds that several of the Beta Leonid meteors of early March were included in Lindblad's data. By comparing the orbits of these two streams it can be seen that they do have distinct differences.

The Author has plotted the radio meteors, photographic meteors and available visual radiants and determined the daily motion of this shower as +0.9° in α and –0.4° in δ.

Orbit

The orbit as determined by Zdenek Sekanina after analyzing the data gathered during the 1968-1969 session of the Radio Meteor Project is as follows:

	ω	Ω	i	q	e	a
S1976	256.4	1.9	2.4	0.853	0.238	1.119

The Author has identified 14 photographic meteors in MP1961 and BK1967, and has determined the following average orbital elements:

ω	Ω	i	q	e	a
252.7	352.8	3.7	0.728	0.628	1.955

It should be noted that two of the 14 photographic meteors actually possess perihelion distances and eccentricities very close to that given for the radio meteor orbit.

Delta Mensids

Observer's Synopsis

The duration of this stream extends from March 14-21. At maximum on March 18 (λ=358°), an hourly rate of 1 to 2 meteors can be seen emanating from a radiant of α=55°, δ=–80°.

History

This shower was discovered during 1969, by observers operating the radio meteor system at Adelaide Observatory in Australia. In analyzing the data, G. Gartrell and W. G. Elford found two fairly distinct, but similar streams active while the system was in operation during the period March 16 to 22. The first stream, designated 3.04, possessed a radiant of α=51°, δ=–81°, while the second stream, designated 3.05, was located at α=50°, δ=–78°.[11]

The orbits of these two streams were found to be nearly identical except for their semimajor axes—stream 3.04 had an average value of 2.13 AU, while stream 3.05 was determined as 10.0 AU. Gartrell and Elford indicated these streams provided a very important key to meteor stream formation, when they

[11]GE1975, p. 596.

are compared to comet Pons of 1804. As can be seen below, the orbits of all of these objects are very similar except in their values of the semimajor axes. The authors said the orbital similarities "give further evidence that the low eccentricity orbits could be the result of evolution rather than direct formation from low eccentricity comets." They added that if the orbit of comet 1804 is parabolic as indicated, then "only 170 years have been available for 3.04 to contract from a>10 a.u. to the present value of 2.13 a.u."[12]

No apparent records appear to exist concerning past observations of this shower; however, southern hemisphere observers have been making occasional observations of this stream during the 1970's and 1980's. According to Jeff Wood, director of the meteor section of the National Association of Planetary Observers (Australia), this shower has a duration spanning March 14 to 21. At maximum on March 18, 1 to 2 meteors per hour can be detected from an average radiant of α=55°, δ=–80°.

Orbit

Two orbits were determined for this stream at Adelaide during 1969. They were published in GE1975 and were designated 3.04 (average date March 18) and 3.05 (average date March 19). Stream 3.04 was based on 11 meteors, while 3.05 was based on 10 meteors.

	ω	Ω	i	q	e	a
3.04	347	178	55.3	0.98	0.53	2.13
3.05	346	178	58.3	0.98	0.87	10.00

The orbit of comet Pons (1804) is listed in Brian G. Marsden's *Catalog of Cometary Orbits* (1983) as follows:

	ω	Ω	i	q	e	a
Pons 1804	332.0	178.8	56.5	1.071	1.0	∞

Gamma Normids

Observer's Synopsis

The duration of this shower extends over the period of March 11 to 21. Maximum occurs on March 16 (λ=356°), from an average radiant of α=245°, δ=–49°. The maximum ZHR reaches 5-9.

History

Ronald A. McIntosh (Auckland, New Zealand) discovered this meteor shower on March 10.1, 1929. He plotted seven meteors which indicated a radiant of

[12]GE1975, pp. 613-614.

α=241.5°, δ=–43°.[13] Confirmation came in 1932, when Murray Geddes (New Plymouth, New Zealand) plotted six meteors on March 7.1, from α=242.7°, δ=–54.7°. Geddes plotted another five meteors on March 12.0—the radiant then being α=240°, δ=–52°.[14] McIntosh summarized these radiants in his 1935 paper "An Index to Southern Meteor Showers." The duration was given as March 7-12, and the weighted average radiant was α=241°, δ=–53°.[15] It was referred to as the "Scorpiids."

This stream were virtually ignored until 1953, when radar equipment used by A. A. Weiss (University of Adelaide, South Australia) accidentally detected activity on March 15-16. Although the radiant position was estimated as α=250°, δ=–50°, Weiss said it could not "be fixed precisely because of low activity and also because of the marked deficiency of large meteors in this stream."[16] He elaborated by noting that the number of radar echoes with a duration of one-half second or longer were practically no greater during the shower than on non-shower days. Weiss also indicated that the radiant's culmination after sunrise would make visual observations difficult.

Curiously, C. D. Ellyett and C. S. L. Keay (Christchurch, New Zealand) made an attempt to confirm this shower during March 1956. The equipment was set at the same sensitivity as Weiss' during March 8-14, and it was set at a higher sensitivity during March 15-23, but neither session revealed the shower. The authors concluded the shower "is variable in activity from year to year."[17]

The next observation of the Gamma Normids came during March 16-22, 1969, while G. Gartrell and W. G. Elford operated the radio meteor system at Adelaide, South Australia. Two associations were noted which possessed radiants close to that of this stream. The first was based on three meteor orbits and possessed a radiant position of α=250°, δ=–43° on a mean date of March 20. The second association was considered less reliable since it was based on only two meteors. Its radiant was α=253°, δ=–41° on a mean date of March 19.[18]

Michael Buhagiar (Perth, Western Australia) published a list in 1981, which gave details of meteor showers observed by himself during 1969-1980. Radiant number 339 (called the "Beta Arids") was given a duration of March 15-21. Maximum was said to have occurred on March 17, from α=245°, δ=–50°. The maximum hourly rate was given as 4.[19]

---

[13]Olivier, Charles P., *FOR*, No. 5, p. 21.

[14]Olivier, Charles P., *FOR*, No. 15, pp. 43-44.

[15]M1935, p. 714.

[16]Weiss, A. A., *AJP*, **8** (1955), pp. 157-158.

[17]Ellyett, C. D., and Keay, C. S. L., *AJP*, **9** (1956), p. 479.

[18]GE1975, pp. 596 & 619.

[19]Buhagiar, Michael, *WAMS Bulletin*, No. 160 (1981).

Observers of the Western Australia Meteor Section (WAMS) have contributed greatly to observations of this shower in recent years. During 1979, the Gamma Normids were observed over the period of March 16-18. Maximum came on March 17, when a ZHR of 8.45±1.60 was detected from α=248°, δ=−49°. In 1980, observations were made during March 14-15. At maximum on March 15, the ZHR was 8.90±2.30 and the radiant was α=242°, δ=−50°.[20]

The WAMS made very extensive observations during 1983. The earliest Gamma Normid activity came on the night of March 10/11, when the ZHR was about 1.5±0.3. After another low ZHR of 1.6±1.0 on March 11/12, a sharp rise to a ZHR of 9.6±2.3 came on March 13/14, followed by a rate of 4.6±0.6 on March 14/15. Thereafter, rates were 2.2±0.8 on March 15/16, 0.5±0.1 on March 17/18, and 0.7±1.1 when last seen on March 18/19. Based on 63 meteors, the average magnitude was determined as 2.68, while 9.5% had trains. For the meteors of magnitude 2 or brighter, 64% were white, 24% were yellow, 8% were orange, and 4% were blue.[21]

Orbit

The two orbits determined by Gartrell and Elford (GE1975) from radio-echo data obtained during 1969 are listed below. Although the respective radiants are very similar, the differences in semimajor axes cause some significant differences in the orbital elements. Orbit 3.15 is based on three meteors, while 3.46 is based on two. As mentioned earlier, the latter orbit is considered uncertain.

	ω	Ω	i	q	e	a
3.15	97	179	137.4	0.66	0.43	1.176
3.46	49	179	145.4	0.85	0.72	3.125

Eta Virginids

Observer's Synopsis

Observations of this shower indicate a duration of February 24 to March 27. Maximum is not prominent, but seems to fall on March 18 (λ=358°), from a radiant of α=185°, δ=+3°. The maximum hourly rate reaches about 1 to 2. A possible southern branch of this stream seems to exist about 10° to the south.

History

The Eta Virginids appear to be a fairly diffuse branch of the Virginid complex of February to April. Although visual radiants are not particularly abundant in the

[20]Wood, Jeff, Personal Communication (October 15, 1986).

[21]Wood, Jeff, *WGN*, 12 (February 1984), p. 8.

literature, this stream has produced its share of photographic meteors and has appeared in each of the sessions of the Radio Meteor Project conducted at Havana, Illinois, by Zdenek Sekanina during the 1960's.

The first detection came during 1961-1965, when 11 meteors were detected during March 12-27. The nodal passage came on March 18 from a radiant of α=184.9°, δ=+2.7°.[22] The second detection came during the 1968-1969 survey, when 26 meteors were detected during February 24-March 14. Nodal passage came on March 10 from an average radiant of α=182.2°, δ=+13.8°.[23]

Sekanina's data indicates radiants separated by about 10° between the two sets of data. A look at the orbital elements reveals that this difference is entirely due to the inclination, since all the other elements are nearly identical. The Author has examined the individual radio meteor orbits used in Sekanina's two analyses and finds a diffuse stream, which might possess a radiant diameter of 10-12°. There is also a possibility that this stream may itself be composed of two filaments—one being of high inclination, while the other is of low inclination. This latter hypothesis could explain the differences in duration and dates of maximum activity among the two radio-meteor streams detected by Sekanina. Support is also apparent among the recent observations of the Western Australia Meteor Section (WAMS).

Observations of both Eta Virginid showers were made by members of the WAMS during 1979 and 1980. In the former year the early stream was observed during February 23-March 5, with a maximum ZHR of 2.20±0.52 coming on March 5 from a radiant of α=184°, δ=+1°. The second stream was detected during March 16-24, with a maximum ZHR of 3.33±1.66 coming on March 20 from α=183°, δ=+2°. In 1980, the early Eta Virginids were seen during February 23-March 10, with a maximum ZHR of 2.27±0.34 coming on March 9 from α=184°, δ=+1°. The second shower was observed during March 14-23, with a maximum ZHR of 3.22±1.07 coming on March 21 from α=187°, δ=0°.[24]

When all things are considered, it is interesting that the shower that occurs in the second half of March is very similar to the stream detected by Sekanina during 1961-1965. The February-March activity noted by the WAMS does not, however, match the 1968-1969 radar data of Sekanina in the declination of the radiant. In fact, there is about a 10° difference at the time of maximum.

The Author has conducted a mathematical determination of the radiant ephemeris of this stream and revealed a daily motion of +0.9° in α and −0.4° in δ. Confirmation of this was given when the 37 radio meteor orbits were grouped according to inclination and then subjected to a least-squares fit.

---

[22]S1973, pp. 255 & 258.

[23]S1976, pp. 275 & 292.

[24]Wood, Jeff, Personal Communication (October 15, 1986).

In 1980, Sam S. Mims (Baton Rouge, Louisiana) suggested a relationship between this stream (he simply referred to it as the Virginids) and a comet discovered by Dunlop (Parramatta) on September 30, 1833. The comet was only observed 15 times in 16 days, by a few observers in South Africa and Australia, so that the orbit is considered as somewhat uncertain. Nevertheless, during 1888, Schulhof showed that the eccentricity could be as small as 0.8. Mims pointed out that the orbits of the comet and meteor stream are close except for the longitude of perihelion, which is about 50° off. He suggested that, if the comet was indeed of short period, it may "have been perturbed (by Jupiter) before 1833...." Although realizing that several assumptions would have to be accepted to support some of his suggestions, Mims added, "it is interesting to think that we could learn much more today about a comet observed for only a short period in the early 19th century...."[25]

One final point about this shower's activity is that a third radiant with a position about 10° to the south may be present throughout the month of March. The strongest support exists in the data collected by Sekanina during 1961-1965, when a radiant referred to as the "Southern Eta Virginids" was described as having a duration extending from March 9 to April 9. The date of the nodal passage was given as March 22, at which time the radiant position was $\alpha=178.9°$, $\delta=-8.2°$.[26]

Orbit

The orbits determined by Sekanina during the two sessions of the Radio Meteor Project are as follows:

	ω	Ω	i	q	e	a
S1973	282.4	357.3	3.7	0.496	0.707	1.691
S1976	281.8	349.3	11.3	0.501	0.703	1.690

As noted above, Sam S. Mims has suggested this stream is related to comet Dunlop (1833). The orbit of this comet is given in Brian G. Marsden's *Catalog of Cometary Orbits* (1983) as

	ω	Ω	i	q	e	a
1833	259.6	324.9	7.3	0.458	1.0	∞

Although the orbit is given here as parabolic, primarily due to the comet's being observed for only 16 days, the original investigation by Schulhof in 1888, revealed the possibility of both a hyperbolic and an elliptical orbit, with the latter fitting the observations better. Schulhof indicated that it was not impossible that the eccentricity could be as small as 0.8.[27]

---

[25]Mims, Sam S., *MN*, No. 48 (January 1980), pp. 3-4.

[26]S1973, pp. 255 & 258.

[27]Schulhof, *Bulletin Astronomique*, **5** (1888), pp. 248 & 532.

Pi Virginids

Observer's Synopsis

The duration of this meteor shower extends from February 13 to April 8, with maximum occurring sometime between March 3 and 9 (λ=342°-348°). The average radiant during maximum is α=182°, δ=+3°, while the ZHR seems to peak at 2-5.

History

The discovery of the Pi Virginids should be attributed to Cuno Hoffmeister, who, in his 1948 book *Meteorströme*, included a meteor shower meeting this stream's basic description among the nine visual radiants making up his "Virginid Complex." This was the earliest of the active Virginid showers, with maximum occurring at a solar longitude of 344.5°, from α=178.8°, δ=+0.6°.[28] Nine visual radiants had been collected during 1908 to 1938, which provided the data for the Pi Virginids.

Visual radiants from this stream are not prominent in the records. The Author has noted that the overall confusion of activity in the entire Virginid region during March and April generally makes visual observations useless—especially in the 19th century, when Denning's stationary radiant theory was in full swing. At that time, Denning noted weak, but continuous activity during March and April from a position matching that of the Pi Virginid maximum; however, since, the radiant of this stream moves steadily southeastward, one can only conclude that these "stationary" Virginid radiants were the product of several different streams of the Virginid complex. Careful, accurate plots are one answer to studying showers of the Virginid complex. Another accurate method of studying these streams, and, especially the Pi Virginids, is by the use of radar.

Radar studies revealed definite traces of the Pi Virginids during the 1960s. The first study was that of B. L. Kashcheyev and V. N. Lebedinets, both of the Kharkov Polytechnical Institute (USSR). When the radar began operating on March 14, the first members of the Pi Virginids were detected and observations of stream members continued until March 23. The average radiant was given as α=188°, δ=+1°.[29] In the following year, C. S. Nilsson (University of Adelaide, South Australia) conducted a radar survey which ultimately revealed three meteors during the period of March 13-16. These meteors came from a radiant of α=189°, δ=–4°. On this occasion, the radar equipment only operated during March 11-16, so that later activity would have easily been missed. It is also

---

[28]H1948, p. 138.
[29]KL1967, p. 188.

interesting that Nilsson noted many other meteors in the vicinity of this radiant, but he stated that the equipment was not capable of resolving the individual radiants displaying activity.[30]

Both sessions of the Radio Meteor Project, conducted by Zdenek Sekanina at Havana, Illinois, during the 1960s, detected the Pi Virginids. The first session occurred during 1961-1965. It indicated the shower's duration was February 23-March 13, with a nodal passage on March 5.8 (λ=344.7°), and an average radiant of α=182.4°, δ=+7.4°.[31] The second session was conducted during 1968-1969. On this occasion, the duration was determined as February 13-April 8, while the nodal passage came on March 9.0 (λ=347.9°). The average radiant was α=184.1°, δ=+0.5°.[32]

Recent observations of this shower have been made by members of the Western Australia Meteor Section. In 1979, shower members were detected during March 2-5, with a maximum ZHR of 2.28±1.02 coming on March 3, from α=182°, δ=+7°. In 1980, shower members were detected during March 2-10. A maximum ZHR of 5.38±3.11 came on March 3, from α=183°, δ=+9°.[33]

Orbit

Orbits have been determined for this stream on four occasions—all during the 1960's and all during radio-echo surveys. Kashcheyev and Lebedinets' 1960 survey was based on 9 meteors. Nilsson's 1961 survey orbit was based on 3 radio meteor orbits. Sekanina's 1961-1965 survey was based on 9 meteors and his 1969 survey was based on 60 meteors.

	ω	Ω	i	q	e	a
KL1967	297	356	6	0.36	0.82	1.94
N1964	304.3	354.5	2.9	0.26	0.89	2.38
S1973	299.7	344.7	9.8	0.325	0.826	1.869
S1976	303.9	347.9	2.9	0.289	0.839	1.792

Theta Virginids

Observer's Synopsis

The duration of this shower extends from March 10 to April 21. A maximum of 1-3 per hour seems to occur around March 20, from a radiant of α=194°, δ=–2°. The radiant possesses a daily motion of +0.90° in α and –0.31° in δ.

---

[30]N1964, pp. 226-228 & 239.

[31]S1973, pp. 255 & 258.

[32]S1976, pp. 275 & 292.

[33]Wood, Jeff, Personal Communication (October 15, 1986).

History

The first person to recognize that this stream produces an active annual shower was Cuno Hoffmeister. In his 1948 book, *Meteorströme*, his analysis of visual radiants revealed a large group of ecliptic showers visible during March and April, which he referred to as the Virginid complex. He demonstrated that this large, diffuse stream was composed of nine distinct radiants, with distinct dates of maximum. Although it had long been known that abundant activity was present from the Virginid area during March and April, Hoffmeister's study was the first to attempt to identify the individual streams. The strongest supported shower of this complex belongs to the Theta Virginids and was based on 12 individual radiant determinations. It was most active on March 15 (λ=355.7°), from α=192.4°, δ=–1.5°.[34]

The first apparent observation of the Theta Virginids was made by Robert P. Greg and A. S. Herschel during the interval of 1850-1867, when it became apparent that a radiant was active during March 5-17, from α=190°, δ=+1°.[35] Ronald A. McIntosh also recognized this stream in his classic paper "An Index to Southern Meteor Showers." Although relying on only two visual radiants, McIntosh noted a shower, which he referred to as the "46 Virginids," was active during March 19-31 from an average radiant of α=194°, δ=–3°.[36]

Recent observations of the Theta Virginids were made by the Western Australian Meteor Section. In 1977, meteors were observed during March 18-19. Maximum came on the latter date, when the ZHR was 3.17±2.24 and the radiant was at α=196°, δ=–1°. In 1979, the shower was observed during March 16-31. Maximum came on the 20th, when the ZHR peaked at 2.61±0.87 from an average radiant of α=196°, δ=–1°. During 1980, observations of the shower were made during March 19-23. At maximum on March 21, the ZHR reached 1.39±0.46, while the average radiant was determined as α=196°, δ=–1°.[37]

The orbit of this stream was finally determined by Zdenek Sekanina, while evaluating the radio-echo data collected during the two sessions of the Radio Meteor Project. The 1961-1965 data revealed a stream, called the "Northern Virginids," which possessed a duration of March 9-April 5. The nodal passage came on March 20.6, at which time the radiant was at α=201.6°, δ=–3.3°.[38] Curiously, Sekanina's computer analysis revealed two streams among the 1968-1969 data, which possessed nearly identical orbits (see below). The first stream, designated the "Southern Virginids," possessed a duration of March 10-April 17.

---

[34]H1948, p. 138.

[35]D1899, p. 260.

[36]M1935, p. 713.

[37]Wood, Jeff, Personal Communication (October 15, 1986).

[38]S1973, pp. 255 & 258.

The stream's nodal passage came on March 16.4, at which time the radiant position was α=196.0°, δ=–1.0°. The second stream, called the "Northern Virginids," possessed a duration of March 10-April 21. Its nodal passage was given as April 2.1, with the average radiant being at α=210.5°, δ=–8.4°.[39]

The Author examined the raw orbital data from both the 1961-1965 and 1968-1969 surveys and subjected it to a fairly restrictive D-criterion of 0.10. The result was 31 radio meteors that are definite members of the Theta Virginids. These meteors indicate the radiant's daily motion is +0.90° in α and –0.31° in δ. This motion seems to connect the three radiant positions given above in Sekanina's data. At first glance, this might indicate multiple peaks in the activity levels; however, the Author notes that, in 1969, there were two critical periods when the radar equipment was inactive. These periods were March 20-22 and March 29-April 6, and, since these were situated during the time the stream should have been producing its maximum activity, the two 1969 streams should probably be combined.

Orbit

Three orbits were included in Sekanina's two papers which may relate to the Theta Virginids. The first orbit represents that of the stream detected during the 1961-1965 survey, which was referred to as the "Northern Virginids." The other orbits are from the 1968-1969 survey and possess nearly identical durations, though the dates of nodal passage and the radiant positions were different. As discussed earlier, the motion of the radiant almost perfectly connects the radiants of each stream.

ω	Ω	i	q	e	a
315.7	359.5	8.4	0.222	0.833	1.329
310.6	355.3	6.4	0.288	0.759	1.196
310.3	11.9	4.8	0.278	0.785	1.295

The Author's examination of the radio-meteor orbits from both the 1961-1965 and 1968-1969 revealed 31 members of the Theta Virginids.

ω	Ω	i	q	e	a
312.7	4.7	6.4	0.252	0.807	1.306

Additional March Showers

March Aquarids

This daytime meteor shower was first detected by C. S. Nilsson (Adelaide Observatory) during 1961. Radar equipment was operated during March 11-16

[39]S1976, pp. 276 & 292.

and three meteors were detected during March 12-16 from an average radiant of α=339.5°, δ=–7.6°. Nilsson suggested the stream was closely related to the Northern Iota Aquarid stream (see August).[40]

The stream was next detected during March 16-22, 1969, by G. Gartrell and W. G. Elford (Adelaide Observatory). Seven meteors were detected from an average radiant of α=338°, δ=–8°. The authors concluded that, although there was some discrepancy between the ascending node of the March stream and that of the July stream, "The correspondence of the longitudes of perihelion is excellent." They added that since the July stream was apparently broad, a link with the March stream "may still be acceptable."[41]

Radio-echo equipment at Adelaide Observatory has produced two available orbits for this stream.

	ω	Ω	i	q	e	a
N1964	59.7	353.7	2.5	0.298	0.86	2.128
GE1975	42	359	1.8	0.18	0.89	1.695

The Author has examined the original 39,145 radio meteor orbits obtained during the two sessions of the Radio Meteor Project. Although Zdenek Sekanina gives no orbit corresponding to this stream for either February, March or April, it seems the March Aquarids were present. The data indicates a very diffuse stream that begins in February and ends in April. Both of the above orbits could easily be represented among these radio meteor orbits and this might indicate that two or more filaments are present.

Leonids-Ursids

Although visual radiants of this stream are a distinct rarity, it is interesting that its strongest support for existing is based on several photographic meteors detected during the 1950's.

In all, seven meteors were detected by cameras operating in the United States and Czechoslovakia, with individual details subsequently being reported in H1959, MP1961 and C1977. The indicated duration covers March 18 to April 7, while the average radiant is α=175.7°, δ=+23.0°. The resulting orbit is

ω	Ω	i	q	e	a
240.0	2.0	9.4	0.793	0.709	2.721

Zdenek Sekanina's 1968-1969 session of the Radio Meteor Project also detected this stream. A very short duration of only 1.1 day was indicated, centered on March 10.6. The radiant was located at α=166.6°, δ=+28.1°.[42] The subsequent average radio-echo meteor orbit is given on the next page.

[40]N1964, pp. 226 & 240-241.
[41]GE1975, p. 606.
[42]S1976, pp. 275 & 292.

	ω	Ω	i	q	e	a
S1976	241.8	349.5	8.5	0.814	0.518	1.691

There is a possibility that a radiant noted by Alexander S. Herschel during March 2-28, 1860-1881, might be associated with this stream. The average radiant position was given as α=162°, δ=+24°.[43]

Rho Leonids

Although visual observations of this shower seem virtually nonexistent, support for this stream appears in two radar studies conducted during the 1960's, as well as five photographic meteors detected over the period of 1937 to 1954.

The first official detection of this stream was made by B. L. Kashcheyev and V. N. Lebedinets (Kharkov Polytechnical Institute) during a radar survey conducted in 1960. The radar was not operated continuously, but five meteors from this stream were detected during March 14-15, from an average radiant of α=172°, δ=+3°.[44] The stream was referred to as the "Sigma Leonids," and the orbit was given as

ω	Ω	i	q	e	a
83	175	0.1	0.64	0.67	1.95

The stream was next detected by Zdenek Sekanina during the 1968-1969 session of the Radio Meteor Project. The data revealed a duration of February 13 to March 11. The nodal passage came on March 3.4 (λ=342.3°), at which time the radiant position was given as α=160.9°, δ=+7.3°.[45] The orbit is as follows:

ω	Ω	i	q	e	a
84.7	162.3	0.5	0.618	0.711	2.138

The Author has examined the photographic data of the last few decades and has found five meteor orbits published in W1954 and MP1961, which were detected between 1937 and 1954. The indicated duration was February 17-March 13, while the average radiant was α=156.9°, δ=+5.3°. The following orbit was determined.

ω	Ω	i	q	e	a
81.8	159.7	3.8	0.606	0.832	3.607

[43] D1899, p. 257.
[44] BL1967, p. 188.
[45] S1976, pp. 275 & 292.

Chapter 4: April Meteor Showers

Tau Draconids

Observer's Synopsis

The duration of this stream extends from March 13-April 17. Maximum occurs over the period of March 31-April 2, from an average radiant of α=285°, δ=+69°. Visual observations reveal low activity from a radiant 2°-5° across.

History

This stream was first recognized in a 1973 study involving 2401 photographic meteor orbits obtained during the Harvard Meteor Project of 1952-1954. The study was conducted by Allan F. Cook, Bertil-Anders Lindblad, Brian G. Marsden, Richard E. McCrosky and Annette Posen. They identified four meteor orbits, which formed the "Delta Draconids." Possessing a duration covering March 28-April 17, the average radiant position was given as α=281°, δ=+68°.[1]

Shortly after the above photographic data was published, Zdenek Sekanina published the third paper in his four-part "Statistical Model of Meteor Streams" series. Covering the period of 1961-1965, it listed 72 minor meteor streams, one of which was the "Tau Draconids." This stream was given a duration of March 24-April 12. The nodal passage came on April 1.7 (λ=11.5°), at which time the radiant position was α=291.6°, δ=+71.3°.[2] The last paper of that series was published in 1976 and covered the period of 1968-1969. The Tau Draconids were again detected—this time with a duration extending from March 12-April 12. The date of the nodal passage was given as March 27.0, while the average radiant was α=286.4°, δ=+69.1°.[3]

The first visual observations of this shower were actually occurring while Sekanina's second session of the Radio Meteor Project was in progress. On March 25.92, V. K. Leichenok, N. S. Malikov, L. M. Afanas'eva and T. A.

---

[1]C1973, pp. 1-5.
[2]S1973, pp. 255 & 258.
[3]S1976, pp. 276 & 292.

Kopycheva plotted more than 10 meteors from a radiant of α=280.0°, δ=+73.0°. The radiant diameter was determined as 5.0°.[4]

As a follow-up to the 1969 observations, the Russian observers strove to confirm the many radiants they had found by conducting another extensive visual survey in 1973. On March 26, S. V. Safonov plotted three meteors from a 1.5°-diameter radiant at α=272°, δ=+72°. During March 27-29, N. V. Smirnov plotted six meteors from a 2.0°-diameter radiant at α=285°, δ=+71°.[5]

Orbit

The orbit determined in C1973 was based on four photographic meteors and is as follows:

ω	Ω	i	q	e	a
171.1	13.7	37.5	0.996	0.724	2.770

The orbits determined by Sekanina during the two sessions of the Radio Meteor Project are listed below. The first is from S1973, while the second is from S1976.

ω	Ω	i	q	e	a
166.3	11.5	30.9	0.985	0.533	2.109
169.0	5.8	33.0	0.988	0.542	2.156

Lyrids

Observer's Synopsis

The Lyrids are typically visible between April 16 and 25. Maximum occurs during April 20-21 (λ=31.4°), from an average radiant of α=272°, δ=+33°. Although the maximum ZHR is about 10, there have been instances during the last 200 years when rates were near or over 100 per hour. The average magnitude of the meteors is near 2.4 and the speed is described as rapid. About 15% of the meteors leave persistent trains.

History

Interest in this meteor shower was very slow to develop due to the relative infancy of meteor astronomy. A storm of about 700 meteors per hour had been observed by numerous people in the eastern part of the United States during April 19-20, 1803, but no further attention was given to this shower until 1835. That year closely followed the discovery that the Leonids of November were an annual shower, and as astronomers struggled to identify other annual meteor

[4]KM1971, p. 100.
[5]SK1975, p. 99.

showers, Dominique François Jean Arago (1786-1853) conjectured that April 22 might be a date of frequent meteor activity.[6]

Much of the leg work for confirming Arago's remark should be credited to Edward C. Herrick (New Haven, Connecticut), who, during 1839, not only carried out coordinated observations of this meteor shower, but also collected accounts of the activity in 1803. Herrick also uncovered further appearances on April 9.6, 1095, April 10, 1096, and April 10.6, 1122.[7] His visual observations (made in conjunction with Francis Bradley) proved that weak, but definite, activity was present during 1839, with the radiant for April 19 being α=273°, δ=+45°.[8] Despite this apparent confirmation, the Lyrids were again ignored until April 19-20, 1864, when Professor Alexander Stewart Herschel observed 16 meteors from a radiant of α=277°, δ=+35°.[9] This observation preceded a new wave of interest in meteor showers in general—an interest that again encouraged observations of the Lyrids.

During 1866, the annual Perseid shower had been linked to periodic comet Swift-Tuttle (1862 III) and the Leonids were linked to the newly discovered periodic comet Tempel-Tuttle (1866 I). As 1867 began, astronomers were still busy seeking further evidence linking meteor showers to comets. In Vienna, Professor Edmond Weiss was busy calculating probable close encounters between Earth and comet orbits. One comet orbit, that of Thatcher (1861 I), was found to come within 0.002 AU of Earth's orbit on April 20. As Weiss searched through various publications for evidence of this shower's presence, he came across several references to observed showers around April 20.[10] Later that same year, Johann Gottfried Galle mathematically confirmed the link between comet Thatcher and the Lyrids and he successfully traced the history of the shower back to March 16, 687 BC.[11]

Observations of the Lyrids increased during the late 1860s and early 1870s, and, as has been the case with so many meteor showers, William F. Denning played an important role in the understanding of this shower. By 1885, he had obtained evidence that the radiant of this shower moved about one degree eastward each day. By 1923, the evidence had become so convincing, that Denning published the following radiant ephemeris:[12]

---

[6]Olivier, Charles P., *Meteors*. Baltimore: The Williams & Wilkins Company (1925) p. 64.
[7]Olivier, pp. 63-64.
[8]Herrick, Edward C., *AJS*, **35** (1839), pp. 362-363.
[9]D1899, p. 272.
[10]Weiss, E., *AN*, **68** (1867), p. 382.
[11]Galle, J. G., *AN*, **69** (1867), p. 33.
[12]D1923B, pp. 43-56.

Lyrid Ephemeris

Date	RA(°)	DEC(°)
April 10	259	34
April 12	262	34
April 14	264	33
April 16	266	33
April 18	269	33
April 20	271	33
April 22	274	33
April 24	276	33
April 26	278	33
April 28	281	34
April 30	284	34

Denning admitted to only having seen Lyrids between April 14 and 26, but was convinced further, very weak traces of activity might be present outside of that range—hence, the extended radiant ephemeris. Visual studies have thus far failed to provide convincing evidence of this extended activity; however, during the period of 1961-1965, the Radio Meteor Project, under the directorship of Zdenek Sekanina, detected probable members of this stream up through May 3.[13]

Aside from the abnormal activity of 1803, the maximum hourly rates have remained relatively consistent from year to year, though there have been other unexpected outbursts. Denning pointed out that in 1849 and 1850, observers in New Haven and India, respectively, noticed "unusual numbers" of meteors on April 20.[14] Denning himself observed a maximum hourly rate of 22 during his observations of 1884,[15] H. N. Russell (Greece) found a rate of 96 on April 21, 1922,[16] Koziro Komaki (Nippon Meteor Society, Japan) saw 112 meteors (most were Lyrids) in 67 minutes on April 22, 1945,[17] and several observers in Florida and Colorado noted rates of 90-100 on April 22, 1982.[18]

Several observers have attempted to estimate the orbital period of this meteor stream from the visual observations above. Herrick concluded from his historical study of Lyrid activity that the stream possessed an orbital period of 27 years. Based on the activity observed in 1803 and 1850, Denning concluded that the Lyrids possessed an orbital period of 47 years,[19] but his prediction of

---

[13]S1970, p. 477.
[14]Denning, W. F., *Observatory*, **20** (April 1897), pp. 174-175.
[15]Denning, W. F., *Observatory*, **7** (1884), p. 217.
[16]Hindley, Keith B., *JBAA*, **79** (1969), p. 479.
[17]Olivier, Charles P., *FOR*, No. 67 (1947), p. 22.
[18]Marsden, Brian G., *IAU Circular*, No. 3691 (April 28, 1982).
[19]Denning, W. F., *Observatory*, **20** (April 1897), pp. 174-175.

possible enhanced activity in 1897 was met by rates not exceeding 6 per hour. After the outburst in 1982, many researchers remarked that the period was about 60 years, based on the showers of 1803, 1922 and 1982. Unfortunately none of these suggested orbital periods fit the observations perfectly, and it might be possible that the Lyrid orbit contains several irregularly spaced knots of material that could make it impossible to arrive at an accurate period based on visual observations.

Using the more precise methods of radar and photographic techniques, several attempts have been made to determine the period of the Lyrid stream. A collection of photographic orbits published by Fred L. Whipple in 1952, revealed two "reliable" Lyrid meteors with periods differing by 300 years![20] In 1971, Bertil-Anders Lindblad published a Lyrid stream orbit, which had a period of 131 years, that was based on 5 meteors photographed during 1952 and 1953,[21] and, in 1970, Sekanina published a Lyrid stream orbit based on radio meteors which had an average period of 9.58 years.[22]

The discrepancy in the orbital period of the Lyrids is primarily due to a lack of data. The number of meteors obtained from the major lists of photographic meteors totals 12, with only 6 being considered reliable (and, incidentally, giving a period of 139 years—close to Lindblad's despite sharing only 2 meteors). Comet Thatcher's period of 415 years is probably much more reliable today than the computed orbital period of the Lyrids.

The Lyrids are known to possess a sharp peak of maximum activity—a feature generally exhibited by meteor streams which are young or not prone to serious planetary perturbations. Since the inclination of the comet's orbit is 79.8°[23] and since evidence exists showing activity as long ago as 687 BC, then the latter scenario seems most appropriate. Typically, the time of maximum occurs around solar longitude 31.6°, with other well-documented visual observations falling within the range of 31.4° to 31.7°. The earlier mentioned study of photographic orbits by Lindblad gave a value of 31.6°, while Sekanina's radar study gave 32.0°. All of these tend to indicate a much more pronounced peak of maximum activity than is generally present for other meteor streams.

In 1969, Keith B. Hindley pointed out that the close agreement of the maximums of both visual and photographic meteors "indicates that there is no evidence which could be interpreted as the result of the action of dispersive

---

[20]W1954, pp. 201-217.

[21]Lindblad, Bertil-Anders, *SCA*, **12** (1971), pp. 14-24.

[22]S1970, p. 476.

[23]Marsden, Brian G., *Catalog of Cometary Orbits*. New Jersey: Enslow Publishers (1983), p. 13.

forces of the Poynting-Robertson type."[24] As support for this statement, Hindley added that occasional observations extending back to 687 BC indicate there has been little or no motion in this stream's orbital nodes for at least 2600 years!

The Author generally supports Hindley's view of the great age of the Lyrid stream after noting that, despite a lack of serious planetary perturbations, Lyrid activity is visible every year. Thus, particles within the stream have spread completely around the orbit—though, admittedly, somewhat unevenly. It is, however, curious that such a large difference exists between the determined orbital period of the Lyrids when photographic and radar data are considered. The photographic data indicates a period over 100 years, while the radar data indicates a period of about 10 years. Such a discrepancy can only be explained by either some minor presence of the Poynting-Robertson effect or a serious error in the determination of the radar orbital data.

Recent studies of the characteristics of the Lyrids, have revealed many interesting features. In 1972, observations by members of the Moscow planetarium during April 16-20, revealed an average magnitude of 3.3.[25] Hindley's analysis revealed an average magnitude of 2.09 in 1969. Further average magnitudes, as well as the percentages of meteors leaving persistent trains are as follows:

Lyrid Magnitudes and Trains

Year(s)	Ave. Mag.	# Meteors	% Trains	Observer(s)	Issue
1960-1976	2.65	277	—	N. McLeod	*MN*, No. 32
1971-1984	2.90	246	—	N. McLeod	Personal Comm.
1974	2.43	107	15.9	5 in Calif.	*MN*, No. 21
1977	2.86	58	32.8	N. McLeod	*MN*, No. 37
1977	2.70	61	16.4	F. Martinez	*MN*, No. 37
1982	3.62	79	—	N. McLeod	*MN*, No. 58
1985	2.12	55	—	B. Katz	*MN*, No. 70
1985	2.77	240	10.4	DMS	*Radiant*, **7**

Observers in the above table are Norman W. McLeod, III (Florida); Felix Martinez (Florida); Bill Katz (Canada); Dutch Meteor Society.

Although some of the variation in average magnitude can be attributed to observing conditions, several of the observers were under skies with limiting zenithal magnitudes of 6.5, so that the difference could also be due to a mass variation within the stream. Such was suggested by V. Porubcan and J. Stohl in

[24]Hindley, Keith B., *JBAA*, **79** (1969), pp. 477-480.
[25]Demidova, I. G., *SSR*, **8** (1974), pp. 52-53.

1983, after analyzing visual observations obtained by observers at Skalnate Pleso Observatory in 1945, 1946, 1947 and 1952. They noted that an abnormally strong maximum of 40 meteors per hour in 1946, was characterized by an increase in the number of fainter stream members.[26] Similarly, the very strong return of 1982 was also characterized by an increase in faint stream members, with the average brightness dropping by almost one full magnitude from what other years typically possess.

Recent estimates of the shower's *normal* hourly rate seem to show no noticeable change from what observers were reporting 80-90 years ago, with ZHRs typically between 8 and 15. Recent estimates have been 13.5 in 1969, 17 in 1974,[27] and 13.1 in 1985.[28]

The duration of this shower is fairly short. Four amateur astronomers from southern California (Alan Devault, Terry Heil, Greg Wetter and Bob Fischer) observed the Lyrids during April 20 to 24, 1974, and concluded that the shower remained above 1/4 its maximum rate for 3.6 days.[29]

Orbit

A photographic orbit, based on six reliable photographic meteor orbits obtained from W1954, JW1961 and C1964, was obtained by the Author, while a radar orbit comes from the 1961-1965 session of the Radio Meteor Project, conducted by Zdenek Sekanina.

	ω	Ω	i	q	e	a
Photo	214.2	32.0	79.6	0.920	0.966	26.804
Radar	215.3	32.0	76.9	0.922	0.796	4.511

The orbit of comet Thatcher (1861 I) is

ω	Ω	i	q	e	a
213.5	31.2	79.8	0.92	0.98	55.68

April Piscids

Observer's Synopsis

This daylight shower is detectable during the period April 8 to 29. It reaches maximum around April 20 (λ=29°) from an average radiant of α=7°, δ=+5°. Due to inconsistencies in the quantity of meteors detected in four separate radar surveys, this stream may be periodic in nature.

---

[26]*Meteoros*, 14 (May 1984), p. 29.

[27]*MN*, No. 21 (June 1974), p. 3.

[28]Veltman, Rudolf, *Radiant*, 7 (July-August 1985), pp. 79-81.

[29]Fischer, Bob, *MN*, No. 21 (June 1974), pp. 2-3.

History

This meteor stream was discovered by B. L. Kashcheyev and V. N. Lebedinets (Kharkov Polytechnical Institute, USSR) during a radar survey in 1960. Thirty-four radio meteors were detected during April 15-25, with the average radiant being α=7°, δ=+3°. The date of the stream's nodal passage was given as April 20/21.[30]

This stream was again detected during 1961, while C. S. Nilsson (University of Adelaide, South Australia) was conducting a radio survey of southern hemisphere meteor streams. Only three meteors were noted during the interval of April 13 to 29. The indicated date of the nodal passage was April 18/19, at which time the radiant was located at α=6.5°, δ=+4.3°. Nilsson said his stream actually did not qualify as a group, due to the excessive "scatter in the values obtained for the right ascension;" however, he noted a close agreement between the orbits of this stream and a stream detected in August, which the Author has identified with the Northern Iota Aquarids.[31]

It is surprising that this stream was not recognized by Zdenek Sekanina in either of the two sessions of the Radio Meteor Project. In an attempt to discover why, the Author examined the 39,145 radio meteors orbits determined by Sekanina. The radio equipment at Havana, Illinois, operated during April of 1962-1965 and in 1969. Thirteen probable members of the April Piscids are present in the sample. These meteors indicate a duration extending from April 8 to April 26. The date of the nodal passage is determined as April 19/20 (λ=29.7°), at which time the radiant is at α=7.4°, δ=+7.2°. What is most interesting is the yearly distribution: 5 meteors in 1962, 1 in 1963, 4 in 1964, 2 in 1965, and only 1 in 1969.

When Sekanina's data is compared with that obtained in the Russian and Australian surveys discussed earlier, it appears that the Russian data was based on an uncharacteristic return of this stream, with 34 meteors being detected. In fact, the 1968-1969 session of the Radio Meteor Project involved the most sensitive equipment ever used in radar meteor surveys and April was well covered. The fact that only 1 meteor was detected in 1969, indicates this daylight stream may be periodic.

Orbit

Two radar studies had determined orbits for this stream. During 1960, B. L. Kashcheyev and V. N. Lebedinets determined the first orbit based on 34 meteors (KL1967). During the following year, C. S. Nilsson isolated 3 meteors (N1964).

---

[30]BL1967, p. 188.

[31]N1964, pp. 226-228 & 241.

	ω	Ω	i	q	e	a
KL1967	45	30	0.5	0.22	0.82	1.32
N1964	48.6	28.3	5.8	0.282	0.76	1.176

Although this stream was not recognized by Sekanina in either of the two sessions of the Radio Meteor Project, the Author examined the 39,145 radio meteor orbits obtained during these sessions and has found 13 probable members of this stream. The average orbit follows:

ω	Ω	i	q	e	a
49.1	29.7	4.7	0.263	0.809	1.380

The orbit of the Northern Iota Aquarids of August, as determined by Nilsson during August 16-24, 1961, is as follows:

	ω	Ω	i	q	e	a
N1964	310.4	146.0	7.9	0.301	0.75	1.205

Pi Puppids

Observer's Synopsis

The duration of this shower extends from April 18 to April 25. The short-duration maximum occurs during April 23-24, from a radiant of α=112°, δ=−43°. The shower is associated with the periodic comet Grigg-Skjellerup. Although this comet was officially discovered in 1902, it was only recently perturbed by Jupiter into a close-approach orbit with Earth. Activity was first noted in 1972, and visual hourly rates of 18 to 42 meteors per hour were noted during the comet's perihelion returns of 1977 and 1982. Activity levels are typically very low or nonexistent in other years.

History

During 1971, H. B. Ridley was examining the predicted orbital elements for the coming return of periodic comet Grigg-Skjellerup, when he noticed that Earth would make a close approach to the comet's orbit on April 23.02, 1972. The separation between the orbits was calculated as only 0.004 AU, while the encounter was to occur only 50 days after the passage of the comet. The predicted radiant was α=107.5°, δ=−45°.[32]

Observations made during the predicted appearance of this meteor shower revealed a very poor display. During the period of April 16-23, 17 observers in the United States obtained average hourly rates of only 1.9, with a maximum of about 4 per hour being observed by B. Edwards (Jacksonville, Florida) during a

[32]Marsden, Brian G., *IAU Circular*, No. 2371 (November 22, 1971).

three hour interval on April 18/19.[33] Observers in West Australia were met with even weaker activity, as 7 observers compiled 70 hours of searching during April 21-24, only to detect three possible shower members during an eight-hour interval on the night of April 22/23.[34] A more positive Southern Hemisphere observation was made by W. J. Baggaley (University of Canterbury, New Zealand), who utilized radio equipment in his search. He detected an "increase in the rate of radio-meteor echoes over the normal sporadic activity on the four days 1972 April 21, 22, 23 and 24...." He added that the observed activity flux was consistent with a radiant at α=107.5°, δ=–45°, though the rates were considered too low for an accurate radiant determination.[35]

Comet Grigg-Skjellerup was next expected at perihelion in 1977. Predictions for a possible meteor shower in this year had actually been first made by G. Sitarski during 1964. He said that on April 23.3, 1977, Earth would cross the comet's orbit just 12 days after the comet, and that activity would probably emanate from α=109.6°, δ=–44.3°.[36] Once again, observers in the United States were not successful in observing activity, but circumstances were different in West Australia.

Jeff Wood, A. Saare and G. Blencowe, observing in Perth, West Australia, individually observed maximum rates of 18-24 meteors per hour during a three-hour interval centered on April 23.5, 1977. Numerous meteors were plotted, which revealed a radiant of α=112°, δ=–43°, and the ZHR was calculated as 36.47±2.61. The overall duration of the activity was given as April 23-25. The meteors were typically bright and slow.[37]

A weak return of the Pi Puppids was observed by the West Australian meteor observers during 1979—a time when the comet was actually nearing its greatest distance from the sun. Activity was noted from April 21 to 24, with a maximum ZHR of 3.54±1.77 coming on April 23. The average radiant was given as α=112°, δ=–43°.[38]

The comet's next perihelion passage came on May 14, 1982. A very strong return of the Pi Puppids was observed on the night of April 23/24. The first detection of increased activity was made by A. Gozalos Beltran (Cochabamba, Bolivia) when 58 meteors were detected during a period of 1 hour 35 minutes. He described the meteors as being predominantly yellow. A strong return was also noted in West Australia. Individuals reported 25-42 meteors per hour, with

---

[33]*MN*, No. 11 (June 1972), p. 6.

[34]*MN*, No. 12 (August 1972), p. 5.

[35]Baggaley, W. J., *The Observatory*, **93** (February 1973), pp. 23-26.

[36]Marsden, Brian G., *IAU Circular*, No. 3055 (March 30, 1977).

[37]Wood, Jeff, Personal Communication (October 15, 1986).

[38]Wood, Jeff, Personal Communication (October 15, 1986).

the ZHR reaching 22.8 on April 24.49. By April 24.56, the ZHR had dropped to 7.1. The West Australian observers reported that 56.5% of the meteors were yellow, while 19.6% were orange. Trains were observed among 16.1% of the meteors and the average magnitude of 447 meteors was 1.97.[39] It is interesting that one year later, West Australian observers detected a maximum ZHR as high as 12.7 on April 23/24, and estimated the average meteor magnitude as 2.33.[40]

Orbit

This meteor stream is definitely associated with periodic comet Grigg-Skjellerup and is very new, as is evidenced by the near total lack of activity in years when the comet is not at perihelion. Therefore the difference between the orbits of the comet and meteor stream is negligible. Grigg-Skjellerup last passed perihelion in 1982 and possessed the following orbit:

ω	Ω	i	q	e	a
359.3	212.6	21.1	0.989	0.666	2.959

April Ursids

Observer's Synopsis

The duration of this shower is fairly long—extending from March 18 to May 9. At maximum on April 19, the average radiant is at α=149°, δ=+55°.

History

The April Ursids were observed extensively during the last third of the 19th century and into the 20th century, but their current appearance seems primarily limited to occasional bright meteors and telescopic activity.

The April Ursids were first observed by Eduard Heis (Münster) during April 16-30, 1849-1861, when he determined the radiant position as α=150°, δ=+61°.[41] However, the most extensive set of observations was made by Giuseppe Zezioli (Bergamo, Italy), who plotted meteors from this shower during both 1868 and 1869. According to Giovanni Virginio Schiaparelli's analysis, Zezioli observed 10 meteors from α=163°, δ=+47° on April 10, 1869, another 10 from α=168°, δ=+47° on April 14, 1868-1869, and 12 meteors from α=142°, δ=+53° on April 25, 1868.[42] All of these radiants possessed D-criterions of 0.04 to 0.10.

---

[39]Simmons, Karl, *MN*, No. 58 (July 1982), pp. 7-8.
[40]*MN*, No. 63 (October 1983), p. 9.
[41]Newton, H. A., *AJS (2nd Series)*, **43** (1867), p. 286.
[42]D1899, pp. 254 & 258.

Although a few additional radiants were detected during the 1870's, only occasional fireball and stationary meteor observations were made during the next 30 years. Finally, during April 14-18, 1915, William F. Denning plotted four meteors from α=161°, δ=+58°.[43] Interestingly, neither *Meteorströme*, the American Meteor Society, or any other source has revealed any further visual radiant from this stream.

Individual meteors continue to appear from this stream. Six photographic meteors were detected in three surveys during the period of 1950 to 1969. The indicated duration is April 7 to 23, with an average radiant of α=173.7°, δ=+59.0°. Several fireballs have also been noted, which prompted Charles P. Olivier to include a radiant from this stream in his "Catalogue of Fireball Radiants." Possessing an average position of α=167°, δ=+63°, radiant number 5053 had an estimated endurance of six days centered on April 25.[44]

The most convincing modern-day support for this stream's existence comes from the 1968-1969 session of the Radio Meteor Project. Zdenek Sekanina detected a stream active over the period of March 18-May 9. The radar stream's nodal passage was given as April 18.7 (λ=28.2°), at which time the radiant position was α=149.3°, δ=+54.9°.[45]

Orbit

Based on 6 meteors obtained from W1954, MP1961 and CF1973, the following orbit was obtained:

ω	Ω	i	q	e	a
188.4	25.3	10.7	1.000	0.532	2.137

During the 1968-1969 session of the Radio Meteor Project, conducted at Havana, Illinois, Sekanina isolated 21 meteors with the following orbit:

	ω	Ω	i	q	e	a
S1976	183.5	28.2	9.4	0.993	0.473	1.865

Alpha Virginids

Observer's Synopsis

Typical of streams lying near the ecliptic, the Alpha Virginids show evidence of a long duration—spanning from March 10 to May 6—and a diffuse radiant. Maximum hourly rates typically reach between 5 and 10 during April 7 to 18, with the average radiant being α=204°, δ=–11°. The meteors are generally slow.

---

[43]D1923B, p. 51.

[44]Olivier, Charles P., *FOR*, No. 146 (1964), p. 13.

[45]S1976, pp. 277 & 293.

History

This shower was apparently first detected by A. S. Herschel during April 10-17, 1895, when 8 meteors were plotted from a radiant of α=209°, δ=–7°.[46] Aside from 6 fireballs that were noted from this radiant between 1896 and 1915, no other shower activity was noted until April 12-15, 1915, when William F. Denning saw a "rich shower" from α=209°, δ=–10°.[47]

The Alpha Virginids represent the most active discernible shower that occurs from the Virginid complex of meteor streams. The shower possesses one of the longest known durations for any meteor stream and may actually represent the main core of the Virginid stream from which most of the other weaker showers originated.

Studies of the Virginid complex have revealed several unusual characteristics. The British Meteor Society (BMS) conducted a study during 1972 which revealed an excess of meteors between magnitude 0 and 5 as compared to the sporadic background.[48] Also pointed out was the fact that "different sub-centres are active to a different extent each year...." The BMS also showed that the shower's sub-centers rarely produced hourly rates greater that 8-10, but rates of 5 per hour were more common. Robert A. Mackenzie (director of the British Meteor Society) wrote in 1980, that the Virginid complex involved "16 main radiant centres whose periods of activity overlap."[49] In the BMS Radiant Catalogue, Mackenzie mentions that the Alpha Virginids possess an maximum ZHR of 10.

Jeff Wood said observers of the Western Australia Meteor Section obtained excellent observations of this shower in 1979 and 1980. In the former year the ZHR reached 4.11±1.01, while the 1980 maximum reached 6.09±0.59. Wood summarized the Australian observations by pointing out that the Alpha Virginids typically produce 5-10 meteors per hour, with a maximum that more commonly falls on April 11. The duration of activity extends from March 30 to April 17.[50] Other meteor showers in the "Southern Hemisphere Meteor Stream List," supplied by Wood, and distributed by the meteor section of the National Association of Planetary Observers of Australia, possess durations that fall either before or after that quoted for the Alpha Virginids and even have radiants that lie within a few degrees; however, they possess different dates of maximum activity and are, therefore, listed as separate showers.

---

[46]D1899, p. 262.

[47]D1923B, p. 52.

[48]Mackenzie, Robert A., *Solar System Debris*. Dover: The British Meteor Society (1980), pp. 33-34.

[49]Mackenzie, p. 33.

[50]Wood, Jeff, Personal Communication (October 24, 1985).

An excellent example of the Virginid complex' numerous irregularly active sub-centers is demonstrated by the following list of nine visual active showers published by Cuno Hoffmeister in 1948:[51]

Radiants of the Virginid Complex

Solar Longitude	RA(°)	Dec(°)	# of radiants
344.3	178.8	+0.6	9
355.7	192.4	–1.5	12
1.6	186.0	–0.4	6
9.3	209.4	–0.4	5
12.6	199.3	–10.7	3
16.6	197.6	–6.1	6
24.3	205.0	–10.7	7
32.0	216.2	–10.8	4
44.7	219.2	–17.8	4

The radiant at solar longitude 24.3° represents the actual Alpha Virginid maximum as derived by Hoffmeister.

This stream was detected during the 1968-1969 session of the Radio Meteor Project. Zdenek Sekanina determined the duration as March 10-May 6. The date of the nodal passage was given as April 8.7, at which time the radiant was at α=203.6°, δ=–11.7°. This survey revealed the Alpha Virginids to be in a very low inclination (1.5°) orbit and this seems to explain the Author's puzzling findings when searching for photographic members of this stream.

Altogether, the Author identified 12 probable members of the Alpha Virginids among the 2,529 meteors listed in MP1961. The puzzle that developed involved the argument of perihelion and ascending node of the meteor orbits, as when the meteors were averaged it was unclear whether these two orbital elements should be 104.7° and 200.3°, or 284.7° and 20.3°, respectively. Since 75% of the meteor orbits indicated the latter set of elements, the Author would tend to trust these as indicating the true orbit, despite the fact that the former set agrees well with the radar data.

A study of this stream's evolution was published in 1973, by E. I. Kazimirchak-Polonskaya and A. K. Terent'eva.[52] Adopting the orbit of Harvard meteor number 7333 (included in the photographic meteor orbit listed below) as representing the orbit of this stream, 10 groups were created at various positions along the stream's orbit and then subjected to perturbations by 7 planets (Venus to Neptune) over the interval 1860-2060.

---

[51]H1948, p. 138.

[52]Kazimirchak-Polonskaya, E. I., and A. K. Terent'eva, *SA*, **17** (Nov.-Dec. 1973), pp. 368-376.

Examining the three most interesting groups, the Russian study demonstrated how Jupiter's influence caused the stream to variously approach and move away from Earth's orbit, as well as how the radiant moves by as much as 30° in both right ascension and declination. For group "I" of the study, it was found that the stream is within 0.085 AU of Earth's orbit during the period of 1895 to 1991, with the date of maximum moving between solar longitudes 18° and 25° (April 8 and 15) and the radiant executing "looped motions over a 3°x5° area...." Group "VI" was found to possess a very complicated motion which caused it to approach and recede from Earth's orbit three times during the interval examined. A diagram of the maximum activity's radiant was given for each year and it demonstrated how maximum could variously occur between solar longitudes 1.5° and 51.6° (March 22 to May 12). The authors pointed out that these old dates of maximum would persist—eventually causing activity to be present from this stream for a period of 51 days. Between 1860 and 2051, the radiant producing maximum activity for this group would be expected to move within an elliptical area with dimensions of 15°x31°. For the third group perturbations caused changes that generally lay between the two groups already discussed.

Overall, the authors concluded that "perturbations by Jupiter represent the principal factor governing the evolution of the meteor streams belonging to its family and the evolution of their radiants. The magnitude and character of these perturbations has a strong influence on the size of the shower radiation area and the length of time that the shower remains visible."[53] The calculated duration of the activity coincides very well with present observations, as does the large, diffuse area of the radiant.

Orbit

During 1969, Zdenek Sekanina conducted a radar study of meteor streams from Havana, Illinois. He detected the following orbit from meteors detected within the period of March 10 to May 6:

ω	Ω	i	q	e	a
106.7	198.4	1.5	0.477	0.691	1.541

The Author located 12 photographic meteors in MP1961 which appear to be members of the Alpha Virginids; however, due to the fact that the ecliptic separates the meteors into two groups, it is unclear whether the shower is occurring at the stream's ascending node or descending node. The two possible orbits are listed below. As can be seen, the former orbit agrees well with that obtained by Sekanina, but the Author notes that 75% of the meteors indicate the latter orbit.

[53]Kazimirchak-Polonskaya, E. I., and A. K. Terent'eva, p. 375.

ω	Ω	i	q	e	a
104.7	200.3	3.0	0.471	0.756	1.930
284.7	20.3	3.0	0.471	0.756	1.930

Gamma Virginids

Observer's Synopsis

This meteor shower occurs during the period of April 5 to 21 (λ=15° to 31°). Its maximum occurs during April 14 to 15, from an average radiant of α=185°, δ=–1°. The daily motion is about 0.70° in α and +0.16° in δ. Hourly rates are weak—perhaps less than 5 per hour.

History

The discovery of this meteor shower should be attributed to A. S. Herschel, who plotted 12 meteors from a radiant of α=192°, δ=+4° during April 10-11, 1864.[54] During April 16 to 25, 1895, W. Doberck plotted 7 meteors from a similar radiant at α= 190.5°, δ=+7°.[55]

Aside from a few scattered fireballs and radiants between 1895 and 1963, the decade containing the most observations of this stream was the 1930s. Cuno Hoffmeister, in his famous book *Meteorströme*, listed four radiants (numbers 1628, 2103, 3156 and 5266) observed during 1931 to 1934.[56] As a group, these radiants spanned across the period of solar longitudes 16.8° to 31.5°. Also during the same period, R. A. McIntosh observed two radiants which he combined into a radiant he designated number 91 in his "Index to Southern Hemisphere Meteor Showers". Occurring during the period April 12 to 16, the radiant moved from α=184°, δ=0°, to α=188°, δ=0°.[57]

The Author has accumulated 26 meteor orbits from the photographic catalogs and Sekanina's raw data obtained during the two sessions of the Radio Meteor Project. The average position of the radiant is α=185.0°, δ=–0.7°. The indicated daily motion is +0.70° in α and +0.16° in δ. When the shower first becomes active on April 5 (λ=15°) the radiant is at α=180.7°, δ=–1.7°, and when the last traces occur on April 21 (λ=31°) the radiant is at α=191.8°, δ=+0.9°. These values are, however, tentative as the radar meteors tend to fall from the southern part of the radiant, while the photographic meteors fall from the northern part. This setup may be due to the differences in the average semimajor

[54]D1899, p. 260.
[55]D1899, p. 260.
[56]H1948, pp. 214, 219, 230 & 251.
[57]M1935, p. 713.

axes of the two types of meteors or just a curiosity that develops due to a lack of data.

Among the several meteor showers emanating from the Virgo-Libra region during April, this meteor stream stands out from the rest due to its large perihelion distance of about 0.82 AU (compared to 0.1 to 0.4 AU for the rest of these streams). This stream is similar to the other streams as it occurs especially close to the ecliptic. Subsequently, both photographic and radar meteors tend to possess two orbits—each separated by 180° in the ascending node and argument of perihelion. The orbits represented below belong to the most commonly occurring orbit. The Gamma Virginids are sometimes included in lists of showers belonging to the Virginid complex of meteor showers.

Orbit

A photographic orbit, based on 16 meteors gathered by the Author from JW1961, MP1961, M1976 and C1977, revealed the following orbit for the Gamma Virginids:

ω	Ω	i	q	e	a
238.6	21.3	3.1	0.813	0.658	2.377

During 1969, a radar study conducted by Zdenek Sekanina, from facilities at Havana, Illinois, revealed 10 meteors during the period April 7 to 21, with the following average orbit:

ω	Ω	i	q	e	a
240.8	25.2	0.5	0.829	0.517	1.717

Additional April Showers

Librids

The primary sources for data supporting this stream's existence come from Ronald A. McIntosh's 1935 paper entitled "An Index to Southern Hemisphere Meteor Showers," and the two sessions of Zdenek Sekanina's Radio Meteor Project.

McIntosh listed two very similar radiants in his "Index." The first was designated number 134 and was based on nine previously observed radiants. Its duration was given as April 13-21, during which time the radiant moved from α=233.5°, δ=–19° to α=238.5°, δ=–22°. The second radiant was designated number 141 and was based on three previously observed radiants. Given a duration of April 17-19, the average radiant was α=236°, δ=–15°.[58]

---

[58]M1935, p. 714.

Zdenek Sekanina's Radio Meteor Project revealed three radiants during the 1960's. The 1961-1965 session revealed a stream Sekanina called the "Librids." The duration was given as March 24-May 7, the date of the nodal passage was April 17.6 (λ=27.2°), and the average radiant was given as α=232.1°, δ= –16.0°.[59] Sekanina's 1968-1969 session of the Radio Meteor Project revealed two fairly similar radiants in Libra. The first was called was called the "Librids" and was said to confirm the 1961-1965 stream. Its duration was March 10-April 21, the date of nodal passage was April 5.5 (λ=15.2°), and the average radiant was α=224.3°, δ=–12.8°.[60] The second stream detected in 1968-1969 was called the "Theta Librids." Its duration was given as March 11-May 5, with the nodal passage occurring on April 18.4 (λ=27.9°), and the average radiant being α=236.3°, δ=–18.4°.[61]

The orbits of the three radio meteor streams follow. The first orbit represents the "Librids" of 1961-1965, while the latter two orbits represent the "Librids" and "Theta Librids" of 1968-1969.

ω	Ω	i	q	e	a
324.1	27.1	5.6	0.159	0.887	1.408
326.7	15.2	5.8	0.191	0.794	0.926
332.0	27.9	2.7	0.101	0.920	1.269

The only additional visual radiant noted in the literature, aside from McIntosh's, was detected by William F. Denning during April 16-21, 1887, when five slow and long meteors were plotted coming from a radiant at α=235°, δ=–15°.[62]

Delta Pavonids

The discovery of the Delta Pavonids is attributed to Michael Buhagiar (Perth, West Australia). During the period of 1969-1980, he succeeded in observing this shower in six different years. The duration of activity was determined as April 3-8, while the date of maximum was established as April 6. The average radiant position was α=303°, δ=–63°. The hourly rate was described as variable, but did reach a high of 10.[63] Buhagiar suggested an association with comet Grigg-Mellish. The activity curve begins with a slow rise to maximum, which occurs at a solar longitude of 16.5°, and then rapidly declines.[64]

---

[59] S1973, pp. 256 & 259.
[60] S1976, pp. 276 & 292.
[61] S1976, pp. 277 & 293.
[62] D1899, p. 266.
[63] Buhagiar, Michael, *WAMS Bulletin*, No. 160 (1981).
[64] Mackenzie, Robert A., *Solar System Debris*. Dover: The British Meteor Society (1980), p. 39.

The variability of the activity from this shower is best evidenced by observations made by the Western Australia Meteor Section in 1980 and 1986. According to Jeff Wood (director of the WAMS) the maximum ZHR reached 1.88±0.19 on April 5, 1980. The duration was given as April 4-8, while the average radiant position was given as α=305°, δ=–65°.[65] During 1987, 35 members of the WAMS observed for 369 man-hours during late March and early April. The highest ZHR came on March 29/30, when rates were 7.3. The next highest rates were 5.0 on March 21/22 and 4.7 on April 7/8. The meteors possessed an average magnitude of 3.12, while 12.9% left trains.[66]

Comet Grigg-Mellish (1907 II) was once believed to possess a period of 164 years, but recent investigations by Brian G. Marsden (Smithsonian Astrophysical Observatory, Massachusetts) makes the short-period nature of this comet "highly improbable."[67] The orbit is currently given as parabolic.

ω	Ω	i	q	e	a
328.8	189.7	110.1	0.924	1.0	∞

April Virginids

This meteor shower probably represents one of the weaker branches of the Alpha Virginid stream. Its orbital elements are very similar (though separated by 180° in ω and Ω), but its inclination is about 13° higher.

The primary support for this stream's existence is the fact that it was detected by Zdenek Sekanina in both sessions of the Radio Meteor Project. The 1961-1965 survey revealed meteors during April 5-10, with the nodal passage coming on April 7.3, and the average radiant being α=213.4°, δ=–1.1°.[68] The 1968-1969 survey only detected meteors on April 7 and 8, since the radar had been shut down during March 29 to April 6. The date of the nodal passage was given as April 7.8, while the average radiant was determined as α=213.1°, δ=+3.6°.[69] The first orbit that follows is from the 1961-1965 survey, while the second is from 1968-1969.

ω	Ω	i	q	e	a
298.7	17.0	14.1	0.344	0.801	1.731
291.5	17.5	15.3	0.434	0.711	1.501

The Author has collected eight photographic meteors from MP1961. The indicated duration is April 1-16. The nodal passage falls on April 11, at which

---

[65]Wood, Jeff, Personal Communication (October 15, 1986).
[66]*MN*, No. 79 (October 1987), p. 6.
[67]Marsden, Brian G., *Catalog of Cometary Orbits*. New Jersey: Enslow Publishers (1983), p. 49.
[68]S1973, pp. 255 & 258.
[69]S1976, pp. 276 & 292.

time the radiant is at α=215.6°, δ=–0.9°. The following average orbit was determined:

ω	Ω	i	q	e	a
294.1	20.5	14.0	0.398	0.759	1.651

A comparison of the photographic and radar orbits indicates no significant difference in the shapes and sizes of the two types of orbits, although the nodal passage of the photographic meteors does occur about 3 days later than that given for the radar meteors. Noting this similarity, the Author combined 15 of the radio meteor orbits and the 8 photographic orbits in an attempt to determine the daily motion of the radiant. The result was +0.77° in α and –0.17° in δ.

Only three probable visual observations of this stream were made prior to Sekanina's discovery. The first observation was made by Cuno Hoffmeister on April 6, 1933 (λ=15.5°), when the radiant was given as α=212°, δ=+5°.[69] The next observation was also made by Hoffmeister, this time on April 7, 1934 (λ=17.5°), when the radiant was given as α=210°, δ=–4°.[70] The third observation was made by Franklin W. Smith (Glenolden, Pennsylvania) on April 8, 1934—less than 24 hours after Hoffmeister's second observation. The radiant was given as α=211°, δ=+2°.[71]

In an article published in 1973, E. I. Kazimirchak-Polonskaya and A. K. Terent'eva showed the evolution of several theoretical particles within the Alpha Virginid stream.[72] According to their tables showing details of this orbital evolution, the Alpha Virginids can attain a high inclination of 14° (during the 200 years examined), but usually did not spend too many years in that configuration due to fairly frequent approaches to within 0.32 AU of Jupiter. This Soviet study demonstrates that any meteors left in such an orbit would not be plentiful. Such is the case with the April Virginids, as only seven visual radiants appear in the major visual lists of radiants published from 1899 to the present.

---

[69]H1948, p. 228.
[70]H1948, p. 219.
[71]Olivier, Charles P., *FOR*, No. 63 (1943), p. 41.
[72]Kazimirchak-Polonskaya, E. I., and A. K. Terent'eva, pp. 369-371.

Chapter 5: May Meteor Showers

Eta Aquarids

Observer's Synopsis

This shower is visible during the period of April 21 to May 12. It reaches maximum on May 5 (λ=44°), from an average radiant of α=336°, δ=–1°. During the period of greatest activity hourly rates usually reach 20 for observers in the northern hemisphere and 50 for observers in the southern hemisphere. The radiant's daily motion is +0.96° in α and +0.37° in δ.

History

Hints that a shower might be active at the end of April and in early May began in 1863, when Professor Hubert A. Newton, examined the dates of ancient showers and suggested a series of periods which deserved the attention of observers. One of those periods was April 28-30, and included observed showers in 401 A.D., 839 A.D., 927 A.D., 934 A.D. and 1009 A.D.[1]

The Eta Aquarids were officially discovered in 1870, by Lieutenant-Colonel G. L. Tupman (Mediterranean Sea). On April 30, 15 plotted meteors indicated a radiant of α=325°, δ=–3°, and on May 2-3, 13 plotted meteors indicated a radiant of α=325°, δ=–2°. At a later date, William F. Denning examined the records of the Italian Meteoric Association and identified 45 meteors that were plotted during April 29 to May 5, 1870, from an average radiant of α=335°, δ=–9°.[2] Finally, the shower's first confirmation came on April 29, 1871, when Tupman plotted 8 meteors from α=329°, δ=–2°.

Observations of the Eta Aquarids were rare, but, during 1876, Professor Alexander Stewart Herschel discovered something which at least began to generate a greater interest in the shower. He conducted a mathematical survey to find which comets were most apt to produce meteor showers. Comet Halley was found to be closest to Earth on May 4, at which time a radiant was predicted to

---

[1]Newton, H. A., *AJS (Series 2)*, **36** (1863), pp. 148-149.
[2]D1899, p. 283.

occur at α=337°, δ=0°. Herschel immediately noted that Tupman's observed radiants of 1870 and 1871 were very near these predictions.[3]

The Eta Aquarids remained a poorly observed shower due to a lack of active meteor observers in the southern hemisphere. Only occasional hints of an active shower were reported, since northern observers had to face the beginnings of twilight shortly after the radiant rose above the eastern horizon. Nevertheless, H. Corder detected activity on the morning of May 4, 1878, with 3 plotted meteors revealing a radiant at α=334°, δ=–1°.[4] During this same year, Herschel examined all available observations and noted that the shower's radiant seemed to move further eastward as each day passed.[5]

Denning finally managed to observe this shower during April 30 to May 6, 1886. A total of 11 plotted meteors revealed a radiant at α=337°, δ=–2.5°. From these observations, he stated that the radiant seemed 5° to 7° in diameter. He added that the apparent closeness of his radiant to that predicted by Herschel placed the identity of this shower to Halley's comet "beyond doubt."

Fortunately, several good meteor observers appeared in the southern hemisphere during the 1920's, and the knowledge of primarily southern meteor showers increased dramatically. One of the most prolific observers was Ronald A. McIntosh (Auckland, New Zealand) and he published one of the more significant studies of the Eta Aquarids during 1929. McIntosh stated that his observations of that year showed activity during April 22 and May 13, which he said presented "a good illustration of the dispersive action of the planets during the centuries that the parent comet has been in existence."[6] His first radiant was determined on May 3.2 (α=334.0°, δ=–1.5°), while the last came on May 12.19 (α=342.7°, δ=+2.5°). He stated that maximum definitely came in early May, though bad weather prevented it from being pinpointed; however, hourly rates remained between 10 and 20 during the period of May 2 to 11. The radiant diameter was consistently about 5° across, and McIntosh's orbital calculations showed excellent agreement with the orbit of Halley's Comet.

During 1935, McIntosh published his investigation of the radiant motion of the Eta Aquarids. Using observations made by Murray Geddes (New Zealand) and himself during 1928 to 1933, he precisely determined the radiant's daily motion as +0.96° in α and +0.37° in δ.[7] He also plotted the observed activity of this stream and developed an activity curve that revealed the shower to begin with rates of 1 per hour on April 28, then rapidly rise to a flat maximum of 10

---

[3]Herschel, A. S., *MNRAS*, **36** (1876), p. 210.
[4]Corder, H., *The Observatory*, **2** (1878-1879), p. 103.
[5]Herschel, A. S., *MNRAS*, **38** (1878), p. 379.
[6]*MNRAS*, **90** (November 1929), p. 158.
[7]*MNRAS*, **95** (May 1935), p. 604.

per hour during May 3 to 6, and finally slowly decline to rates of 1 per hour by May 16.

Beginning in 1947, the Eta Aquarids joined the ranks of the first streams to be detected by radio-echo techniques. During May 1 to 10, an average radiant of $\alpha=339°$, $\delta=0°$ produced an hourly rate of 12.[8] Little additional data was gathered about this stream by the Jodrell Bank observers during the remainder of the 1940's and throughout the 1950's. In fact, the stream was largely ignored since the radio equipment was rarely operated during the early half of May. Fortunately, observers using the radar equipment at Springhill Meteor Observatory (Ottawa, Canada) and, later, at Ondrejov Observatory in Czechoslovakia, were able to supply some of the most extensive series of data ever accumulated on this stream.

The sensitive radar equipment at Springhill observed the Eta Aquarids during 1958 to 1967. Hourly rates were typically between 350 and 500 at maximum. An analysis of this data, as well as visual data accumulated during the years 1911 to 1971, was published in 1973 by A. Hajduk. He noted that there was an "instability of meteor frequencies in individual returns," which he attributed to "variations of the stream density along the orbit."[9] Hajduk noted that "no regular periodicity in the shower activity can be identified."

Overall, the Springhill data covered the period of May 1 to 10, and a fact revealed by Hajduk was the complexity of the activity rates. Using an average compiled for the period 1958-1967, it was noted that two apparent radar maxima occurred: one on May 4 and the other on May 7. These figures represented all radio echoes, but a further study of only the long-duration echoes (lasting ≥ 1 second) revealed the same two dates of maxima, except the decline between the two dates was not as pronounced. Also present was a further rise to maximum that came on May 10.[10]

The above figures represent a 10-year average and, although they show some interesting characteristics for the activity levels of the Eta Aquarids, the annual activity levels given in the same paper are even more interesting—especially when they are compared to the unusual peaks and valleys noted in the activity curves of the Orionids (see Chapter 10). Hajduk's study of the Orionids led him to conclude that the abnormal activity levels were due to Earth's encounter with filaments within the stream. The same explanation was also given as the reason the Orionids occasionally possess secondary maxima or primary maxima on a date other than that usually accepted as the date of maximum. The same is also true for the Eta Aquarids. In fact, of the 10 years

---

[8]CHL1947, pp. 373-377.

[9]Hajduk, A., *BAC*, **24** (1973), pp. 9-11.

[10]Hajduk, A., p. 10.

examined by Springhill Observatory, only 3 years represented what might be considered a normal activity curve. Some examples of unevenly distributed matter within the Eta Aquarid stream are as follows:

***1962:** During May 1 and 2, radio-echo rates increased from 303 to 328 per hour. On May 3 they had declined to 133, and by May 4 they were up to 468.

***1964:** Between May 1 and 4, hourly rates steadily increased from 366 to 415. They quickly dropped to 302 by May 6, but jumped back to 445 by the 7th.

***1965:** A double maximum was prominent in the radio-echo records for this year. A normal activity curve peaked at 370 per hour on May 4 and steadily declined to 287 by May 9. On May 10, rates of 349 had appeared.

***1966:** A fairly prominent plateau of 432 to 440 per hour was present during May 3 to 7, accept for a decline to 399 on May 6. A maximum of 498 came on May 9.

Hajduk's study not only revealed interesting details about this stream, but also about the Orionids of October—long known as the Eta Aquarids' sister stream. Although there is a distinct similarity between the characteristics of the meteors and activity levels of both streams, an interesting feature displayed in the Springhill data seems directly due to the distances each stream lies from Earth's orbit. Using the orbit of Halley's comet as representing the center of the associated meteor stream, Hajduk noted that the Eta Aquarids occur when Earth is 0.065 AU from the stream's core, while the Orionids occur when Earth is 0.15 AU away. According to the Springhill data, there is a smaller variation between the annual activity rates for the Eta Aquarids than exists for the Orionids.[11]

The evolution of this stream was discussed during 1983, by B. A. McIntosh (Herzberg Institute of Astrophysics, Ottawa, Canada) and Hajduk (Astronomical Institute of the Slovak Academy of Sciences, Bratislava, Czechoslovakia). They published the details of a proposed model of the meteor stream produced by Halley's comet. Using a 1981 study published by Donald K. Yeomans and Tao Kiang, which examined the orbit of Halley's comet back to 1404 BC,[12] McIntosh and Hajduk theorized that "the meteoroids simply exist in

---

[11] Hajduk, A., p. 11.

[12] Yeomans, D. K., and Kiang, T., *MNRAS*, **197** (1981), pp. 633-646.

orbits where the comet was many revolutions ago."[13] Further perturbations have acted to mold the stream into a shell-like shape containing numerous debris belts. These belts are considered as the explanation as to why both the Orionids and Eta Aquarids experience activity variations from one year to the next.

A good example of how observing conditions vary between the northern and southern hemispheres occurred during 1971. Observations in the United States and Japan revealed the shower to have peaked on May 5/6 with a ZHR of 13. During the same period, Australian observers noted a peak ZHR of 85.[14] Although this example may be somewhat extreme, average rates tend to be about 20 per hour in the northern hemisphere, and 50 per hour in the southern hemisphere, according to organizations in the United States, England, Japan, Australia and New Zealand.

Eta Aquarid Magnitudes and Trains

Year(s)	Ave. Mag.	# Meteors	% Trains	Observer(s)	Issue
1971-1984	3.04	315	—	McLeod	Personal Comm.
1978	3.07	394	23.9	WAMS	*MN*, No. 45
1984	3.05	104	67.3	Lunsford	Personal Comm.
1984	2.90	128	30	Bolivia	*MN*, No. 66
1984	2.96	25	32	Swann	*MN*, No. 66
1985	3.04	2872	29.4	WAMS	*MET*, **16**, No. 3
1986	2.46	—	32.0	WAMS	*MET*, **16**, No. 4
1986	2.68	55	67.9	Lunsford	Personal Comm.
1986	3.19	108	32	Swann	*MN*, No. 74
1986	2.33	406	28.3	Bolivia	*WGN*, **14**, No. 4
1987	2.40	92	54.3	Lunsford	Personal Comm.

Observers in the previous table were Norman W. McLeod III (Florida), Western Australia Meteor Section, Robert Lunsford (California), Asociacion Boliviana de Astronomia (LaPaz, Bolivia), David Swann (Texas).

During the 1985-1986 apparition of Halley's Comet, several meteor organizations around the world put their members on alert to check for possible increased activity in the Eta Aquarids (and the Orionids). Reports from groups in Australia, New Zealand, Bolivia, North America, and Japan generally indicate that no enhanced activity from this stream was present. It is interesting, however, that the data given in the previous table does seem to indicate an increase in the average magnitude of the Eta Aquarids. Thus, this stream's meteors may have been slightly larger than normal during 1986.

[13]McIntosh, B. A., and Hajduk, A., *MNRAS*, **205** (1983), p. 931.
[14]*MN*, No. 7 (August 1971), p. 5.

Orbit

During the years 1961-1965, the Radio Meteor Project detected eight radio-echo meteors from this stream in the period of May 3-8. The orbit determined by Zdenek Sekanina is as follows:

	ω	Ω	i	q	e	a
S1970	79.5	44.9	161.2	0.468	0.834	2.823

Epsilon Arietids

Observer's Synopsis

This daylight shower possesses a duration extending from April 25 to May 27. At maximum on May 9 it has an average radiant of α=44°, δ=+21°. This stream might be associated with the Southern Taurid stream of November.

History

The discovery of this daylight meteor shower should be attributed to C. S. Nilsson (University of Adelaide, South Australia). During the interval of May 19-27, 1961, his radio equipment detected six members of this stream, which indicated a radiant of α=58.8°, δ=+23.7°. Nilsson commented that W. G. Elford had "reanalyzed the data using the stream search program of Southworth and Hawkins and suggests that the May day-time shower...is due to the S. Taurid stream...." Nilsson commented that the agreement was good when the angular elements of the orbits were considered, but said the eccentricity was "slightly low."[15] It should be pointed out that the radio equipment did not operate during May 1-18, so it is possible that the shower could have been active sooner than indicated.

The Epsilon Arietids were next detected in 1969, during the second session of the Radio Meteor Project. Zdenek Sekanina analyzed the data obtained by the equipment at Havana, Illinois, and noted that meteors were detected during the interval of April 25 to May 22. The established date of the nodal passage was given as May 8.5, at which time the radiant was α=43.6°, δ=+20.9°.[16] It should be noted that the radar did not operate during May 24-June 1.

Orbit

This daylight shower's orbit has been determined on two separate occasions: the first instance was in 1961, by Nilsson, and the second was in 1969, by Sekanina. The orbits are listed below.

---

[15]N1964, pp. 226-228 & 245.
[16]S1976, pp. 277 & 293.

	ω	Ω	i	q	e	a
N1964	89.5	62.3	2.7	0.604	0.71	2.083
S1976	90.0	47.4	2.8	0.592	0.708	2.026

May Arietids

Observer's Synopsis

This daylight meteor shower is active during the period of May 4 to June 6. Maximum occurs around May 16 from an average radiant of α=37°, δ=+18°.

History

The May Arietids were first detected in 1960 while the radio equipment at the Kharkov Polytechnical Institute was being operated by B. L. Kashcheyev and V. N. Lebedinets. Activity was detected during the period of May 5-27. The stream's nodal crossing came on May 15 (λ=54°), at which time the radiant was at α=41°, δ=+23°.[17]

The stream was next detected in 1961. C. S. Nilsson had been operating the radio equipment of the University of Adelaide, South Australia, during May 19-28. Meteors from this stream were detected during the entire period. The date of the nodal crossing was determined as May 23 (λ=62.1°), at which time the radiant was at α=46.5°, δ=+19.1°.[18]

The last radar survey to successfully detect this stream was the 1968-1969 session of the Radio Meteor Project. Zdenek Sekanina found the duration to be May 7 to June 6, with a nodal crossing on May 15.6 (λ=54.3°). The average radiant was at α=36.5°, δ=+17.8°.[19]

Orbit

The orbit of this stream has been determined on three occasions. The first orbit represents that of Kashcheyev and Lebedinets and is based on 16 meteors detected in 1960. The second orbit was determined by Nilsson during 1961 and is based on 11 meteors. The last orbit was obtained from radio meteors detected during 1969 by Sekanina.

	ω	Ω	i	q	e	a
KL1967	74	54	6	0.44	0.77	1.94
N1964	64.8	62.1	2.9	0.391	0.75	1.563
S1976	60.8	54.3	3.4	0.363	0.763	1.528

---

[17]BL1967, p. 188.
[18]N1964, pp. 226 & 228.
[19]S1976, pp. 277 & 293.

Omicron Cetids

Observer's Synopsis

This meteor shower occurs exclusively during daylight hours during the period May 7 to June 9. Maximum may occur during the period of May 14 to 25, with a maximum hourly radio-echo rate of 18. The average position of the radiant near maximum is α=28°, δ=–3°. The meteor shower seems to occur annually, but may be prone to periodic or irregular increases in activity.

History

The Omicron Cetids were first detected on May 14, 1950, when A. Aspinall and Gerald S. Hawkins of Jodrell Bank radio observatory detected meteors at a rate of 18 per hour coming from α=27.5°, δ=–3.5°. Despite a 24-hour-a-day operation, this stream was again detected only on May 16 (α=28.5°, δ=+0.0°), 21 (α=29.5°, δ=–5.0°) and 23 (α=34.0°, δ=+1.0°), with hourly rates ranging from 18 to 22 per hour. The radiant diameter was less than 3° on each day, except for the 21st, when it was estimated as 10° across.[20]

The Omicron Cetids were again detected at Jodrell Bank observers during May 14 to 17, 1951 (activity levels of 18 to 25 per hour), but were missed during 1952. Since the stream had generally produced activity levels greater than those observed for the Eta Aquarids, this prompted Mary Almond, K. Bullough and Hawkins to conclude that if the stream was present it "must have been less active than in 1950 or 1951."[21]

The stream was possibly observed again at Jodrell Bank on May 17, 1953. Bullough gave the radiant as α=11°, δ=–2° and the hourly rate as 8. The radiant was less than 3° across.[22] No further observations were made at this radio observatory during the period 1954 to 1958.

The next observation of this stream was made during 1961, by radio equipment at Adelaide Observatory in Australia. C. S. Nilsson noted three Omicron Cetids during the period of May 23-28, from an average radiant of α=35.5°, δ=+1.0°.[23]

The true extent of this stream was finally realized during the two sessions of the Radio Meteor Project conducted during the 1960's by Zdenek Sekanina at Havana, Illinois. During 1962-1965, 11 meteors were detected during May 19-June 9, from an average radiant of α=20.6°, δ=+1.2°. The probable date of

---

[20]Aspinall, A., and Hawkins, G. S., *MNRAS*, **111** (1951), pp. 20-21.
[21]ABH1952, p. 16.
[22]B1954, p. 79.
[23]N1964, pp. 226-228.

maximum was given as May 27.9.[24] During 1969, 11 meteors were detected during the period of May 7-21, with the average radiant being α=21.5°, δ=–4.0°. The probable date of maximum was then given as May 9.0.[25]

Orbit

From observations obtained during 1950, at Jodrell Bank, Mary Almond was able to compute the following orbit:

	ω	Ω	i	q	e	a
A1951	211	238.0	34	0.11	0.91	1.3

Additional orbits were later computed by Nilsson and Sekanina, using observations obtained during radio-echo studies conducted during the 1960's.

	ω	Ω	i	q	e	a
N1964	212.5	244.9	33.8	0.127	0.91	1.408
S1973	200.2	246.1	36.3	0.066	0.937	1.055
S1976	213.9	227.9	32.6	0.122	0.925	1.623

May Librids

Observer's Synopsis

This stream possesses a very short duration extending from May 1 to 9. At maximum around May 6, the radiant is at α=233°, δ=–18°. The ZHR lies between 2 and 6.

History

This shower seems to have first been noticed on May 5, 1929, by three Texas members of the American Meteor Society. Oscar E. Monnig and Robert Brown were the first to detect the radiant. Monnig plotted 3 meteors from α=230°, δ=–16°, while Brown plotted 2 meteors from α=229°, δ=–16°. The mean time of both observations was given as May 5.32, and the radiants were given a poor rating of reliability. The third Texas observer was Blakeney Sanders. He plotted 3 meteors from α=234°, δ=–18°, at a mean time of May 5.35. The radiant was also considered to be of poor reliability.[26]

The year of 1933 brought three more observations of this shower—this time from observers in Germany. The first was made by N. Richter on May 2, when his meteor plots revealed a radiant at α=231°, δ=–20°. Just 24 hours later,

---

[24] S1973, pp. 256 & 259.
[25] S1976, pp. 277 & 293.
[26] Olivier, Charles P., *FOR*, No. 5 (1929), p. 25.

Cuno Hoffmeister observed a radiant at α=235°, δ=–19°. Finally, on May 6, Richter observed a radiant at α=226°, δ=–16°.[27]

In Ronald A. McIntosh's 1935 paper entitled "An Index to Southern Meteor Showers," two radiants were recognized which might offer further observations of this shower. Radiant number 132 was called the "Gamma Librids" and was based on two previously observed radiants. The duration was given as May 9-11, while the average radiant was α=232°, δ=–15°. Radiant number 133 was called the "42 Librids." Its duration was May 3-9, while the average radiant was given as α=233°, δ=–23.5°. This shower was based on three previously observed radiants.[28]

In the three decades following the 1930's, only three indications of this shower's existence appear in various publications. Paul Anderson (Beechwood, Michigan) plotted four meteors from α=233°, δ=–18° on May 7.24, 1942.[29] The cameras of the Harvard Meteor Project photographed a magnitude 0.1 meteor on May 1, 1954, which possessed a radiant of α=230°, δ=–19°.[30] The 1961-1965 session of the Radio Meteor Project detected several meteors from this stream during May 3-7. Zdenek Sekanina, director of the project, determined the date of nodal passage as May 5.8 (λ=44.8°), while the radiant was given as α=233.0°, δ=–19.5°.[31]

The dearth of visual observations of this shower has changed in the last two decades as two well-organised groups of amateur astronomers in the Southern Hemisphere have begun systematic observations. J. E. Morgan, director of the Royal Astronomical Society's New Zealand meteor section, compiled a list of the active showers observed by his group during the 1970's. One radiant, designated as number 126, was called the "Lambda Librids" and was observed during May 4-15. The average radiant was given as α=236°, δ=–19.5°, while the maximum observed ZHR was 6. In Western Australia, Jeff Wood heads the meteor section of the National Association of Planetary Observers. During 1979, his group observed a radiant they called the "K Librids." Meteors were detected during May 5-6, with a maximum ZHR of 3.30±0.72 coming on May 5, from α=233°, δ=–19°. During 1980, Wood's group observed a radiant referred to as the "Iota Librids." Activity was detected during May 3-4, with a maximum ZHR of 2.15±0.10 coming on May 3 from α=232°, δ=–20°.[32]

---

[27]H1948, pp. 229-230.

[28]M1935, p. 714.

[29]Olivier, Charles P., *FOR*, No. 84 (1951), p. 2.

[30]MP1961, p. 49.

[31]S1973, pp. 256 & 259.

[32]Wood, Jeff, Personal Communication (October 15, 1986).

Orbit

This stream was detected during the 1961-1965 session of the Radio Meteor Project. Eleven radio meteors indicated the following orbit:

	ω	Ω	i	q	e	a
S1973	113.7	224.8	0.3	0.410	0.739	1.570

In MP1961, the following orbit was given for a meteor photographed by the Harvard Meteor Project on May 1, 1954:

ω	Ω	i	q	e	a
117	220	0	0.35	0.80	1.80

Northern May Ophiuchids

Observer's Synopsis

This ecliptic stream possesses a long duration extending from April 8 to June 16. Maximum occurs during May 18-19 from an average radiant of α=253°, δ=–15°. The maximum ZHR is about 2-3, while the radiant diameter is near 3°.

History

The first observation of this meteor shower was made by H. Corder during May 12-14, 1896. He had plotted five "slow" meteors from a radiant of α=248°, δ=–15°.[33]

During 1935 this shower experienced the beginning of recognition as an annual shower when Ronald A. McIntosh listed it in his paper "An Index of Southern Meteor Showers." Designated number 160 and called the "Phi Ophiuchids," this shower was given a duration of May 7-14, with the average radiant being given as α=248.5°, δ=–15°.[34]

The publication of Cuno Hoffmeister's *Meteorströme* during 1948 contained even more support for the Northern May Ophiuchids. Hoffmeister noted a collection of radiants which indicated a radiant of α=250°, δ=–18° reaching maximum at a solar longitude of 54° (May 15).[35] This radiant was based on seven individual radiants determined by German observers during 1914 to 1937. One of the radiants, designated number 693, was composed of 16 meteors whose intersection indicated a radiant diameter of 3°.[36]

The greatest amount of data to be accumulated on the Northern May Ophiuchids came from the two sessions of the Radio Meteor Project. Zdenek

[33]D1899, p. 267.
[34]M1935, p. 714.
[35]H1948, p. 83.
[36]H1948, p. 205.

Sekanina found a duration extending from May 3 to June 4 during the 1961-1965 survey. The date of the nodal passage was determined as May 17.6 (λ=56.2°), at which time the radiant position was α=252.3°, δ=−16.9°.[37] During the 1968-1969 survey, Sekanina found a duration of April 8 to June 16. The nodal passage was determined as May 19.0 (λ=57.6°), at which time the radiant was at α=256.2°, δ=−12.8°.[38]

The Western Australia Meteor Section has obtained good observations of the shower in recent years. According to the section's director, Jeff Wood, observations conducted in 1979 revealed "Eta Ophiuchids" during May 16-27. A maximum ZHR of 2.46±0.56 came on May 19, at which time the radiant was at α=256°, δ=−13°.[39]

Orbit

Both sessions of Sekanina's Radio Meteor Project determined orbits for this stream. The first orbit listed below represents the orbit from the 1961-1965 data, while the second orbit is from the 1968-1969 data.

	ω	Ω	i	q	e	a
S1973	306.2	56.2	7.4	0.282	0.835	1.707
S1976	308.3	57.6	13.5	0.283	0.807	1.464

Southern May Ophiuchids

Observer's Synopsis

This ecliptic stream possesses a long duration extending from April 21 to June 4. Maximum occurs between May 13 and 18 from an average radiant of α=252°, δ=−23°.

History

The first observations of activity from the Southern May Ophiuchids were made by two members of the American Meteor Society (AMS). Ronald A. McIntosh (Auckland, New Zealand), a long-time Southern Hemisphere member of the AMS, plotted nine meteors from a radiant of α=249.5°, δ=−26° on May 15, 1926,[40] while G. W. Ridley (Alameda, California) plotted three meteors from a radiant of α=249°, δ=−26.5° on May 18, 1931.[41]

---

[37] S1973, pp. 256 & 259.
[38] S1976, pp. 277 & 293.
[39] Wood, Jeff, Personal Communication (October 15, 1986).
[40] Olivier, Charles P., *FOR*, No. 15 (1933), p. 42.
[41] Olivier, Charles P., *FOR*, No. 15 (1933), p. 25.

The 1930's marked the true beginnings of notable activity from this stream, with the bulk of the observations coming from Cuno Hoffmeister and other German observers. In his book *Meteorströme*, Hoffmeister lists nine radiants for the year 1933 alone which appear to represent activity from this stream. The observations covered the period of May 19 to June 2 (λ=58.1° to 72.3°).[42] These observations indicate the radiant is rather diffuse, but there is a definite tendency for it to move eastward at a rate of about one degree daily. These observations do not show a clear trend as to whether the declination moves to the north or south. Additional radiants were given for 1936 and 1937.

The first apparent recognition of this stream as a potential annual producer of meteors was in 1935, when Ronald A. McIntosh's paper "An Index to Southern Meteor Showers" was published. Although a strong indication of the shower was not present, four fairly similar radiants were listed which could all refer to the Southern May Ophiuchids. Radiant 157 was based on four previously observed radiants and was given a duration of May 6-15. The average radiant position was given as α=246°, δ=–26°. Radiant number 159 was said to be active during May 17-24. It was based on two previously observed radiants and possessed an average position of α=248°, δ=–22.5°. Radiant 161 was based on only one previously observed radiant. It was said to be active on May 15 from α=249.5°, δ=–25°. Radiant 170 was based on three previously observed radiants and was given a duration of May 19-22. The average radiant position was α=258°, δ=–23°.[43]

The Southern May Ophiuchids were well represented in the 1952-1954 Harvard Meteor Project. From a list of 2529 photographic meteors published in MP1961, the Author has isolated seven meteors (two of which had precise orbits computed in JW1961) which appear to be probable members of this stream. The indicated duration is May 5 to June 4, and the date of nodal passage is May 17 (λ=55.4°). The average radiant position was at α=251.6°, δ=–24.8°.

The 1960's were a decade when the true extent of the Southern May Ophiuchid stream was finally realized. Details were obtained during both sessions of the Radio Meteor Project. From the 1961-1965 data, Zdenek Sekanina established the duration as May 3-June 4. The nodal passage was determined as May 13.0 (λ=51.8°), at which time the radiant was α=250.1°, δ=–22.4°.[44] From the 1968-1969 session, the duration was determined as April 21-June 3, while the date of the nodal passage was given as May 18.5 (λ=57.1°). The average radiant was given as α=254.0°, δ=–25.0°.[45]

---

[42]H1948, pp. 229-231.
[43]M1935, p. 714.
[44]S1973, pp. 256 & 259.
[45]S1976, pp. 277 & 293.

Recent observations of activity from this stream have been obtained by members of the Western Australia Meteor Section. According to the group's director, Jeff Wood, the Southern May Ophiuchids were detected in 1979 during the period of May 17-26 (they were referred to as the "Theta Ophiuchids"). A maximum ZHR of 2.27±0.55 was detected on May 19 from an average radiant of α=254°, δ=–25°. Possible activity was also detected in 1980 when meteors were observed during May 23-24 from an average radiant of α=256°, δ=–22°. The shower was labelled the "Xi Ophiuchids." A maximum ZHR of 1.78±0.07 was observed on May 23.[46]

Orbit

This stream was detected during both sessions of the Radio Meteor Project. The first orbit is based on data gathered by Sekanina during the years 1962-1965, while the second orbit is based on material gathered in 1969.

	ω	Ω	i	q	e	a
S1973	132.9	231.8	0.6	0.243	0.833	1.458
S1976	130.0	237.1	3.5	0.264	0.824	1.502

The Author has collected seven photographic meteor orbits from MP1961 and JW1961, and determined the following average orbit for the Southern May Ophiuchids.

ω	Ω	i	q	e	a
129.7	235.4	5.4	0.220	0.919	2.708

May Piscids

Observer's Synopsis

The duration of this daylight stream extends over the period of May 4-27. The probable date of maximum activity falls on May 12 (λ=51°), at which time the radiant is at α=13°, δ=+22°. The maximum hourly rate is near 8, while the radiant diameter is about 3°.

History

This daylight stream seems to have first been detected by observers at Jodrell Bank (England), as J. A. Clegg, V. A. Hughes and Professor Alfred Charles Bernard Lovell listed a radiant of α=7°, δ=+20° that was detected on May 4, 1947. Although the radar was operated during May 1 to 30, no additional activity from this stream was noted.[47]

---

[46]Wood, Jeff, Personal Communication (October 15, 1986).
[47]CHL1947, p. 374.

The Jodrell Bank equipment again detected this stream on May 12, 1955 (λ=50.8°). T. W. Davidson found 33 meteors which indicated a radiant of α=12°, δ=+24° and a radiant diameter of 3°. The hourly rate was given as 8. The radiant was called the "Upsilon Piscids," and Davidson noted it was near a radiant detected at Jodrell Bank in 1951.[48]

The next radar survey to detect this stream was conducted by B. L. Kashcheyev and V. N. Lebedinets (Kharkov Polytechnical Institute, USSR) in 1960. During the period of May 4-27, 17 meteors were observed. The date of the nodal passage was given as May 12 (λ=52°), at which time the radiant was at α=17°, δ=+19°.[49]

The last radio meteor survey to detect the May Piscids was the 1968-1969 session of the Radio Meteor Project. Zdenek Sekanina noted that activity was detected during May 7-9, from an average radiant of α=11.9°, δ=+19.0°. The date of the nodal passage was given as May 8.4 (λ=47.3°).[50] The equipment was shut down during May 10-18, so that any extension in the duration of the shower would have been missed.

Orbit

The first orbit represents that determined by Kashcheyev and Lebedinets during 1960, while the second was established by Sekanina from data obtained during 1969.

ω	Ω	i	q	e	a
32	52	30	0.11	0.93	1.64
35.9	47.3	29.1	0.147	0.896	1.412

Additional May Showers

Epsilon Aquilids

This meteor shower is apparently only visible with some optical aid, as no visual or photographic detection has ever been made.

The discovery of the Epsilon Aquilids should be credited to B. L. Kashcheyev and V. N. Lebedinets (Kharkov Polytechnical Institute, USSR), who detected 17 radio meteors from this stream during May 4-27, 1960. They determined the date of the nodal passage as May 17, at which time the radiant was located at α=276°, δ=+13°.[51]

---

[48]D1956, pp. 121 & 125.
[49]BL1967, p. 188.
[50]S1976, pp. 277 & 293.
[51]KL1967, p. 188.

This stream was again detected in 1969, during the second session of the Radio Meteor Project. Despite Zdenek Sekanina giving the duration of the Epsilon Aquilids as May 19-21, it should be noted that the radar system at Havana, Illinois, had been shut down during May 10-18, so earlier members might have been missed. Sekanina gave the date of nodal passage as May 20.3, at which time the radiant was at α=284.1°, δ=+15.5°. Both the date of nodal passage and the radiant might have been altered had the radar been in operation during mid-May.

The orbits determined during the two radar surveys just discussed are as follows:

	ω	Ω	i	q	e	a
KL1967	312	55	56	0.35	0.64	0.96
S1976	318.3	58.8	59.6	0.354	0.594	0.873

The orbits are very similar and, with their semimajor axes both being less that 1 AU, they seem to hint that the body responsible for their formation is moving in an Aten-class asteroid orbit.

As noted earlier, no apparent visual observations of this stream appear in any records of the last 150 years. The primary sources checked by the Author are D1899, K1916, M1935, H1948 and over 6000 radiants of the American Meteor Society. Photographic sources checked included W1954, MP1961 B1963, C1964, BK1967 and C1977.

Chapter 6: June Meteor Showers

Arietids

Observer's Synopsis

This is the strongest daylight meteor shower of the year. The duration extends from May 22 to July 2, with maximum activity occurring on June 8 (λ=76.8°) from α=44.5°, δ=+23.6°. The hourly rate is near 60 at maximum. The radiant's mean daily motion is about +0.5° in α and +0.3° in δ, while the radiant diameter remains a fairly consistent 3° in diameter.

History

This meteor shower was discovered during 1947, by operators of the radio equipment at Jodrell Bank (England). Activity was first noted from this stream on May 30 and continued until June 17. The radiant was not accurately determined, but was noted to fall within the range of α=45° to 55°, δ=+25° to +35°.[1]

The Arietids have been detected in every major radar study since 1947. The orbit of this stream was first determined by Mary Almond during 1951.[2] Although the average orbital inclination was found to be 18°, Almond used the "smoothed radiant positions for the first and last dates of observations" and found the stream's orbit to begin with an inclination of 3° and end with a value of 34°, thus "the line of apsides swings gradually forward, and the main centre of the stream rises farther away from the ecliptic day by day."

The question of the stream's inclination has remained ever since 1951, with radar studies in Australia, the United States, and the Soviet Union variously revealing values of 19° to 38°; however, in 1975 the results of an Australian radar survey conducted during 1969 were published, which perhaps shed new light on this problem.

---

[1]CHL1947, pp. 374-375.

[2]Almond, Mary, *MNRAS*, **111** (1951), pp. 37-44.

G. Gartrell and W. G. Elford detected six associations during mid-June 1969 that possessed radiants and velocities very close to those accepted for the Arietids. The problem that existed was that the orbital inclinations of these streams varied from 2.4° to 65.3°. They pointed out that the inclination spread noted by other workers was definitely confirmed; however, "no evidence of the progressive increase in inclination with passage through the stream...was found in this survey."[3] Thus, numerous ringlets of material seem to be the reason leading to the confusion of the Arietid stream's orbital inclination.

Just as confusing as the determination of the stream's inclination has been the determination of the stream's daily motion of the radiant. In 1951, A. Aspinall and Gerald S. Hawkins produced a smoothed radiant ephemeris, based on Jodrell Bank data, that indicated a daily motion of +0.74° in α and +0.92° in δ.[4] Using more precise observations obtained at Jodrell Bank during 1950-1953, K. Bullough determined the radiant's daily motion as +0.48° in α and +0.30° in δ.[5] Still another determination from Jodrell Bank data, this time covering the period 1950-1955, was made by T. W. Davidson in 1956. He found the radiant's daily motion to be +0.47° in α and +0.39° in δ.[6] A radar survey conducted in the Soviet Union during 1960, by B. L. Kashcheyev and V. N. Lebedinets, determined the daily motion to be +0.7° in α and +0.1° in δ.[7] Although no formal explanation seems to have been offered as to why the radiant's daily motion has been so difficult to determine, it may be possible that the inclination variances noted previously are directly responsible—especially since the radiant's motion in declination seems the hardest to establish. Thus, since Gartrell and Elford's data seem to indicate several filaments working simultaneously to produce the Arietid activity, the same filaments might be contributing to the confusion of the determination of the shower's daily motion.

Other details concerning this stream have not been easy to obtain. From the numerous Jodrell Bank studies (many of which have been cited in this section) the diameter of the radiant appears to be about 3°, while the maximum hourly rate ranges from 54 to 76. From observations made in the United States and Australia during 1971, it appears that meteors from this shower can be visually detected coming up from the horizon during the hours immediately after sunset and immediately before sunrise.[8] During June 6/7, Karl Simmons estimated combined rates of the Arietids and Zeta Perseids (later in chapter) reached 1 to 2

---

[3]GE1975, p. 601.

[4]Aspinall, A., and G. S. Hawkins, *MNRAS*, **111** (1951), p.21.

[5]B1954, p. 75.

[6]D1956, p. 120.

[7]KL1967, p. 189.

[8]*MN*, No. 7 (August 1971), p. 6.

meteors per hour. During one hour on the morning of June 2, 1973, John West (Bryan, Texas) observed four Arietids.[9]

Various researchers have arrived at some interesting conclusions concerning links between this stream and other solar system bodies. In 1951, while obtaining the first determination of this stream's orbit, Almond concluded that another shower should be encountered as Earth crossed the stream's orbital plane on July 28. The estimated radiant position was α=336°, δ=–11°, which falls within 15° of the position of the Southern Delta Aquarid meteor stream. After examining both stream orbits, Almond concluded that, although the orbits "are now different, it seems probable that they may have had a common origin in the past."[10]

In articles published during 1973 and 1976, Zdenek Sekanina suggested several possible associations of meteor streams with comets and asteroids. For the Arietids, he noted that the Apollo asteroid Icarus (1566) possessed an orbit with similar characteristics. A D-criterion calculation by Sekanina, comparing the meteor stream orbit to the orbit of Icarus, revealed a value of 0.245 for the earlier data and 0.286 for the latter data.

Orbit

The exact orbit of this stream has not been easy to determine since its discovery due to the large spread in inclination; however, some of the orbits determined from large meteor samples are as follows:

	ω	Ω	i	q	e	a
KL1967	29.9	76.6	18.7	0.10	0.94	1.67
S1973	29.5	78.0	27.9	0.094	0.946	1.750
GE1975	28	81	17.4	0.08	0.96	2.778
S1976	25.9	76.9	25.0	0.085	0.938	1.376

The orbit from KL1967 was based on 380 meteors, that from S1973 was based on 41 meteors, that from GE1975 was based on 32 meteors, and that from S1976 was based on 48 meteors. The very small inclination orbit determined by GE1975 (2.4°) was based on 3 meteors, while the high inclination orbits of N1964 and GE1975 (33°-65°) were based on 6-8 meteors.

The orbit of the Apollo asteroid Icarus (1566) is as follows:

ω	Ω	i	q	e	a
31.0	87.6	22.9	0.187	0.827	1.078

The orbit of the Southern Delta Aquarid stream is as follows:

ω	Ω	i	q	e	a
151.9	307.3	29.9	0.083	0.955	1.853

---

[9]*MN*, No. 18 (Mid-October 1973), p. 2.

[10]Almond, p. 41.

June Boötids ("Pons-Winneckids")

Observer's Synopsis

This shower is currently active during June 27 to July 5 and possesses a maximum of activity that falls on the 28th. The greatest activity levels only reach 1 to 2 per hour, but the stream is noted for an especially strong display in 1916, and good displays in 1921 and 1927. At maximum, the radiant is diffuse—probably greater than 5° in diameter—with the average position being at α=223°, δ=+58°. The shower is notable in that its meteors are primarily faint, with an average magnitude near 5; however, bright meteors do occur regularly.

History

The entrance of this stream into the lists of active meteor showers occurred as a result of intense activity that was noted on the night of June 28, 1916. Observers in England were treated to a very active display of meteors beginning soon after sunset and, one of the most experienced observers lucky enough to be outside that evening was William F. Denning.

After a cloudy day, the skies of England began clearing shortly before sunset. Denning went outside at 10:25 p.m. (local time) and very quickly noted that a meteor shower was in progress. "Large meteors came in quick succession from a radiant in the region between Boötes and Draco," Denning wrote.[11] He described the meteors as "moderately slow, white with yellowish trains, and paths rather short in the majority of cases. Several of the meteors burst or acquired a great intensification of light near the termination of their flights, and gave flashes like distant lightning."

Denning enlisted the help of a friend in observing the spectacle, while he plotted as many meteors as possible. Observations ended after 1:15 a.m. when clouds again moved in. Denning concluded that the greatest part of the display occurred prior to midnight. He also stated that the exact location of the radiant was impossible to pinpoint. As he projected his plotted paths backward he noted the "directions were from a wide region or area of about 12° to 15° diameter." What Denning considered to be the main radiant was given the position of α=231°, δ=+54°.

Though clouds prevented observations on June 29, Denning was able to observe between 10:15 and 11:30 p.m. on the 30th, but only one meteor was noted that could have been from the shower of the 28th. This meteor was also detected and plotted by Mrs. Fiammetta Wilson (Totteridge). From both plots, Denning was able to determine the radiant to have been at α=223°, δ=+41°.

---

[11]Denning, W. F., *MNRAS*, **76** (1916), pp.740-743.

Shortly after observing this strong shower, Denning wondered if its sudden appearance might be attributable to a comet. After searching through lists of cometary orbits, he concluded that periodic comet Pons-Winnecke was probably responsible. "The radiant was placed in the correct region and the date agreed. Moreover, the comet passed through perihelion on September 1 last year."

Charles P. Olivier (then director of the American Meteor Society) received several charts during the latter half of 1916 which clearly indicated that, although the great shower of June 28 had been missed in the United States, observers had independently noted activity from a similar area of the sky during May, June, and July—John Koep and Philip Trudelle (both of Chippewa Falls, Wisconsin) observed these meteors during the period of May 19 to June 5, while Raymond Lambert (Newark, New Jersey) detected a radiant of α=206.7°, δ=61.2° on July 4.[12]

Olivier tabulated the known 1916 observed radiants and computed both parabolic and elliptical orbits for each. He upheld Denning's belief of a relationship with Pons-Winnecke and added that the minor activity seen in the United States during the period of late May to early July, and the strong activity noted on June 29, were probably produced by the same stream.

A later investigation into the link with this meteor shower and Pons-Winnecke was published in 1932.[13] Franklin W. Smith extended Olivier's mathematical work to check the similarity between a radiant ephemeris based on the observations and an ephemeris based on the orbit of Pons-Winnecke. Franklin took a typical radiant from the Koep and Trudelle data and noted that it would drift toward the *general* position of the June 28 activity (as noted earlier, no obvious center of activity was noted on the 28th); however, the link between the orbits of the observed activity and that of Pons-Winnecke was far from perfect. Smith theorized that "the meteors had been separated from the actual orbit of the comet by dispersive forces, which would be expected in this case because of the direct motion of the comet, its short period, and the low inclination of its orbit."

Indeed, the forces being exerted upon the comet by Jupiter were great. In fact, from the beginning of the 20th century until about 1940, the comet had been locked into a nearly perfect 2:1 resonance with Jupiter. The main effect of this resonance was the rapid increase in Pons-Winnecke's perihelion distance. From the time of the comet's discovery in 1819, until shortly after the 1869 perihelion passage, the comet's perihelion distance remained about 0.2 AU inside of Earth's orbit. During the next 46 years the perihelion distance quickly moved

---

[12]Olivier, Charles P., *MNRAS*, **77** (November 1916), p. 73.

[13]Smith, Franklin W., *MNRAS*, **93** (December 1932), p. 156-158.

away from the sun, and in 1915 it was only 0.04 AU inside of Earth's orbit. By 1921, the perihelion was 0.03 AU *outside* of Earth's orbit, and by 1964 it was located 0.22 AU outside of Earth's orbit. The orbit of this comet has remained in a fairly stable state since 1964.

A search for signs of activity prior to the 1916 appearance has revealed some interesting data. In the same article cited earlier, Denning pointed out that "in 1860 and 1861 June 30 Mr. E. J. Lowe observed 'many meteors.'" During 1899, Denning published his "General Catalogue of Radiant Points of Meteoric Showers and of Fireballs and Shooting Stars Observed at More Than One Station"[14] which did list two possible early detections of this radiant. The first was detected during June 26 to July 11, 1872, by observers of the Italian Meteoric Association. The plots were examined by Denning, who determined a radiant of α=216°, δ=+47°. It was based on 10 meteors. The second radiant was seen by Denning during June 14-28, 1887, from a radiant of α=213°, δ=+53°. It was based on 4 plotted meteors. This latter radiant possesses very similar circumstances present when the 1916 shower was noted, namely Pons-Winnecke had passed perihelion at the beginning of September of the previous year.

Following 1916, two notable, though weaker, appearances of this shower occurred during the next two perihelion dates of the parent comet. During 1921, observations in the United States and England revealed predominantly weak activity, with only Robert M. Dole (Wilmington, North Carolina) and Denning being able to secure enough plotted meteors to reveal radiants. On June 29.17 Dole plotted 7 meteors from α=213.6°, δ=+47.2°, while on June 30.10, he plotted 8 meteors from α=213.2°, δ=+47.0°.[15] On June 28, Denning plotted 7 meteors from α=228°, δ=+58°.[16] But, while these observers were barely collecting enough data for radiant determinations, Kaname Nakamura (Kyoto, Japan) noted increased numbers in early July.[17]

Nakamura searched for meteors of Pons-Winnecke on several occasions during late June and early July, in weather that varied from clear to mostly cloudy skies. Beginning his observations on June 25, Nakamura observed his first meteors from this shower on the 26th. Although the number of meteors from this shower increased by the 27th and 28th, the greatest rates were said to have occurred on July 3, when 153 meteors were seen in 35 minutes. Cloudy skies were present on the 4th, but observations on the 5th revealed meteors still

---

[14]Denning, W. F., *MRAS*, 53 (1899), p. 263.

[15]Olivier, Charles P., *Publications of Leander McCormick Observatory, University of Virginia*, V, Part 1.

[16]Denning, W. F., *MNRAS*, 84 (November 1923), p. 52.

[17]Yamamoto, Issei, *The Observatory*, 45 (March 1922), p. 82.

falling at a rate of 91 per 41 minutes at one point. During the period of June 26 to July 11, Nakamura was able to plot enough meteors to obtain 9 daily positions, with the radiant moving slowly southeastward. On June 28, the radiant was determined to have been at α=212.5°, δ=+49°, while the position on the night of maximum activity levels was α=212°, δ=+47°. Nakamura's radiants are very similar to those determined by Dole.

Nakamura had been described as possessing "very sensitive eyes," and his daily estimations of the mean magnitude of these meteors showed the shower to have begun at 5.4, slowly brightened to 3.5 on July 1, then varied between 4.5 to 5.0 during July 3 to 11. Responding to Yamamoto's letter, Denning showed some doubt about the sensitivity of Nakamura's eyes unless "Nakamura is able to discern meteors of 6th, 7th and 8th magnitudes."[18] Despite Denning's views interesting observations were made during the next appearance of this shower.

During late June and early July 1927, several members of the meteor section of the Russian society Mïrovédénïé, observing at Tashkent, noted increased activity which reached hourly rates of 500 on June 27.[19] According to their director, Vladimir A. Maltzev, "about 90 per cent of the meteors were fainter than the 5th magnitude, which leads to the conclusion that our observations are confirming those made in Japan in Kyoto in 1921." According to N. Sytinskaja, the observations revealed maximum to have fallen on June 27.21, while the radiant's daily motion was noted to be +1.0° in α and –0.3° in δ.[20] Although the number of meteors seen in the United States was much lower than that in Russia, Dole (East Lansing, Michigan) detected 145 meteors during June 26-30. The radiant moved from α=212.5°, δ=+55° on the 27th, to α=218°, δ=+59.5° by the 30th.[21] He remarked that "many brilliant individual meteors" were seen, but, overall, they were "very faint." Curiously, the radiant motion indicated by Dole's observations indicate a northeastward motion, while the Russian data shows a southeastward motion. Similarly, Nakamura's 1921 observations also revealed a southeastward motion, while Franklin Smith's calculations revealed a northeastward motion for the radiant. An explanation for this discrepancy among observers is not easy to explain, but could be linked to the apparent diffuse nature of the radiant, as well as the faintness of the meteors.

Recent activity from this stream indicates it has weakened considerably since the 1920s. In 1968, Edward F. Turco wrote that observations had revealed recent rates of 3 to 5 per hour, "with meteors being on the fairly dim side."[22] In

---

[18]Denning, W. F., *The Observatory*, **45** (March 1922), p. 83.

[19]King, A., *The Observatory*, **50** (November 1927), p. 361.

[20]King, A., *The Observatory*, **52** (January 1929), p. 29.

[21]King, A., *The Observatory*, **51** (January 1928), p.25.

[22]Turco, Edward F., *Review of Popular Astronomy*, **62** (October 1968), p. 7.

1981, David Swann (Dallas, Texas) wrote that on six occasions during the period 1964 to 1971, he observed this stream and detected rates of 1 to 2 meteors per hour. Concerning the individual meteors, Swann noted that he had "never noticed any trains, even though I have seen several bright shower members."[23] He added that, considering the low number of meteors currently coming from this radiant, "the possibility of seeing a fireball from this stream seems good."

Other observers have recently managed to determine radiants from the shower's meager activity levels. On June 28.6, 1964, D. Conger (Elizabeth, West Virginia) observed 5 meteors from α=226°, δ=+59°,[24] while on June 28, 1970, Pennsylvania observers Gary Becker (Allentown) and Mark Adams (Warrington) found radiants at α=223°, δ=+59° and α=224°, δ=+58°, respectively.[25]

It should be noted that, since 1916, attempts have been made to observe this shower in several different years. The entire month of June has gained the attention of potential observers of this shower, but only 1916 displayed activity extending several weeks prior to June 28. Based on the then-published theories of a link with Pons-Winnecke, this indicates that the close proximity of the Earth and comet orbits during June 1916 caused Earth to encounter a broad, diffuse cloud of material as early as May 19. The June 28 activity is, however, a very sharply condensed knot of material. Since 1916, it appears that only material from this well-condensed ring has been encountered, though the weakness of the subsequent displays indicates Earth must only be moving through its outer fringes. On the other hand, an alternative theory might be available for the late May-early June activity of 1916, and this will be discussed later in this chapter under the Tau Herculids.

Although the June Boötids are present in both visual and photographic records, their presence in radar records is not as pronounced. Admittedly, Australian and New Zealand radar surveys would be too far south to enable an identification of this stream. On the other hand among northern hemisphere radar surveys, only Sekanina's Radio Meteor Project conducted in two stages during 1961-1965 and 1969 show signs of this stream.

In Sekanina's early survey, a stream identified as the "July Draconids" was found to be active on July 1 and 2, from an average radiant of α=239.5°, δ=+68.8°. The discrepancy of the radar radiant from that determined visually was due to the radar orbit possessing an orbital inclination of 30.3°. The stream was based on only 5 meteors, and this may represent an excellent example of why researchers should place a low value on data supported by small samples.

---

[23]Swann, David, *MN*, No.53 (April 1981), p. 2.

[24]Olivier, Charles P., *FOR*, No. 155, p. 16.

[25]*MN*, No. 3 (October 1970), p. 2.

At the same time, it has been shown that the June Boötids possess a fairly diffuse radiant due to the rapid alterations of the parent comet's orbit, so that this small sample may have represented a random collection of high-inclination members of this stream. It should be noted that during the comet's transition from possessing a perihelion distance lying within Earth's orbit to possessing a perihelion outside of Earth's orbit, perturbations by Jupiter also acted on the orbital inclination. When discovered in 1819, the inclination was only 10°, while the present orbit is at 22°. Since the June Boötids displayed an obvious diffuse radiant during 1916, 1921 and 1927, which appears due to an orbital inclination spread of about 8°, such a diffusion would have been continued, if not enhanced, by the closest approaches with Jupiter that followed. Thus, Sekanina's orbit seems based on a collection of meteors orbiting near the edges of the main orbit of the June Boötid stream.

Sekanina's 1969 survey more conclusively defined the orbit of the June Boötids. During the period of June 2 to July 19, 54 meteors were detected from a stream Sekanina identified as the "Alpha Draconids." The average radiant was determined as α=207.4°, δ=+64.0° and the orbit was shown to cross the ecliptic on June 22.[26] This orbit closely matches that determined for the visually observed radiants detected from 1921 to the present, as well as the 1927 orbit of Pons-Winnecke. A second stream detected in 1969, was referred to by Sekanina as the "Bootids-Draconids." Possessing a duration extending from July 1 to 4, from an average radiant of α=233.7°, δ=+52.2°, its orbit crossed the ecliptic on July 2. This orbit closely matches that determined by Olivier from the English observations of June 28, 1916.

Orbit

Listed below are the elliptical orbits computed by Olivier in 1916. They reveal a fairly large difference between the orbits of the May-June activity (designated below as "A") and the June 28 activity (designated "B"). As a comparison, the Author has computed elliptical orbits representing the 6 visual radiants seen on June 28/29 of 1921, 1964 and 1970 (designated "C"). All orbits use the semimajor axis of Pons-Winnecke (1915) for calculation purposes.

	ω	Ω	i	q	e	a
(A)	216.9	67.9	17.8	0.927	0.716	3.261
(B)	188.7	96.9	18.7	1.009	0.690	3.261
(C)	178.0	96.6	21.0	1.017	0.688	3.261

A possible explanation of the difference between the orbit of "A" and those of "B" and "C" is given by the Author later in this chapter under the Tau Herculids.

---

[26]S1976, pp. 265-321.

The two orbits listed below are of Pons-Winnecke during its 1915 and 1927 perihelions.

	ω	Ω	i	q	e	a
P-W 1915	172.4	99.8	18.3	0.971	0.702	3.261
P-W 1927	170.4	98.5	18.9	1.039	0.686	3.306

The radar orbits determined by Sekanina from observations made during 1962-1965 (orbit "D") and 1969 (orbits "E" and "F"), from Havana, Illinois, are as follows:

	ω	Ω	i	q	e	a
(D)	171.2	99.0	30.3	1.009	0.616	2.625
(E)	168.0	90.5	21.7	1.000	0.596	2.479
(F)	184.0	99.8	21.3	1.014	0.596	2.511

As can be seen, the orbits "D" and "E" bear a close resemblance to the 1915-1927 orbits of Pons-Winnecke—"D" represents Sekanina's July Draconids, while "E" represents his Alpha Draconids. Orbit "F" (Sekanina's Boötids-Draconids) closely matches the orbit calculated for the 1921, 1964 and 1970 observations (see orbit "C").

Using an orbit similar to "B," Ken Fox (F1986) determined the stream orbit for 1000 years into the past and future. The following were determined for the June Boötids:

	ω	Ω	i	q	e	a
950	179.6	97.6	18.0	1.06	0.68	3.31
2950	182.3	97.9	17.8	0.98	0.70	3.25

Interestingly, this study reveals the present June Boötid stream to be in a fairly stable orbit, even though the observations fail to bare this out.

Tau Herculids

Observer's Synopsis

The duration of this shower extends from May 19 to June 19. Maximum occurs on June 9, from an average radiant of α=236°, δ=+41°, with the average magnitude then being about 4.

History

The discovery of this meteor shower occurred as a result of the discovery of its parent comet. The latter discovery occurred on May 2, 1930, when plates exposed by A. Schwassmann and A. A. Wachmann (Hamburg Observatory), during a regular minor planet survey, revealed the diffuse image of a comet. The saga of the Tau Herculids unfolded a short time later at Kwasan Observatory (Kyoto, Japan).

Observations of Schwassmann-Wachmann 3 were made daily at Kyoto, and, after a few positions had been obtained, Watanabe computed a preliminary parabolic orbit. From this orbit, Shibata deduced a radiant point and predicted a shower would soon occur.[27]

Beginning on the night of May 21, several Japanese observers, most of whom were located at Kwasan Observatory, began watching the skies for the predicted shower. Searches were fruitless until May 24, when a careful watch around Boötes revealed a stationary meteor at α=230°, δ=+48°.[28] The next evening, several more meteors were seen radiating from a point slightly less than a degree east of the previous night's position.

Observations of this shower ceased for a short time following May 25, since Schwassmann-Wachmann 3 had attained naked-eye visibility at the end of May, but by June 3 the Tau Herculids were again under scrutiny and were noted coming from a radiant of α=232°, δ=+46°. Further observations were made on the 6th and 7th as the radiant continued its southeastward trek, but the fairly weak activity observed up to this point suddenly changed on June 9, when 59 meteors (most of which were described as fainter than 4th magnitude) were detected in one hour. Rates were slightly higher on the next evening when 36 meteors were seen in 30 minutes. On the former date, short-trailed meteors were used to determine the radiant position as α=236.25°, δ=+41.5°. Weak displays had again returned when observations resumed on June 12 and 13, and the shower was last detected on June 19, from a radiant of α=244.5°, δ=+39°.[29]

It should be noted that by the first days of June the prediction of a possible strong meteor shower had been published in newspapers around the world, but Kaname Nakamura was the only observer to note a strong display. Even Issei Yamamoto, director of the Kwasan Observatory, noted that "Mr. Nakamura was practically the only observer" among staff members of the observatory. However, Yamamoto did point out that a radiant position of α=237.5°, δ=+41° had been determined by K. Siomi (Hukutiyama, Japan) on the night of June 12-13. Thus, the shower seems to have been confirmed.[30] Unsuccessful, however, were the meteor section members of the British Astronomical Association. They noted that bright moonlight interfered with their observations on June 5, 7 and 9, so that no display of any kind was noted.[31] From this data it seems the Tau

---

[27]Nakamura, Kaname, *MNRAS*, **91** (November 1930), p. 204.

[28]Nakamura, pp. 204-205.

[29]Nakamura, p. 206.

[30]Yamamoto, Issei, *MNRAS*, **91** (November 1930), p. 209.

[31]King, A., *The Observatory*, **54** (July 1931), p. 201.

Herculids possess a very sharply defined maximum, with meteors being predominantly faint.

Attempts to locate past appearances of this shower have revealed few possibilities. Most promising, however, are a series of radiants determined by John Koep and Philip Trudelle (both of Chippewa Falls, Wisconsin) between May 19 and June 5, 1916.[32] The radiants are as follows:

Possible Tau Herculid Radiants of 1916

Date (UT)	RA(°)	DEC(°)	Meteors	Observer
May 22.1	224.5	25.3	8	Koep
May 27.18	230.3	27.4	7	Trudelle
May 27.75	231.0	27.5	7	Koep
May 28.17	232.1	26.8	4	Trudelle
May 30.64	232.7	28.	15	Koep
June 4.2	234.4	27.5	16	Koep
June 5.18	235.8	25.6	9	Trudelle

Admittedly, the declination is nearly 20° too far south in May and about 10° too far south in June; however, as will be shown in the "Orbit" section below, the shower's orbit is closer to the Tau Herculids.

The only other pre-1930 detection of this shower came during June 3-7, 1918, when William F. Denning detected 4 very slow "Theta Coronids" from a radiant of α=230°, δ=+34°.

Following the 1930 shower, attempts at further observations proved fruitless. J. P. M. Prentice (Stowmarket, England) watched on May 20, 22, 23 and 24, 1931. His total observing time amounted to 11 hours and 20 minutes, but no activity was noted. Stars to magnitude 6 were visible some of the time.[33] Additional attempts to reobserve this shower—usually during the years of the comet's predicted perihelions—occurred on several occasions during the 20 years following 1931, but no traces of the shower were ever noted. However, using photography during the early 1950's, the Harvard Meteor Project revealed the stream to still be producing meteors.

The first detection of the photographic Tau Herculid meteors among the 1950's data was made by Richard B. Southworth and Gerald S. Hawkins in 1963. They detected just two meteors, but the similarity of the derived orbit to Schwassmann-Wachmann 3 was close.[34] A further look at possible photographic meteors associated with this comet occurred in 1971, when Bertil-

[32]Olivier, Charles P., *MNRAS*, **77** (November 1916), p. 73.

[33]King, p. 201.

[34]Southworth, R. B., and Hawkins, G. S., *SCA*, **7** (1963), pp. 274 & 280.

Anders Lindblad (Lund Observatory, Sweden) identified 14 meteors from the Harvard Meteor Project. The average orbit "suggests good agreement in all orbital elements, and the proposed comet-meteor relation may now be considered very probable."[35]

The last extensive search for meteors of this stream came in 1974, following a prediction by K. Kono.[36] With Schwassmann-Wachmann 3 coming toward a mid-March 1974 perihelion, Kono predicted that a shower might occur on June 1.0 from α=146°, δ=+54° (the comet's orbit having underwent some changes since its 1930 apparition). Harold Povenmire noted that observations for 2 hours on June 1 by members of the Florida Fireball Network revealed only one meteor from the radiant, though moonlight was interfering.[37]

Other recent observations of this shower continue to show extremely weak activity:

> **1976**—During 19 hours and 20 minutes of observing during June 1 to 25, Bert Matous (Grandview, Missouri) saw only 4 Tau Herculids.[38]
>
> **1977**—John West (Bryan, Texas) saw 2 Tau Herculids in 4 hours on May 26 and 28, C. Smith (Bryan, Texas) saw 3 in 3 hours.[39] Matous saw 17 in 12 hours 30 minutes during June 6-11, while Norman McLeod (Florida) recorded 5 in 15 hours and 9 minutes during May 21/22-June 11/12.[40]
>
> **1979**—F. Roy (Pte. Gatineau, Quebec, Canada) saw 3 meteors in 2 hours 15 minutes on June 3/4.[41] (Schwassmann-Wachmann 3 was recovered for the first time since 1930, and passed perihelion on September 2.)
>
> **1982**—West saw 4 meteors in 4 hours on May 22/23 and 27/28, while M. Zalcik (Edmonton, Alberta, Canada) observed 2 in 3 hours on May 23/24.[42]
>
> **1984**—David Swann (Dallas, Texas) saw no meteors during 2 hours on May 29/30, one during 2

---

[35]Lindblad, Bertil-Anders, *SCA*, **12** (1971), pp. 19 & 23.

[36]Marsden, Brian G., *IAU Circular*, No. 2672 (May 21, 1974).

[37]Marsden, Brian G., *IAU Circular*, No. 2676 (June 5, 1974).

[38]*MN*, No. 33 (Mid-October), p. 3.

[39]*MN*, No. 37 (August 1977), p. 11.

[40]*MN*, No. 38 (Mid-October, 1977), p. 7.

[41]*MN*, No.47 (October 1979), p. 7.

[42]*MN*, No. 58 (July 1982), p. 9.

> hours on May 31/June 1 and one during 1 hour 30 minutes on June 1/2. Bill Katz (Willowdale, Ontario, Canada) observed during 6 hours on June 3/4. Tau Herculids were noted each hour with the highest hourly rate being six and the lowest being two.[43] This marked some of the highest rates in recent years. (The comet was approaching a January 11, 1985, perihelion.)

From an examination of the apparent observations of this comet, it seems that a few meteors have managed to venture to the opposite end of the comet's orbit; however, this still amounts to less than one meteor per hour near maximum. On the other hand, it seems activity is still strongest during years when the comet is near perihelion. Since the meteor stream is probably fairly young, it may be possible that a sharply defined and fairly strong maximum might occur during years the comet reaches perihelion, as in 1930 and 1984. The 1916 activity came roughly two years after a perihelion passage of Schwassmann-Wachmann 3.

Orbit

Orbit "A" represents an elliptical orbit computed from Nakamura's positions of 1930, using the semimajor axis of Schwassmann-Wachmann 3. Orbit "B" represents the orbit obtained from 11 photographic meteors collected from MP1961, JW1961 and B1963. Also listed is the elliptical orbit computed by Olivier in 1916 for the May-June activity of the June Boötids (designated below as "C"). The latter orbit was based on the semimajor axis of Pons-Winnecke (1915).

	ω	Ω	i	q	e	a
(A)	202.1	75.9	22.2	0.985	0.681	3.090
(B)	195.7	78.5	18.1	0.998	0.617	2.606
(C)	216.9	67.9	17.8	0.927	0.716	3.261

Following are the orbits of two comets. The first represents that of Schwassmann-Wachmann 3 during its discovery apparition of 1930, while the second is that of Pons-Winnecke during its 1915 perihelion.

	ω	Ω	i	q	e	a
SW 3 1930	192.3	77.1	17.4	1.011	0.673	3.090
PW 1915	172.4	99.8	18.3	0.971	0.702	3.261

The orbits of both comets are very similar, but the main differences are in ω and Ω—with Schwassmann-Wachmann 3 actually fitting the May-June 1916 data better than Pons-Winnecke.

---

[43] *MN*, No. 66 (July 1984), p. 7.

Using an orbit midway between "A" and "B," Ken Fox (F1986) determined the stream orbit for 1000 years into the past and future. The following were determined for the Tau Herculids:

	ω	Ω	i	q	e	a
950	169.1	90.2	23.1	1.07	0.60	2.68
2950	321.5	322.3	13.6	0.91	0.66	2.69

This study reveals a very interesting variation of the Tau Herculid activity over 2000 years. About 1000 years ago the shower's maximum occurred during the latter half of June from α=220.6°, δ=+72.4°, while 1000 years in the future maximum will occur in mid-August from α=154.7°, δ=−29.4°. Even more interesting is the similarity between the orbit of 950 and the present orbits given under the June Boötids, especially orbit "E"—the radar orbit obtained by Sekanina during 1969. Perhaps the June Boötids, and maybe even Pons-Winnecke, are somehow related to the Tau Herculids and Schwassmann-Wachmann 3.

June Lyrids

Observer's Synopsis

This shower is active during June 10 to 21, producing predominantly blue and white meteors at a maximum hourly rate of 8 per hour on June 15 (λ=84.5°). At maximum the radiant is located at α=278°, δ=+35°. The average observed magnitude of this shower is near 3, while about 32% of the meteors leave trains.

History

This meteor shower was discovered on the evening of June 15, 1966, by Stan Dvorak (California) while camping out in the San Bernardino mountains. His attention had been drawn to the region of Lyra by a very bright meteor that moved swiftly to the northeast through that constellation. Another meteor was noted a short time later and Dvorak began plotting additional meteors. After 1 1/2 hours he had managed to plot 16 meteors, of which 13 appeared to originate from a hitherto unknown radiant located at α=278°, δ=+30°.[44] Just a few hours later, F. W. Talbot (Cheshire, England) independently discovered the radiant at α=275.5°, δ=+30°, and noted an hourly rate near 9.[45]

Moonlight from a waning moon interfered with observations in 1967, but, in June 1968, confirmation of this shower's existence came from Richard Nolthenius (Hacienda Heights, California). During one hour on the 15th, he

---

[44]Dvorak, Stan, *Sky & Telescope*, **32** (October 1966), p. 237.

[45]Hindley, Keith B., *JBAA*, **79** (1969), p. 480.

detected 8 June Lyrids, while a similar hour on the 17th revealed 7.[46] Observations of this shower have continued annually ever since; however, to date, the most elaborate study of this shower was made from observations obtained during 1969.

Observations from 46 observers, totalling 172 man-hours, were gathered and analyzed by Keith B. Hindley of the British Astronomical Association.[47] The observations covered the period of June 11.5 to 21.0 (λ=80.2° to 89.2°) and the total number of June Lyrids observed was 363. Against a fairly constant sporadic meteor rate of 8.7 per hour, the June Lyrids displayed a broad maximum of about 6 per hour during June 13 to 17, with a sharp peak of 9 per hour on June 16.0 (λ=84.5°). The average magnitude was found to be 2.0 and 32% of the shower's meteors showed persistent trains. Colors displayed a prominence of white, as did sporadic meteors, however, there was also a large number of blue meteors. The exact percentages of colors seen among the June Lyrids were as follows: 33%, blue; 52%, white; 9%, yellow; and 6%, orange-red.

The average radiant, as determined by Hindley, was found to be α=278°±2°, δ=+35°±3°. From this position, a parabolic orbit was calculated, which revealed a close, but "not convincing" similarity to comet Mellish (1915 II). This comet possesses a slightly hyperbolic orbit with an eccentricity of 1.0002.[48] A radar orbit obtained by Zdenek Sekanina from observations made at Havana, Illinois, in 1969, actually seemed to strengthen the similarity between the June Lyrids and 1915 II, except for the fact that the radar orbit revealed a period of 2.94 years.[49] These orbits are compared later.

Although observations continued after 1969, there seemed an indication that the meteor rates of 8 or 9 per hour had vanished. In 1971, about 215 hours of observations were acquired by 26 observers from Canada and the United States. The shower's ZHR irregularly varied during the period of June 10 to 24, with maximum peaks of 1.3 to 3.5. The radiant of the stream was determined from 37 meteors plotted by an Ottawa observing group from observations made during June 14 to 17, with the result being α=278.3°, δ=+41°.[50]

In 1972, the shower again showed a poor return. *Meteor News* combined the observations of 20 observers, made during June 9 to 22, and found the maximum ZHR to be only 2.3—a value that came on June 15/16. What the

---

[46]Hindley, p. 480.

[47]Hindley, pp. 481-484.

[48]Marsden, Brian G., *Catalog of Cometary Orbits.* New Jersey: Enslow Publishers, 1983, p. 18.

[49]S1976, pp. 278 & 294.

[50]*MN*, No. 8 (Mid-October 1971), p. 6.

observations also revealed were activity levels between 1.2 and 1.3 during June 12 to 15, which suddenly jumped to 2.3, then rapidly fell to only 0.3 by the night of June 18/19.[51] Thus, the shower's date of maximum and variation of activity levels closely reflected those noted in 1969, but the ZHRs were typically down by about 6 or 7 meteors per hour!

Observations in 1974 indicated a resurgence of activity. The shower's discoverer, Dvorak, observed on four nights during mid-June, with the following average hourly rates (not ZHRs) being noted: June 14/15, 2.9; June 15/16, 6.5; June 16/17, 6.2; June 17/18, 2.3. Dvorak added that the meteors moved swiftly, with the majority being bluish-green. Also, in 1974, Nolthenius, Alan Devault and Bob Fischer (all of California) observed during 7 nights between June 9 and 22. Overall, they observed 32 June Lyrids, with the average magnitude being determined as 3.09. Slightly less than half of their total number of June Lyrids were seen on the shower's night of maximum (June 15/16). The average number of meteors seen from this radiant was 4 per hour.[52]

During 1975, observations were quite scarce. Between June 6 and 15, Norman W. McLeod III (Florida) saw only 2 June Lyrids, while, during the period June 9 to 14, Mark Adams (Pennsylvania) noted 5 members, with 4 coming in less than 3.5 hours on the final date of observation. Finally, Paul Jones (St. Augustine, Florida) observed for 3 hours on the night of maximum and detected 20 June Lyrids.[53]

Interest in the June Lyrids seems to have waned in the latter half of the 1970s and into the 1980s, with only a few individuals continuing to monitor the shower annually—many of them rarely observing around the time of the shower's established date of maximum. As is evident from the previously listed observations, activity from this radiant can be virtually nonexistent on dates other than June 15 and 16, due to the June Lyrids' very pronounced peak of activity.

In 1979, observers in Texas observed on June 15/16 and 17/18, with meteors being seen to reach nearly 2 per hour on the latter date. In Quebec, F. Roy observed during June 16/17, 18/19 and 20/21 and noted 12 meteors in about 3.5 hours[54]—certainly marking the most activity seen from this radiant since 1974. During 1980, the enhanced rates still seemed present, with John West (Texas) noting 18 meteors in only 5 hours on June 13/14 and 14/15.[55] Since 1980, the shower's activity has again dropped to maximum rates of about 1 per hour.

---

[51]*MN*, No. 13 (October 1972), p. 9.

[52]*MN*, No.27 (August 1975), p. 2.

[53]*MN*, No. 28 (Mid-October, 1975), p. 4.

[54]*MN*, No. 47 (October 1979), p. 7.

[55]*MN*, No. 51 (October 1980), p. 5.

A search for pre-1966 observations of this shower has not revealed many clues to the shower's history. From observations by Giuseppe Zeziolli, G. V. Schiaparelli isolated 11 meteors observed on June 14, 1869, from a radiant of α=280°, δ=+35°.[56] No convincing June Lyrid radiants were observed by members of the American Meteor Society prior to 1966, nor were any observations present among the extensive visual radiants obtained by A. S. Herschel, Alphonso King, Cuno Hoffmeister or Ernst Öpik.

Aside from the 1969 study by Hindley, studies of the June Lyrids have been rare, with the only other information gathered on the individual meteors involving the average magnitude. John West observed 59 June Lyrids during 1967 to 1982, and found the average magnitude to be 3.02.[57] During 1967 to 1981, Rick Hill (North Carolina) observed 65 June Lyrids and found the average magnitude to be 2.71.[58]

Orbit

Three orbits are given for this stream in the literature. Orbit "A" is a parabolic orbit computed by Keith B. Hindley from observations made by observers in 1969, while orbits "B" and "C" were computed by Allan F. Cook and are considered "likely extremes."

	ω	Ω	i	q	e	a
A	214	85	52	0.91	1.00	∞
B	237	84.5	44	0.83	0.67	2.5
C	231	84.5	50	0.84	0.92	10

During the Radio Meteor Project directed by Zdenek Sekanina in 1969, 11 probable members of the June Lyrids were detected. The following orbit was published in S1976:

	ω	Ω	i	q	e	a
S1976	224.1	85.5	45.3	0.912	0.556	2.054

The following orbit of comet Mellish (1915 II) was obtained from Brian G. Marsden's *Catalog of Cometary Orbits* (1983):

	ω	Ω	i	q	e	a
1915 II	247.8	72.8	54.8	1.005	1.000	∞

Based on orbit "B" above, Ken Fox computed the following orbits of this stream for 1000 years in the past and future:

	ω	Ω	i	q	e	a
950	238.1	91.7	46.0	0.83	0.67	2.50
2950	235.8	77.7	42.1	0.80	0.68	2.50

---

[56] D1899, p. 273.

[57] *MN*, No. 61 (April 1983), p. 6.

[58] *MN*, No. 63 (October 1983), p. 6.

These indicate that 1000 years ago the June Lyrids were reaching maximum in late June from α=286.0°, δ=+36.4°, while 1000 years from now they will be at maximum during early June from α=271.0°, δ=+35.7°. Thus, if the stream is in a short-period orbit, it should have been visible for quite some time. Since it seems to have suddenly appeared in 1966, this may strengthen the long-period comet hypothesis, though 1915 II may not be this comet.

Ophiuchids

Observer's Synopsis

The duration of this shower extends between May 19-July 2, with maximum activity falling on June 20 (λ=88°) from α=263°, δ=–20°. The maximum ZHR reaches 6, while the average magnitude of the meteors seems slightly fainter than 3. Less than 5% of the stream's meteors leave persistent trains.

History

This weak shower has been with meteor observers for some time, with published observations appearing nearly every year during the 20th century. The shower's discovery seems attributable to William F. Denning, who, during June 14-20, 1887, plotted 5 very slow and bright meteors coming from α=268°, δ=–24°.[59] The first detection of a meteor coming from this radiant seems to have been on June 8, 1841, when a fireball was listed by G. von Niessl as having originated from α=266°, δ=–16°.

Interestingly, the stream seems to produce a large number of bright meteors and fireballs, as the following table illustrates.

Ophiuchid Meteors and Fireballs

Date	RA(°)	DEC(°)	Source
1841, June 8	266	–16	D1899
1899, June 2	250	–23	D1912A
1910, May 24	248	–20	D1912A
1910, June 1	250	–22	D1912A
1914, June 15	260	–22	D1916
1931, June 10	250	–25	*FOR*, No. 146

The first formal recognition that this shower was a producer of annual activity was in R. A. McIntosh's "An Index to Southern Meteor Showers."[60]

[59] D1899, p. 272.

[60] M1935, p. 715.

Designated as radiant 180, it was said to occur during June 14 to 21, from α=266°, δ=–18.5°. Another radiant, designated number 184, was named the "3 Sagittarids" and fits the observations slightly better than 180. It was said to occur during June 14-15, from α=268°, δ=–26°. These two radiants indicate an interesting feature displayed by this shower (and nearly all ecliptic streams, for that matter), and that is the existence of both a north and south branch.

The stream's inclination is so small that some meteors originate above the ecliptic, while others originate below it. This has caused both radar and photographic orbits to be variously published with ω and Ω varying by 180° (this difference is demonstrated in the "Orbit" segment below). Despite this apparent difference in the two key angular elements, these orbits are still similar in size and shape, with the D-criterion calculation of Southworth and Hawkins recognizing that such orbits can be closely related. This is very important since such orbital differences might be due more to inaccurate data than to natural orbital evolution. A similar conclusion was suggested by Bertil-Anders Lindblad in 1971, after noticing several ecliptic streams were being split into northern and southern branches by computer analysis.[61] Most photographic meteors and visual meteors of this stream tend to come from the radiant north of the ecliptic.

Recent observations of this shower indicate weak, but consistent activity. During 1971, *Meteor News* published the results of 38 hourly rates reported by 7 observers in Australia and the United States. Between June 15.2 and June 21.2, rates rose slowly to a maximum of 8 per hour, then rapidly fell off to values of 0 to 1 per hour during June 22.2 to 27.2.[62] This rate of 8 per hour might better represent what activity levels would be like in the Southern Hemisphere since observations from Australia were included. Estimates by observers in the United States reveal values of 1 to 2 per hour on the night of maximum during the years of 1972 to 1985. In addition, activity levels do tend to be significantly lower before and after the date of maximum. The British Astronomical Association lists the ZHR of this shower as 6.[63]

The only major study of this stream to reveal other details besides hourly rate estimates was made in 1978 by observers of the West Australia Meteor Section.[64] A total of 70 shower members were detected. The average magnitude was determined to be 3.25 and 4.3% of the meteors showed persistent trains.

This meteor shower is considered one of the candidates for possible association with periodic comet Lexell (1770 I). The Ophiuchid orbit does not perfectly match that of the comet, but the very low inclination would allow very

---

[61]Lindblad, Bertil-Anders, *SCA*, **12** (1971), p. 18.

[62]*MN*, No. 8 (Mid-October 1971), p. 7.

[63]*The Handbook of the British Astronomical Association* (1981).

[64]*MN*, No. 45 (April 1979), p. 12.

small planetary perturbations to vary the ω and Ω, by large amounts. According to Porter,[65] the probable date of maximum activity from a stream associated with the comet would be July 5, at a radiant of α=272°, δ=–21°.

Orbit

During the Radio Meteor Project conducted by Zdenek Sekanina during 1968-1969, a total of 28 meteors indicated the following orbit (north branch?):

ω	Ω	i	q	e	a
279.3	85.5	0.3	0.503	0.774	2.224

During the Southern Hemisphere Radio Survey, conducted by C. S. Nilsson during 1960-1961, the following orbit was revealed (south branch?):

ω	Ω	i	q	e	a
97.0	265.3	5.0	0.52	0.75	2.083

Sekanina's radar orbit was based on 28 radio meteors, while Nilsson's was based on only 3. If the assumption that a lack of precise data is responsible for orbits such as Nilsson's, then Sekanina's orbit should be considered as the proper orbit for this stream. Most of the visual radiants fit Sekanina's orbit better than Nilsson's. Nilsson pointed out that the node "could in fact be descending, not ascending."[66]

The orbit of comet Lexell (1770 I) is listed in Brian G. Marsden's *Catalog of Cometary Orbits* (1983) as follows:

ω	Ω	i	q	e	a
224.9	133.9	1.6	0.674	0.786	3.153

Theta Ophiuchids

Observer's Synopsis

The duration of this shower extends from May 21 to June 16. A relatively flat maximum of 5 days occurs centered on June 10 (λ=77.8°) from α=265°, δ=–28°, with a ZHR of 10.

History

The discovery of this stream seems attributable to Cuno Hoffmeister and is a major segment of his vast "Scorpius-Sagittarius-System" discussed in his book *Meteorströme* (1948). Hoffmeister noted that numerous active areas appeared in this region of the sky beginning in late April and ending in mid-July. Around June 10 (λ=77.8°), a radiant at α=263.2°, δ=–28.3° becomes active. Hoffmeister

---

[65]Porter, J. G., *Comets and Meteor Streams*. London: Chapman and Hall, 1952.
[66]N1964, p. 239.

referred to this radiant as representing the "distinct core" of the Scorpius-Sagittarius system and he described the activity as a "flat maximum at 75° to 80°."[67]

The first true confirmation of this stream was made in 1963, when Richard B. Southworth and Gerald S. Hawkins identified 5 meteors photographed during the Harvard Meteor Project (1952 to 1954) as members of a "new" stream called the "Theta Ophiuchids."[68] The duration of this stream was given as May 21 to June 16, and it was determined that 0.37 meteors could be photographed every hour. The average radiant was given as α=264.0°, δ=–26.8°.

In 1971, Bertil-Anders Lindblad essentially confirmed Southworth and Hawkins' findings—with one extra meteor being found. Lindblad had conducted a computerized search among 865 precise photographic meteor orbits and noted the same duration, but with the average radiant being at α=265°, δ=–27°.[69] Further proof of the stream's existence was obtained by Lindblad during a second computer stream search among 2401 photographic meteors.[70] The duration on this occasion was determined to be June 4-16, while the average radiant was α=266°, δ=–28°.

This shower has appeared in the observational records for a great number of years, with the name changing depending on the observer or researcher handling the data. It has been variously referred to as the "Alpha Scorpiids," "Scorpiids-Sagittariids" and "Delta Sagittariids." The name of "Theta Ophiuchids" was attached during the early 1960's by Southworth and Hawkins and, in effect, the radiant lies midway between Delta Sagittarii and Alpha Scorpii. In addition, it is closer to Theta Ophiuchi at the time of maximum.

The stream plays an important part in the understanding of the structure of meteor showers. This stream, as well as the Virginid complex of April, has typified the structure of an ecliptic meteor stream—a stream continuously affected by planetary perturbations. According to Hoffmeister, this stream (under the name of the "Scorpius-Sagittarius-System") represented the "typical case of a dissolved current. There is in times no defined radiant appreciable, the meteors radiating uniformly from an area up to 30° in diameter. But sometimes very pronounced centers use to appear, vanishing again within 24 hours."[71]

During 1961-1965, Zdenek Sekanina directed the Harvard Radio Meteor Project and found a stream with a duration covering May 7 to June 28, with an average radiant at α=262.5°, δ=–19.8°. Although at first glance this radiant

---

[67]M1948, pp. 139 & 172.

[68]Southworth, R. B., and Hawkins, G. S., *SCA*, **7** (1963), p. 271.

[69]Lindblad, Bertil-Anders, *SCA*, **12** (1971), p. 8.

[70]Lindblad, Bertil-Anders, *SCA*, **12** (1971), p. 19.

[71]H1948, p. 172.

seems to be too far north to be related to the Theta Ophiuchids, it should be noted that the radiant for this stream lies almost precisely on the ecliptic. This causes two apparently different orbits to appear; however, aside from the shift of 180° in both ω and Ω, the orbits are very nearly in the same plane and are the same shape. Thus, this stream possesses some very similar characteristics to those already discussed for the Ophiuchids and it probably would not be too presumptuous to say that both the Ophiuchids and the Theta Ophiuchids share a common origin. A major difference, however, is that while the Ophiuchids usually originate from a point north of the ecliptic, the Theta Ophiuchids tend to originate south of the ecliptic.

Detailed observations of this shower are hard to come by except for occasional radiant determinations. During 1978, members of the West Australia Meteor Section variously reported hourly rates of 3-4 during June 3/4 and 4-10 during June 10/11.[72] The British Astronomical Association lists the ZHR of this stream as 10.[73] Observers in the United States, however, rarely detect more than 1 to 2 meteors per hour on the night of maximum.

Orbit

The orbit of this stream was first established by Hoffmeister and is given as orbit "A", while orbit "B" was determined by Lindblad using 6 photographic meteors found during the Harvard Meteor Program. They are as follows and may refer specifically to a southern branch:

	ω	Ω	i	q	e	a
A	101	262	4	0.46	0.84	2.90
B	108.7	257.0	4.4	0.43	0.839	2.67

The possible northern branch of this stream detected by Sekanina's Radio Meteor Project during 1961-1965, is as follows:

	ω	Ω	i	q	e	a
S1973	293.3	72.4	3.9	0.386	0.804	1.970

Zeta Perseids

Observer's Synopsis

This daylight shower occurs during May 20 to July 5. Maximum occurs on June 13, from an average radiant of α=63°, δ=+26°. Radar surveys have revealed the activity of this shower to be near 40 per hour. The daily motion of the radiant amounts to +1.1° in α and +0.4° in δ.

---

[72]*MN*, No. 45 (April 1979), p. 12.

[73]*The Handbook of the British Astronomical Association* (1986), pp. 84-85.

History

The Zeta Perseids were discovered in 1947 by operators of the radio equipment at Jodrell Bank (England). Only rough details were available at that time, with the duration being determined as June 2-17, and the radiant being estimated as falling within the range of α=52° to 62°, δ=+15°.[74] Jodrell Bank observations continued during 1949 and 1950, with A. Aspinall and Gerald S. Hawkins establishing a duration of June 1-16 and a typical radiant diameter of less than 3°. The average radiant position was given as α=61.6°, δ=+23.8°.[75]

A radar study was conducted by B. L. Kashcheyev and V. N. Lebedinets during 1960, using equipment at the Kharkov Polytechnical Institute (USSR). The analysis revealed a duration of May 4 to June 19, with the nodal passage coming on June 2 (λ=71°), and the average radiant being α=52°, δ=+23°.[76]

During the first session of the Radio Meteor Project, Zdenek Sekanina determined the duration as May 20 to June 21. The date of the nodal passage was given as June 8.9 (λ=77.6°), at which time the radiant was at α=60.2°, δ=+24.8°.[77] During the 1968-1969 session, Sekanina determined the duration as May 22-July 4. The date of the nodal passage was then given as June 12.2 (λ=80.8°), while the average radiant position was determined to be α=63.3°, δ=+27.1°.[78]

Two radar studies were conducted using equipment at the University of Adelaide (South Australia) during the 1960's. Unfortunately they were operated over periods of about a week, so that their results might be considered somewhat misleading when considering the radiant position and orbit. The first operated in 1961, when C. S. Nilsson detected the Zeta Perseids during June 13-16, and determined the radiant as α=64.2°, δ=+25.4°. He mentioned that the equipment had also been operated during May 19-28, and that a radiant at α=44.3°, δ=+19.5° was probably the same as the June shower.[79] The second survey was conducted during June 9-14, 1969 by G. Gartrell and W. G. Elford. It revealed a radiant at α=65°, δ=+27°, and the authors showed that the stream was probably the twin of the Southern Taurids of November.[80]

From observations made in the United States and Australia during 1971, it appears that meteors from this shower can be visually detected coming up from the horizon during the hours immediately after sunset and immediately before

---

[74]CHL1947, pp. 374-375.

[75]Aspinall, A., and Hawkins, G. S., *MNRAS*, **111** (1951), p. 22.

[76]KL1967, p. 188.

[77]S1973, pp. 256 & 259.

[78]S1976, pp. 277 & 294.

[79]N1964, pp. 226-228 & 242.

[80]GE1975, pp. 597 & 604.

sunrise.[81] Daryl Skelsey (Sydney, Australia) observed 1 Zeta Perseid during 2 hours on June 5/6, while Karl Simmons estimated that the combined rates of the Zeta Perseids and Arietids (earlier in chapter) reached 1 to 2 meteors per hour on the morning of June 6/7.

Orbit

The exact orbit of this stream has been determined on several occasions. The following examples show what appears to be a very well established orbit:

	ω	Ω	i	q	e	a
N1964	55.4	83.8	5.7	0.300	0.82	1.667
KL1967	57	71	6	0.31	0.80	1.61
S1973	59.2	77.6	5.3	0.319	0.834	1.918
GE1975	69	81	7.1	0.30	0.82	1.724
S1976	60.5	80.8	6.5	0.365	0.755	1.492

The orbit from N1964 was based on 9 meteors, that from KL1967 was based on 60 meteors, that from S1973 was based on 44 meteors, that from GE1975 was based on 6 meteors, and that from S1976 was based on 56 meteors.

Phi Sagittariids

Observer's Synopsis

This stream possesses a duration extending from June 1 to July 15. It reaches a fairly weak maximum on June 18 (λ=86.6°) from an average radiant of α=278°, δ=-25° and then attains a ZHR of about 5. A recent name attached to this stream was "Scorpiids-Sagittariids."

History

In his 1948 book *Meteorströme*, Cuno Hoffmeister investigated numerous visual radiants collected during 1908 to 1938, and identified what he called the "Scorpius-Sagittarius-System." It has earlier been shown that the Theta Ophiuchids are apparently the main core of this vast system, but the present stream actually represents one of the weaker of the consistent radiants to become active from that system.

The first definite observation of this shower was made by William F. Denning on June 21, 1917, although this observer claimed to have seen meteors from this area of the sky during June 18 to 28, 1887. Denning also showed that

[81] *MN*, No. 7 (August 1971), p. 6.

several meteors and fireballs had been observed from an average radiant of α=282°, δ=–24° during the period of 1902 and 1917.[82]

Phi Sagittariids Radiants

Date (UT)	Designation	RA(°)	DEC(°)	Observer
1931, June 14.94	AMS 2568	275.5	–27.0	Geddes
1931, June 15.96	AMS 2571	276.5	–26.5	Geddes
1931, June 19.02	AMS 2575	272.5	–23.0	Geddes
1931, June 20.06	AMS 2578	277.8	–24.5	Geddes

The observations by Murray Geddes were obtained from the June-July, 1932, issue of *Popular Astronomy*.

The earliest study of this stream was made by Hoffmeister, who, using observations made during the period of 1920-1937, determined the radiant as α=279.8°, δ=–22.8° when the shower reached maximum at a solar longitude 86.6°. Interestingly, Hoffmeister succeeded in detecting this shower on four nights during June 11-15, 1937.

Surprisingly, this stream has not shown any significant orbital dispersion in radar surveys conducted during the 1960's. C. S. Nilsson's radar survey conducted at Adelaide during 1960-1961, revealed 3 meteors originating from an average radiant of α=275.2°, δ=–24.5°, during June 15-18.[83] During the 1961-1965 Radio Meteor Project conducted at Havana, Illinois, and directed by Zdenek Sekanina, 17 meteors were detected during the period of June 1 to July 2, from an average radiant of α=277.8°, δ=–25.3° (the stream was referred to as the Scorpiids-Sagittariids).[84] Sekanina's second Radio Meteor Project conducted during 1968-1969, again revealed this stream. On this occasion, 31 meteors were detected during the period of June 2 to July 15, from an average radiant of α=282.2°, δ=–25.2°.[85] Finally, during 1968-1969, G. Gartrell and W. G. Elford conducted a radar survey at Adelaide Observatory. Based on 6 meteors, the radiant was seen during June 10-13, 1969, at an average position of α=271°, δ=–24°.

Orbit

The following four orbits, acquired from radar surveys conducted during the 1960's, show a remarkable consistency for an ecliptic stream, although a large variation is present concerning the semimajor axis.

---

[82]Denning, W. F., *The Observatory*, 40 (August 1917), pp. 307-308.

[83]N1964, pp. 226 & 232.

[84]S1973, pp. 259 & 267.

[85]S1976, pp. 265-321.

	ω	Ω	i	q	e	a
N1964	113.7	265.0	4.1	0.33	0.90	3.33
S1973	113.6	266.7	2.8	0.361	0.861	2.594
S1976	113.8	270.4	2.5	0.384	0.799	1.908
GE1975	114.0	261.0	0.6	0.39	0.76	1.667

Chi & Omega Scorpiids

Observer's Synopsis

The Chi and Omega Scorpiids are closely related ecliptic streams. The former, or northern stream, possesses a typical long duration which extends from May 6 to July 2. Maximum activity is not well defined, but seems most likely to occur between May 28 and June 5, from an average radiant of α=245°, δ=–12°. The Omega Scorpiids form the southern component and are active during May 19 to July 11. Maximum occurs during June 3-6, from an average radiant of α=243°, δ=–22°. The daily motions of both radiants are very similar and amount to +0.9° in α and –0.2° in δ.

History

The first detection of the Chi Scorpiids should be credited to William F. Denning. During 1886 to 1912, he observed several meteors emanating from α=252°, δ=–10°, during June 2-4. These meteors were described as "very slow."[86]

In 1935, R. A. McIntosh published his "An Index to Southern Meteor Showers."[87] The Chi Scorpiids were listed as radiant number 147. They occurred during May 28 to June 11. McIntosh's compilation of 5 previously observed radiants indicated the radiant moved from α=241°, δ=–13° to α=247°, δ=–18°. Although this indicates a daily motion slightly exaggerated over the mathematically predicted motion of α=+0.93°, δ=–0.14°, the accumulated visual, photographic and radar data indicate the radiant is fairly diffuse—amounting to a diameter of nearly 8°—so that an accurate visual determination would be truly remarkable.

McIntosh's radiant list also included the Omega Scorpiids. Designated as radiant 146, it was the first time activity had been officially recognized from this stream. The shower was active during the period May 29 to June 11, and moved from α=240°, δ=–21° to α=247°, δ=–18°. Like the Chi Scorpiids, this shower's

[86]D1923B, p. 53.
[87]M1935, p. 714.

radiant should track southeasterly, but since it also possesses a fairly diffuse radiant, McIntosh's determination should not be taken at face value.

Observations of the Chi Scorpiids have been fairly continuous during the 20th century, with nearly every major meteor radiant compilation providing evidence. Radiants from some of the more prominent observers follow:

Chi Scorpiids

Date(UT)	Designation	RA(°)	DEC(°)	Observer
1933, May 26-27	258	244	–14	Ernst Öpik
1932, May 31	99	244	–10	Ernst Öpik
1930, June 8.76	----	244.8	–14.3	V. Maltzev[88]
1912, June 9.2	403	248	–14	C. Hoffmeister

Observations of the Omega Scorpiids have been somewhat sporadic during the 20th century, except among observers in the southern hemisphere. Thus, the shower is represented in several major works. Some of the radiants follow:

Omega Scorpiids

Date(UT)	Designation	RA(°)	DEC(°)	Observer
1933, May 23.8	3065	235	–26	C. Hoffmeister
1933, May 28.3	3235	241	–25	C. Hoffmeister
1937, June 3.3	3415	247	–26	C. Hoffmeister
1937, June 4.0	3429	240	–25	C. Hoffmeister
1937, June 5.3	3439	244	–22	C. Hoffmeister
1937, June 7.4	3466	249	–27	C. Hoffmeister
1937, June 7.1	3476	249	–29	C. Hoffmeister
1937, June 11.4	3526	255	–23	C. Hoffmeister
1937, June 13.5	3548	253	–25	C. Hoffmeister
1937, June 15.0	3569	257	–25	C. Hoffmeister

Taken as a whole, the Chi and Omega Scorpiids are very similar to the Ophiuchids (and even the Theta Ophiuchids) discussed earlier in this chapter—there are northern and southern components with orbits differing only in ω and Ω (this variance equals 180° for each of these orbital elements). Unlike the other ecliptic streams, both branches are equally strong and have been listed as showers in their own right.

Further notice of these northern and southern branches was taken by Bertil-Anders Lindblad in 1971, when he detected the Chi Scorpiid stream in a computer study of photographic meteors. He pointed out that the Chi Scorpiid

---

[88]King, A., *The Observatory*, **55** (October 1932), p. 291.

stream possessed two branches, with the southern branch being identical to McIntosh's radiant number 146.[89]

As indicated earlier, observations of the Omega Scorpiids are at their best from the southern hemisphere. According to Jeff Wood, director of the meteor section of the National Association of Planetary Observers, a shower referred to as the "Omega Scorpiids" is visible every year in Australia. Reaching maximum on June 4 from $\alpha=240°$, $\delta=-22°$, it is visible during May 24 to June 13. Wood claims the ZHR typically varies between 5 and 20.[90]

Interestingly, despite the fact that both streams seem well represented in the literature, the Chi Scorpiids seem the most consistent, with their existence being well represented in visual, photographic and radar data. On the other hand, the Omega Scorpiids seem more sporadic in appearance. The visual observations listed earlier are primarily made in 1937, with few additional observations being at hand. Similarly, Lindblad's photographic meteor study revealed a very weak sample of only 3 meteors to base this stream's orbit on. Finally, no radar survey has delineated this stream (in either the northern or southern hemisphere), while its northern counterpart has been revealed on a couple of occasions.

Orbit

Based on 32 meteors detected during the Radio Meteor Project during 1969, Zdenek Sekanina determined the following orbit for the Chi Scorpiid stream:

ω	Ω	i	q	e	a
261.4	72.9	4.1	0.663	0.687	2.116

From photographic meteors obtained from W1954, MP1961 and BK1967, the following orbits were obtained for the northern and southern branches of this stream. The Chi Scorpiids are based on 12 meteors, while the Omega Scorpiids are based on 3 meteors.

	ω	Ω	i	q	e	a
Chi	256.8	74.0	8.4	0.675	0.768	2.909
Omega	74.7	249.3	1.7	0.693	0.757	2.852

June Scutids

Observer's Synopsis

This shower is active during the period of June 2-July 29. Reaching maximum around June 27 ($\lambda=95°$), the average radiant at that time is $\alpha=278°$, $\delta=-4°$. The hourly rate is probably 2-4.

---

[89]L1971B, p. 20.

[90]Wood, Jeff, Personal communication (October 24, 1985).

History

The discovery of this meteor shower should be credited to R. A. McIntosh, since it was listed in his classic work "An Index to Southern Meteor Showers".[91] Designated radiant number 191 and referred to as the "Eta Serpentids I," the radiant was based on 3 visual radiants. Its duration was given as June 25-30, and the radiant was shown to move from α=274°, δ=–6° to α=277°, δ=–3°. Unfortunately, since the radiant was literally buried amidst 320 other radiants, no apparent searches were made to confirm the shower.

More official recognition of this stream came in 1971, when a computerized stream search among photographic meteor orbits by Bertil-Anders Lindblad revealed a radiant at α=278°, δ=–2°, during the period of June 25 to July 3. Referred to as the "Eta Serpentids," this stream was based on only 2 meteor orbits, but Lindblad noted a similarity to radiants previously noted by R. A. McIntosh and William F. Denning (to be discussed below).

The first person to finally establish the shape of the orbit and duration of the shower was Zdenek Sekanina during the Radio Meteor Project conducted at Havana, Illinois, during 1961-1965. During the period between June 17 and July 15, 7 meteors were noted from an average radiant of α=276.3°, δ=–6.3°. The estimated date of maximum was June 24.7. Sekanina confirmed the stream's existence during the 1968-1969 revival of the Radio Meteor Project. On this occasion, 32 meteors were detected during the period of June 2 to July 29. The average radiant was found to be α=280.7°, δ=+1.0°, while the date of maximum was more clearly determined as June 27.0.

A search for older records of this stream reveals only two *possibilities* in the 19th century. Both of these radiants were detected by the Italian Meteoric Association and were evaluated by William F. Denning,[92] who concluded that 20 meteors detected during June 25-30, 1869-1872, indicated a radiant at α=275°, δ=–9°. Another 20 meteors observed during June 26 to July 11, 1872, indicated a radiant at α=273°, δ=–2°. Neither of these radiants (designated 211-2 and 211-3, respectively) provide positive proof of the existence of the June Scutids prior to the 20th century, but they do barely qualify as being associated, according to the D-criterion.

Additional 20th Century observations have been found in Cuno Hoffmeister's book *Meteorströme* (1948), which lists 5 German-observed radiants from this stream. The first observation came on June 29, 1912 (λ=97.0°), when several meteors were plotted from α=281°, δ=–3°. Two apparent observations were made in 1935, when radiants were determined on June 23 (α=272°, δ=–1°) and June 26 (α=277°, δ=–10°). Another observation

---

[91] McIntosh, R. A., *MNRAS*, **95** (June 1935), p. 715.
[92] D1899, p. 272.

was made on June 25, 1936, when meteors were plotted from α=270°, δ=+4°. The final German observation came on July 4, 1937, when a radiant of α=276°, δ=+4° was detected.

The Author has observed meteors from this shower since July 6/7, 1973, and he has managed to continue to observe members of this stream during the period of June 24 to July 7, at a rate of 0.5 to 1 per hour, up to the present time. On June 29, 1986, he was finally able to actually observe on the night of maximum. In one hour, 4 meteors were plotted from α=277°, δ=–3°. Other observations acquired that year allowed the shower's average magnitude to be estimated as 2.5 (limiting zenithal magnitude=5.5).

Orbit

The radar orbit determined in 1969, by Sekanina during the Radio Meteor Project, is based on 32 meteors and is as follows:

ω	Ω	i	q	e	a
278.8	94.9	15.7	0.599	0.560	1.361

Beta Taurids

Observer's Synopsis

This daylight shower is active during June 5 to July 18. It possesses a relatively flat maximum centered on June 30 (λ=98.3°), that originates from an average radiant of α=79.4°, δ=+21.2°. The maximum hourly rate reaches about 25 to the eyes of radar, and the normally 3° diameter radiant suddenly swells to at least 7° around July 2.

History

This shower was first detected by Jodrell Bank observers during June 20-27, 1947, from a radiant near α=79° to 85°, δ=+20° to +30°. The activity rate ranged from 20 to 30 radio meteors per hour.[93] Apparent confirmation of this stream came during June 24 to July 4, 1948, when the average radiant was determined as α=90°, δ=+26° and the hourly rate reached a high of 35 on July 3.[94] These radio-echo studies were of a fairly low precision and, although the Beta Taurids were detected, it was not until 1950, that more precise details of the stream became available.

During June 26 to July 4, 1950, the Jodrell Bank equipment again found the Beta Taurids, this time at an average radiant of α=86.2°, δ=+18.7°.

---

[93]CHL1947, pp. 375-377.

[94]ACL1949, pp. 357-358.

Researchers analysing the data concluded that maximum activity occurred on July 2 (λ=99°) and that the radiant seemed to move "at random from day to day in an area of sky 8° x 14° in Taurus."[95] The daily radiants of this stream remained less than 3° across until July 2 and 3, when the radiant diameter was 7° and 5°, respectively. The diameter was again less than 3° across on July 4. These same observations were also used to determine the first orbit for this stream, which indicated a semimajor axis of 2.2 AU, a perihelion distance of 0.34 AU and an inclination of 6°.[96]

Radar observations continued at Jodrell Bank during the next several years. From observations obtained during the period 1950 to 1952, Mary Almond, K. Bullough and Gerald S. Hawkins established the mean daily motion as +0.8° in α and +0.4° in δ.[97] In 1954, Bullough determined more refined details of the shower by adding observations made in 1953. The daily motion was given as +0.67° in α and +0.44° in δ. He described the Beta Taurids as "a rather weak stream with no peak of activity. The shower is active for about a week centred on a sun's longitude of 97° to 98°."[98] Curiously, the 1953 data revealed a radiant diameter of 3° to 4° throughout the time of activity until July 2, when it increased to 8°. Finally, after 1958 studies had been conducted, G. C. Evans analysed the Jodrell Bank radar studies made between 1950 and 1958 and concluded that maximum came when the solar longitude was 97.7°, and that the average radiant was then α=85.4°, δ=+19.4°. The daily motion was determined to have averaged +0.57°±0.12° in α and +0.16°±0.22° in δ.[99] Evans described the shower maximum as "rather weak and broad."

The Beta Taurids were also detected from the Southern Hemisphere during 1953. C. D. Ellyett and K. W. Roth delineated 21 shower radiants while using radar equipment at Christchurch, New Zealand. Shower number 10 was detected during the period of June 17-26. The radiant ranged from 82° to 85° in α and +20° to +25° in δ. The mean radiant was given as α=84°, δ=+23°.[100]

During 1961, this stream was detected by C. S. Nilsson, while operating radar equipment at Adelaide Observatory in Australia. Only 7 meteors were found, but the equipment was not in operation after June 19, so that this observation was that of the early part of the shower. The duration was June 12 to 18, and the radiant was at α=75.5°, δ=+20.3°.[101] Nilsson pointed out that his

---

[95]AH1951, p. 22.

[96]A1951, p. 38.

[97]ABH1952, p. 20.

[98]B1954, pp. 75 & 94.

[99]E1960, p. 284.

[100]ER1955, p. 396.

[101]N1964, pp. 226 & 232.

data showed significant correlations between both the radiant position and geocentric velocity with time. He added that the "sign of the mean heliocentric latitude...is not significant, so the node marked as ascending is quite possibly a descending node, for at least part of the stream."[102]

It was not until the studies of the Radio Meteor Project, at Havana, Illinois, that the true extent of the Beta Taurids became known. Zdenek Sekanina, director of the project, found that the 1961-1965 data indicated a duration extending from June 12 to July 6, while the 1968-1969 data revealed a duration extending from June 5 to July 18. The first study revealed an orbit similar to those reported previously, while the second study showed both the ω and Ω shifted by 180°—adding further strength to Nilsson's observation that some of the stream's orbit might be at its descending node when crossing Earth's orbit during June and July.

Strong evidence seems to exist indicating the Beta Taurids are the same stream, or at least closely related to the stream, that later produces the Taurids during October and November. The suggestion was first made in 1951, by Mary Almond and has since been supported on numerous occasions by Nilsson, Sekanina and others. In 1964, Nilsson noted that the correlation of radiant position and geocentric velocity with time was also present during the duration of the Taurids. Other researchers have pointed fingers at the stream's long duration, flat maximum and diffuse radiant as other characteristics previously noted in the Taurids. The Taurids are believed to be a very old remnant of periodic comet Encke—the current record holder for having the shortest period.

It should be noted that the suggestion that the Taurids might produce a shower visible during the daylight hours of summer was first mentioned during 1940 by Fred L. Whipple.[103]

Orbit

The exact orbit of this stream has not been easy to determine since its discovery due to the large spread in inclination; however, some of the orbits determined from large meteor samples are as follows:

	ω	Ω	i	q	e	a
A1951	244	278.1	6	0.34	0.85	2.2
N1964	255.1	264.2	3.7	0.46	0.79	2.17
S1973	239.2	274.5	2.2	0.325	0.825	1.853
S1976	52.3	102.0	0.3	0.274	0.834	1.653

Based on 20 photographic meteors gathered for the Northern Taurids, the following orbit was obtained (see chapter 11 for further details):

---

[102]N1964, p. 245.

[103]Whipple, F. L., *Proceedings of the American Philosophical Society*, **83** (1940), p. 711.

	ω	Ω	i	q	e	a
N. Taurids	295.6	224.2	3.2	0.343	0.843	2.186

The orbit of periodic comet Encke, as obtained from Brian G. Marsden's *Catalog of Cometary Orbits*, is as follows:

	ω	Ω	i	q	e	a
P/Encke	186.0	334.2	11.9	0.340	0.847	2.218

Additional June Showers

June Aquilids

The data supporting this stream's existence is mainly concentrated in three radar studies conducted during the 1960's—N1964, GE1975 and S1976.

During June 13-19, 1961, C. S. Nilsson (Adelaide, South Australia) detected 4 radar meteors from α=293.9°, δ=–8.4°. A second radar survey was conducted by G. Gartrell and W. G. Elford (Adelaide, South Australia) during June 1969. A total of 13 meteors indicated an average radiant of α=289°, δ=–6° centered on June 11. Finally, during June 2 to July 2, 1969, Zdenek Sekanina directed the Radio Meteor Project at Havana, Illinois. Thirty-five meteors indicated a radiant of α=297.1°, δ=–7.1° when the shower reached maximum on June 17.5.

No convincing proof of this shower's visual existence is present in D1899, O1934, M1935 and H1948.

The three established radar orbits obtained from N1964, GE1975 and S1976 are as follows:

	ω	Ω	i	q	e	a
N1964	328.9	84.3	40.1	0.113	0.93	1.613
GE1975	324	80	39.5	0.15	0.90	1.852
S1976	329.5	85.8	39.3	0.114	0.916	1.348

GE1975 suggested the stream might be related to comet 1618 II, the orbit of which follows:

	ω	Ω	i	q	e	a
1618 II	287.4	80.3	37.2	0.390	1.0	∞

Corvids

This shower seems to have been of temporary character as, since its discovery in 1937, no trace of it has appeared in either the visual, photographic or radar surveys.

The Corvids were discovered by Cuno Hoffmeister during a meteor expedition to Southwest Africa in 1937-1938. The shower was first noted on

June 25, just two days after full moon, and on June 26 the ZHR reached a value of 13. Rates declined thereafter and the last remnants were noted on July 3.

Hoffmeister concluded that maximum came on June 27.0 (λ=94.9°). The radiant was determined as α=191.6°, δ=–19.2° on June 28, although it was described as diffuse, with a diameter of nearly 15°. Hoffmeister computed two orbits based on semimajor axes of 2.5 and 3.0 AU and noted "a rather striking resemblance to the orbit of Comet Tempel 3-Swift, except for argument of perihelion...." Interestingly, today's knowledge of asteroids has produced another possibility which seems more attractive than Hoffmeister's—Apollo asteroid 1979VA. Hoffmeister's orbits as well as the orbits of periodic comet Tempel-Swift and 1979VA are as follows:

	ω	Ω	i	q	e	a
I	38.9	274.9	2.5	0.930	0.628	2.5
II	7.7	274.9	3.1	1.013	0.662	3.0
T-S	139.6	264.4	7.1	1.234	0.619	3.240
1979VA	89.60	271.63	2.78	0.982	0.627	2.635

Sagittarids

This stream appears to have been discovered during 1957 and 1958, while radio-echo surveys were being conducted at Adelaide Observatory in South Australia. A. A. Weiss reported the shower to have suddenly appeared with rates of 30 per hour from a compact radiant. The radiant's position was determined as α=307°, δ=–35° in 1957, and α=301°, δ=–36° in 1958. Weiss said no hint of the shower had been noted in 1952, 1953, 1954 and 1956, and that the 1957-1958 appearance was near "the limit of resolution of the equipment." The date of maximum was given as June 11, while the total duration was given as five days.[104]

More sensitive radar equipment was operated by G. Gartrell and W. G. Elford at Adelaide, South Australia, during 1968-1969. Only 4 meteors were detected, so that the stream seemed practically nonexistent. For the mean activity date of June 10, the radiant was established as α=297°, δ=–34°.[105]

Visual activity from this shower seems rare, as some of the most comprehensive southern hemisphere radiant catalogs (i.e. M1935 and H1948) have failed to reveal convincing evidence of the radiant's existence prior to 1957. Recent observations also appear to be rare, although Jeff Wood, director of the meteor section of the National Association of Planetary Observers (Australia), lists what appears to be a probable detection of this shower among the 1980 observations of the Western Australia Meteor Section. The radiant was called the

---

[104]Weiss, A. A., *MNRAS*, **120** (1960), p. 397.

[105]GE1975, p. 597.

"Alpha Microscopiids" and was detected only during June 11 and 12. A maximum ZHR of 1.43±0.13 occurred on June 11 from a radiant of α=305°, δ=–36°.[106] *The BMS Radiant Catalogue* lists this shower as occurring during June 8-16, with a maximum ZHR of 4, and an average radiant of α=304°, δ=–35°.[107]

The orbit of this stream as determined in GE1975, is as follows:

	ω	Ω	i	q	e	a
GE1975	152	259	33.5	0.11	0.90	1.149

[106]Wood, Jeff, Personal Communication (October 15, 1986).

[107]Mackenzie, Robert A., *BMS Radiant Catalogue.* Dover: The British Meteor Society (1981), p. 16.

Chapter 7: July Meteor Showers

Delta Aquarids

Observer's Synopsis

Two distinct showers make up the Delta Aquarid stream. The Southern Delta Aquarids have a duration of July 14-August 18. Maximum hourly rates of 15-20 occur on July 29 (λ=125°) from α=339°, δ=–17°. The radiant's daily motion is +0.9° in α and +0.4° in δ. The duration of the Northern Delta Aquarids extends from July 16 to September 10. Maximum occurs on August 13 (λ=139°), at which time the radiant is at α=344°, δ=+2°. The hourly rates reach a high of 10 and the radiant's daily motion is +0.9° in α and +0.3° in δ.

History

During July and August the Aquarid-Capricornid complex becomes active—a region which contains the northern and southern branches of the Delta Aquarids and Iota Aquarids, as well as several distinct radiants in Capricornus. The strongest activity emanates from the two Delta Aquarid radiants.

Activity was first noticed from the region of the Delta Aquarids in 1870, when G. L. Tupman (Mediterranean Ocean) plotted 65 meteors during July 27-August 6. Tupman found the radiant to have steadily moved during the period of observation, with the position beginning at α=340°, δ=–14°, and ending at α=333°, δ=–16°. Although the motion seems backward with respect to the normal eastward movement of meteor radiants, the Author believes the former radiant represents the true Southern Delta Aquarids, while the latter radiant is either a combination of both the southern and northern streams or a conglomeration of several streams within the Aquarid-Capricornid complex. Following Tupman's discovery, the region became well studied by others, with William F. Denning listing no less than 20 additional observations by experienced observers during the remainder of the 19th Century.[1]

---

[1]D1899, p. 284.

The above observations clearly refer to the Southern Delta Aquarids, and, although the Northern Delta Aquarids were not officially discovered until the 1950's, it should be pointed out that some 19th Century observations do seem present. Denning made three apparent observations between 1879 and 1893, which he erroneously grouped with several additional radiants occurring between May and November to form a stationary shower called the "Beta Piscids." The first observation involved 10 meteors seen during August 21-23, 1879, from an average radiant of α=350°, δ=0°. Another observation was made during August 16-20, 1885, when 7 meteors were seen from α=345°, δ=0°. His final 19th Century observation came during August 13-16, 1893, when 6 meteors were seen from α=347°, δ=0°.[2] Since the observed radiants involved small numbers of meteors, there was no inspiration for other observers to continue observations and there was no hint as to an association with the well-known Delta Aquarid shower.

Observations of the Delta Aquarids continued into the 20th Century, but they continued to primarily refer only to the southern branch. As an example, English observers made quite extensive observations during 1922. A. Grace Cook plotted 13 meteors during July 25, 28 and 31, which revealed an average radiant of α=338°, δ=–12° and, from Ashby, Alphonso King plotted 4 meteors from α=340°, δ=–16° during July 30 and 31.[3] J. P. M. Prentice also detected the radiant, when he plotted 12 meteors from α=341°, δ=–15° during July 30-August 1.[4] Among the 75 additional radiants recognized by these three observers during July and August of that year, only one observation of the Northern Delta Aquarids seems present. That observation was made by Cook during August 17, 19, and 20, when 4 meteors were plotted from α=338°, δ=0°.[5]

The first significant study of the Delta Aquarid stream was published in 1934. Ronald A. McIntosh used the observations made by the New Zealand Astronomical Society during 1926-1933 to determine the daily motion of the radiant. In all, 44 radiants were utilized, with their original observers being McIntosh (Auckland), Murray Geddes (Otekura), F. M. Bateson (Wellington) and A. Bryce (Hamilton).

McIntosh concluded that activity from the shower is continuous from July 22 to August 9, with the radiant moving northeastward from α=334.9°, δ=–19.2° to α=352.4°, δ=–11.8° (average motion +0.96° in α and +0.41° in δ—Author). McIntosh added that a sharp maximum occurs on July 28 (α=340.5°, δ=–17.0°), and a diagram included in his paper revealed the following visual hourly rates: 1

---

[2]D1899, p. 285.

[3]Denning, W. F., *The Observatory*, 45 (September 1922), pp. 301-302.

[4]Denning, W. F., *The Observatory*, 45 (October 1922), p. 332.

[5]Denning, W. F., *The Observatory*, 45 (October 1922), p. 333.

on July 22, 2 on July 25, 3 on July 26, 7 on July 27, 14 on July 28, 9 on July 30, 6 on August 2, and 1 on August 9.[6] As can be seen, McIntosh's interpretation of the New Zealand observations reveals a sharp rise to maximum, followed by a gradual decrease in activity. The radiant ephemeris clearly represents the Southern Delta Aquarids and no mention was made as to a northern branch. During 1935, McIntosh published his classic paper "An Index to Southern Meteor Showers," but among the 320 radiants listed, there are no convincing candidates for the Northern Delta Aquarids.

Cuno Hoffmeister and his fellow German observers obtained good observations of the southern shower between 1908 and 1938. In evaluating the data, Hoffmeister found seven activity centers. Five of these were based on 2-3 observed radiants each, but it was clear to Hoffmeister that maximum occurred on August 3 (λ=130°). He based this statement on two well-established activity centers: one based on seven visual radiants that occurred on August 2 (λ=128.4°) from α=342.4°, δ=–17.7°, and a second was based on ten radiants that occurred on August 6 (λ=132.6°) from α=341.5°, δ=–17.2°.[7] Hoffmeister does seem to have isolated a radiant with the characteristics of the northern branch as well. In a table listing 238 radiants observed on at least 4 occasions, is a radiant at α=349°, δ=+1°. Based on 5 visual radiants, the average date was given as August 13 (λ=139°).[8]

The first radio-echo observations of the Delta Aquarids were made by equipment at Ottawa, Canada, during 1949, when Canadian astronomer D. W. R. McKinley detected both branches of the stream. Unfortunately, orbits were not determined for the two radiants and the northern radiant was not recognized as being associated with the Delta Aquarid shower. In a 1954 paper that appeared in the *Astrophysical Journal*, McKinley revealed how his velocity and radiant determinations on July 26-29, 1949, revealed two distinct radiants: a very strong one at α=339°±2°, δ=–17°±2° and a very weak one at α=340°±5°, δ=0°±5°. The velocities of the two radiants were 40.20±0.1 km/sec and 41.0±0.5 km/sec, respectively.[9]

Radio-echo observations of this shower were also made by equipment operating at Jodrell Bank (England) during 1949-1951. The most reliable data was accumulated during the last days of July 1950. Gerald S. Hawkins and Mary Almond gave the weighted mean date of activity as July 28 (λ=124.5°), at which time the hourly radio-echo rate peaked at 38. The radiant possessed a diameter of 3° and an average position of α=339°, δ=–14°. The 1949

[6]McIntosh, R. A., *MNRAS*, **94** (April 1934), pp. 584 & 587.

[7]H1948, p. 139.

[8]H1948, p. 81.

[9]McKinley, D. W. R., *Astrophysical Journal*, **119** (1954), p. 519.

observation occurred on July 29 (λ=125.8°). Although an hourly rate of 24 was obtained, no further details could be established. Radio-echo rates reached 41 on July 27, 1951 (λ=123.4°).[10] The radiant determination was not considered to be of the highest quality, with the diameter being given as 6° and the position being estimated as α=336°, δ=0°. The only hint that the 1951 shower might have been identical to the Southern Delta Aquarid radiant was its date of maximum activity. The actual position lies closer to the where the radiant of the Northern Delta Aquarids might lie at the end of July.

During 1952, Almond made a specific attempt to determine the velocity of the Delta Aquarid meteors. Using a "more selective beamed aerial," 32 probable members of the stream were detected and a mean velocity of 40.5±2.7 km/sec was revealed. In addition, maximum was found to have occurred on July 28 from α=340°, δ=–15°. When the radiant and velocity were combined they allowed the first accurate determination of the orbit of the Delta Aquarids. From this orbit, Almond noted a strong similarity between the orbit of this stream and that of the Arietids of June (see Chapter 6). Most notable were the similar values determined for the perihelion distance, eccentricity and longitude of perihelion. The discrepancies in the argument of perihelion and ascending node were explained as due to the diffuse nature of both streams. "As the inclination of the orbit plane of the δ Aquarids is 24°," she explained, "the stream would be 0.31 a.u. away from the earth at its second approach on June 9. From the duration of 16 days observed for the daytime Arietids the width of the stream must be at least 0.27 a.u., and there is also evidence that the δ Aquarid shower lasts 18 days in the southern hemisphere. Hence, the system of orbits is so broad, it seems probable that the two showers are connected and are produced by one extended stream."[11]

Using over 2000 photographic meteor orbits determined during the Harvard Meteor Project of 1952-1954, Frances W. Wright, Luigi G. Jacchia and Fred L. Whipple pointed out the mounting evidence supporting the existence of a northern branch of the Delta Aquarids. This marked the first time the northernmost radiant had been recognized as being associated with the Delta Aquarid shower and the authors offered the first hint as to the complex evolution of the stream. They noted that the northern and southern branches were "symmetrical with respect to the ecliptic, or to Jupiter's orbit...." Concerning the suggested link between the Southern Delta Aquarids and the Arietids of June, it was noted that the spread of about 134° between the nodes of the two streams "could have been caused by continual perturbations by Jupiter."[12]

---

[10]HA1952, pp. 222 & 224.

[11]Almond, Mary, *Jodrell Bank Annals*, **1** (December 1952), pp. 24 & 26-27.

[12]Wright, F. W., Jacchia, L. G., and Whipple, F. L., *AJ*, **62** (September 1952), p. 231.

The photographic data accumulated during 1952-1954 was analyzed by several astronomers during the 1960's, but the most complete analysis was made in 1971 by Bertil-Anders Lindblad (Lund Observatory, Sweden). For the Southern Delta Aquarids he isolated 13 meteors which indicated a duration of July 21-August 8. The stream's date of nodal passage was given as July 31, at which time the radiant was at $\alpha=340°$, $\delta=-16°$.[13] A second study by Lindblad utilized 11 precisely calculated meteor orbits. The study began with a D-criterion of 0.15, but he noted that when a more strict value of 0.10 was used the Southern Delta Aquarids remained intact. He concluded that it "indicates a high degree of orbit similarity...."[14] For the Northern Delta Aquarids, Lindblad isolated 9 photographic orbits. The indicated duration was August 5-25. The nodal passage came on August 14 ($\lambda=140.5°$), at which time the radiant was at $\alpha=347°$, $\delta=+1°$.[15]

Although the radio-echo method had actually produced details about the Delta Aquarid streams as early as 1949, radar studies entered a new age of importance during the 1960's. The first major study was conducted in 1960 by researchers at the Kharkov Polytechnical Institute. B. L. Kashcheyev and V. N. Lebedinets obtained 151 radio meteor orbits from the Southern Delta Aquarids during July 14-August 14. When at maximum the average radiant was at $\alpha=341.2°$, $\delta=-16.4°$ and the solar longitude was 126.7°. They gave the daily motion of the radiant as +0.85° in α and +0.35° in δ, and determined an average orbit which revealed a semimajor axis of 2.04 AU. The Northern Delta Aquarids were also detected. The 50 radio-meteors observed revealed a duration extending from July 7-August 14. The date of the nodal passage was August 1 ($\lambda=127.7°$), at which time the radiant was at $\alpha=336.8°$, $\delta=-4.9°$. The radiant's daily motion was given as +0.9° in α and +0.3° in δ.[16] Although many of the details determined for the Northern Delta Aquarids matched those of other researchers made both prior to and after the Kharkov study, several aspects indicate the data may have been contaminated by other meteor showers active at the time. Most notable are the dates of earliest activity and nodal passage, both of which occur at least 10 days prior to that generally accepted for this shower. In addition, the argument of perihelion is the highest ever revealed for this stream's orbit, with a value nearly 15° above that given in most other orbital determinations.

The next significant radar survey was conducted in 1961 by C. S. Nilsson (Adelaide Observatory, South Australia). During July 23 to August 4, 48 radio meteor orbits were obtained, which revealed a maximum on July 28 ($\lambda=125.8°$)

---

[13]L1971B, pp. 16-17.

[14]L1971A, p. 4.

[15]L1971B, p. 16.

[16]BL1967, pp. 188-189.

from an average radiant of α=339.4°, δ=−17.3°. He determined the radiant's daily motion as +0.9° in α and +0.2° in δ, while stream's average orbit possessed a semimajor axis of 2.33 AU. Nilsson tried to be as strict as possible in his evaluation of the data. He said, "there are several distinct radiants... in the vicinity of the main Delta Aquarid radiant. It is not unlikely that some of these have contributed to the data used in previous determinations of the Delta Aquarid radiant."[17] Although Nilsson did not specifically recognize the Northern Delta Aquarids, he did isolate 4 meteors during the period of August 20-23. The average radiant was given as α=352.7°, δ=+6.3°. It should be noted that the radar equipment at Adelaide did not operate during August 5-15, so that the true maximum of the shower would probably have been missed. As can be seen in the "Orbit" section below, Nilsson's orbit bears a striking resemblance to orbits determined by other astronomers for the northern shower.

The most elaborate radio-echo survey to isolate both branches of the Delta Aquarids was the Radio Meteor Project conducted in two sessions during the 1960's. For the 1961-1965 session, Zdenek Sekanina found the Southern Delta Aquarids to have a duration of July 16-August 14, The nodal passage came on July 30.9, at which time the radiant was at α=342.2°, δ=−16.9°. He pointed out that the stream's distribution "could be matched by no single model curve" and he suggested that the stream "might be composed of two constituents: a very compact filament and a more dispersed stream." The Northern Delta Aquarids were given a duration of July 26-August 27. Their nodal passage came on August 13.0, at which time the radiant was α=344.0°, δ=+0.3°. Sekanina considered this stream "to be somewhat looser than the southern branch...and definitely less conspicuous and less populated."[18]

Sekanina's 1968-1969 survey gave the duration of the Southern Delta Aquarids as July 14-August 18. The nodal passage was given as July 29.3, while the average radiant was α=341.8°, δ=−15.9°.[19] For the Northern Delta Aquarids, the duration was given as July 28-September 10. The nodal passage came on August 14.9, at which time the radiant was at α=345.7°, δ=+4.8°.[20]

Following the earliest computations of orbits for the Southern Delta Aquarids, came the studies of physical and evolutionary developments of the meteor stream. In 1963, A. K. Terent'eva examined the structure of the stream. He noted that the small perihelion distance (given as 0.06 AU) would bring the temperature of the individual meteors up to 1100° K, which is the melting point of silicates. Terent'eva suggested this accounted "for the peculiar general

[17]N1964, pp. 227 & 234.

[18]S1970, pp. 476-477, 483 & 486.

[19]S1976, pp. 281 & 296.

[20]S1976, pp. 283 & 297.

appearance of the shower meteors which are sharp, show no wakes, and give off no sparks." Several visual and photographic radiants were studied and the northern and southern radiants were clearly apparent, with their average radiants for July 29 being α=334.5°, δ=–5.4° and α=338.5°, δ=–16.9°, respectively. The daily motion of the Northern Delta Aquarids was given as +0.85° in α and +0.35° in δ, while the Southern Delta Aquarids' motion was +0.88° in α and +0.36° in δ. Finally, it was noted that the orbits of the two streams were "symmetrical relative to the plane of Jupiter's orbit." He added that "This may be the effect of perturbations."[21]

Another very interesting study appearing in 1963, was conducted by S. E. Hamid and Fred L. Whipple. It raises the importance of the Southern Delta Aquarid stream to a very high level as being a link to other meteor showers. It has already been noted that a strong link exists between this stream and the daytime Arietid stream of June, but Hamid and Whipple gave evidence to suggest that the Quadrantids of January also formed from this stream. Taking members of both streams and subjecting them to secular perturbations they found that the orbital planes and perihelion distances were very similar 1300-1400 years ago. "The effects of Jupiter perturbations on i and q are quite remarkable," they said, and "it is possible that the two streams were derived from a single comet...." They added that despite the present differences in the duration and activity levels of the two showers, "the physical characteristics of the meteoroids belonging to the two streams appear to be similar, as judged by their light curves."[22]

Several attempts have been made to identify this shower among ancient displays. Charles P. Olivier suggested the first possible link in his 1925 book *Meteors*. He believed the strong displays of July 19, 714, and July 14, 784, were possible early appearances of the Southern Delta Aquarids.[23] In 1976, Sekanina said the shower of 714 has been classified as a possible early appearance of the Perseids. He added that the most promising early appearance of the Southern Delta Aquarids was a shower that occurred in 1007, in which two independent Japanese sources describe the meteors as flying toward the north—a direction quite inappropriate for a description of the Perseids. Sekanina said this radiant would indicate a nodal regression of 0.8°-1.3°/century.[24]

One very interesting finding about the Southern Delta Aquarids is the diameter of the radiant. Cuno Hoffmeister noted in his 1948 book *Meteorströme*, that "the radiant is at times very diffuse" and he added that activity tended to be

---

[21]Terent'eva, A. K., *SCA*, **7** (1963), pp. 293-295.

[22]Hamid, S. E., and Whipple, F. L., *AJ*, **68** (October 1963), p. 537.

[23]Olivier, Charles P., *Meteors*. Baltimore: Williams & Wilkins Company (1925), p. 44.

[24]S1976, p. 486.

strong within an area 20° in diameter centered on the Southern Delta Aquarid radiant (this, of course, must have included the two branches of the Iota Aquarid stream as well—Author).[25] A similar conclusion of a diffuse radiant has been arrived at by several Northern Hemisphere observers during this century, but for observers south of the equator a different conclusion has been formulated. McIntosh was struck by the fact that the Southern Delta Aquarid radiant seemed fairly small, with New Zealand observers independently determining radiants quite close to one another. From South Australia, Nilsson also noted the "particularly small" radiant diameter based on his radio-echo survey. Such a contrast between northern and southern observations reveals what can happen to a radiant when zenithal attraction comes into play. For the Southern Hemisphere observers the radiant is almost directly overhead, while northern observations tend to occur when the radiant is only about 20° above the horizon.

From the material discussed above, it is obvious that visual activity mainly comes from the Southern Delta Aquarid radiant. Recent observations of this shower reveal quite strong activity levels. Michael Buhagiar (Western Australia) observed 8-12 meteors per hour during 10 hours of observing on July 28/29 and 29/30, 1972, even though a full moon was present.[26] During 1973, visual rates reached 14.6 per hour on July 27/28, according to four observers in the United States, while Buhagiar detected 20 per hour on the same night from Western Australia.[27] During 1974, ZHRs in the United States reached 12.5±3.0 on July 29.4, while Buhagiar and Robert Oates noted hourly rates of 37-44 meteors per hour on July 28 in Western Australia.[28]

Members of the Western Australia Meteor Section (WAMS) have had much success in observing the Southern Delta Aquarids. Section director Jeff Wood said a maximum ZHR of 42.64±9.78 was observed on July 29, 1977, with an overall observed duration extending from July 23-August 14. The radiant position at maximum was given as $\alpha=339°$, $\delta=-15°$. Maximum rates during 1979 were significantly lower, with a ZHR of 16.85±1.21 occurring on July 28, from a radiant of $\alpha=338°$, $\delta=-17°$. The observed duration was July 20-August 5. The shower's 1980 appearance was hampered by a full moon on July 27. Subsequently, hourly rates were significantly lower, with the shower's maximum actually coming on August 3—the night of the last quarter moon. The ZHR reached 7.35±1.19 and the radiant was then $\alpha=343°$, $\delta=-15°$. The duration extended from July 18 to August 10.[29]

---

[25]H1948, p. 139.

[26]*MN*, No. 13 (October 1972), p. 9.

[27]*MN*, No. 18 (Mid-October 1973), p. 2.

[28]*MN*, No. 23 (Mid-October 1974), pp. 5-6.

[29]Wood, Jeff, Personal Communication (October 15, 1986).

Observations of the Northern Delta Aquarid shower by the WAMS have revealed three puzzling facts: very low activity, an uncertain radiant location, and an earlier than usual date of maximum. In 1979 possible meteors from this shower were observed during July 27-August 5. The maximum activity peaked on August 4 when the ZHR reached 3.12±1.10, but the average radiant was then given as $\alpha=328°$, $\delta=-3°$—roughly 7° to the west of the expected position for that date. In 1980, meteors were observed from the shower during August 2-16. A maximum ZHR of 5.67±1.21 came on August 4 (one day after a first quarter moon), from a radiant of $\alpha=341°$, $\delta=-2°$—roughly 6° east of the expected position for that date.[30] Of course both radiants were given dates of maximum activity which definitely contradicts the results of photographic and radar surveys, as well as the apparent visual observations of the past. It is interesting that an earlier maximum has also been noted in the United States by Norman W. McLeod III (Florida). In an article published in *Meteor News* during April 1984, McLeod stated that his observations since 1971 have revealed that maximum occurred at the same time as generally accepted for the Southern Delta Aquarids. He gave the Northern Delta Aquarid radiant as $\alpha=326°$, $\delta=-7.6°$ for July 30, which is about 4° west of the expected position for that date.[31]

The Author believes the minor controversy over the likely date of maximum for the Northern Delta Aquarids might have a simple explanation that points to a more complex structure for the Delta Aquarids as a whole. For the photographic meteors, the orbital inclinations tend to be 6°-8° higher prior to August 10, than after that date. If this means that two different streams actually produce the overall Northern Delta Aquarid activity, then two different maximums might not be out of the question. On the other hand, if it is assumed that the inclination discrepancy among photographic meteors is due to a weak database, then the earlier, less supported maximum might simply reflect how easy it is to observe a weak meteor shower like the Northern Delta Aquarids during late July and early August, than during the time of the Perseid maximum in mid-August.

The characteristics of the Delta Aquarid meteors have not been well studied. Observers tend to lump the Southern Delta Aquarids and Northern Delta Aquarids into one shower and, to make things even worse, the Southern Iota Aquarids and Alpha Capricornids are occasionally thrown in as well. The few instances of specific Southern Delta Aquarid meteor details being obtained are listed in the table at the top of the next page. As will be seen, the average magnitude of these meteors is about 3, while very few of these meteors possess trains.

---

[30]Wood, Jeff, Personal Communication (October 15, 1986).

[31]McLeod, Norman W., *MN*, No. 65 (April 1984), p. 7.

Southern Delta Aquarids

Year(s)	Ave. Mag.	# Meteors	% Trains	Observer(s)	Source
1960-1976	2.70	1903	—	McLeod	*MN*, No. 32
1973	3.33	61	—	Buhagiar	*MN*, No. 18
1973	3.01	—	—	Gates	*MN*, No. 29
1975	3.25	—	—	Gates	*MN*, No. 29
1977	2.96	28	—	McLeod	*MN*, No. 38
1978	3.07	176	9.1	WAMS	*MN*, No. 45

The observers cited are Norman W. McLeod, III (Florida); Michael Buhagiar (Western Australia); Bill Gates (Florida); Western Australia Meteor Section.

The only apparent average magnitude determination for the Northern Delta Aquarids was made in 1977 by McLeod. From 33 observed meteors, he gave the average magnitude as 3.70.[32] Recent studies seem to reveal that both streams' average meteor magnitude fades as each day progresses.[33] This presents a very interesting picture when it is compared to a study published by L. Kresak and V. Porubcan (Czechoslovakia) in 1970. Using all available double-station meteor photographs, they found that the southern stream was essentially compact around the time of maximum, but then proceeded to become more diffuse, especially in right ascension, until the compact northern radiant became active about mid-August.[34]

The photographic meteor rate for the Southern Delta Aquarid shower was determined by American Meteor Society members in Florida during 1976. During July 23-31, of 1970-1971 and 1973-1974, 26 meteors were photographed during 157 hours 38 minutes, revealing a rate of 0.16 per hour.[35]

Orbit

One of the first orbits ever computed for the Southern Delta Aquarids was published by McIntosh during 1934. Based on 44 visual radiants determined by New Zealand observers during 1926-1933, it was a simple parabolic orbit.

ω	Ω	i	q	e	a
159.6	304.7	55.8	0.039	1.0	∞

The first elliptical orbit computed for the Southern Delta Aquarids was computed by Hoffmeister in his 1948 book *Meteorströme*. He indirectly arrived at a value of 34.64 km/sec as the heliocentric velocity of the meteors and determined the following orbit:

[32] *MN*, No. 38 (Mid-October 1977), p. 8.
[33] *MN*, No. 46 (July 1979), pp. 2-3.
[34] Kresak, L., and Porubcan, V., *BAC*, **21** (1970), pp. 153-169.
[35] *MN*, No. 32 (August 1976), p. 7.

ω	Ω	i	q	e	a
147.0	310.0	23.7	0.118	0.926	1.600

The 1960's and 1970's brought a greater understanding of the parameters of the Southern Delta Aquarid orbit, as researchers in the United States, Soviet Union and Australia were able to determine more precise heliocentric velocities for the stream members.

	ω	Ω	i	q	e	a
N1964	152.4	305.8	32.5	0.07	0.97	2.33
BL1967	151.1	306.7	28.4	0.08	0.96	2.04
S1970	151.9	307.3	29.9	0.083	0.955	1.853
S1976	155.4	305.7	28.2	0.069	0.958	1.630

The 1960's and 1970's brought the first orbital determinations for the Northern Delta Aquarids from radar surveys.

	ω	Ω	i	q	e	a
N1964	328.2	148.3	22.4	0.11	0.93	1.563
S1970	324.5	139.8	16.0	0.132	0.927	1.819
S1976	323.2	141.7	19.2	0.169	0.866	1.259

The Author has collected 13 photographic meteors from HS1961, HS1961, JW1961, MP1961 and C1964. The following average orbit was revealed for the Northern Delta Aquarids.

ω	Ω	i	q	e	a
329.2	138.3	21.1	0.095	0.954	2.051

Alpha Lyrids

Observer's Synopsis

This is an apparently strong telescopic meteor shower with a duration extending from July 9-20. When at maximum on July 14 (λ=112°), the radiant is located at α=280°, δ=+38°. At the shower's discovery in 1958, activity rates in binoculars were estimated as 18-33 per hour. Naked-eye rates are about 1-2 per hour. The average magnitude of the meteors is 4.1 and their color appears to be primarily white. The meteors are also described as fast and seem to originate from a radiant 2° in diameter.

History

The Alpha Lyrids were discovered in July 1958, during a Czechoslovakian meteor expedition to Mount Bezovec. Jiri Grygar, Lubos Kohoutek, Zdenek Kviz and Jaromir Mikusek led the team of 44 observers from observatories and astronomy clubs. The purpose of the expedition was to obtain observations of sporadic telescopic meteors, but the observers ended up discovering a new meteor shower.

The Czechoslovakian observers watched the skies between July 10 and 25, but the Alpha Lyrids were only detected during July 10 to 20. Observations made with 25x100 binoculars revealed a maximum hourly rate slightly higher than 18 on July 16, while observations with 10x80 binoculars revealed a maximum rate near 33 on July 15. In all, 839 meteor paths were plotted during 8 nights of actual observations, and the observations revealed two radiants were present, although only the right ascension could be determined precisely: one radiant had α=278.7°, while the other was at α=300.5°. The authors concluded that the latter right ascension corresponded with the antisolar point, but they believed the former right ascension's declination "cannot be too high...."[36]

Fortunately, the telescopic shower had been independently discovered in 1958, by V. V. Martynenko (USSR). Observing at Simferopol, he had observed a very strong telescopic shower, for which he was able to determine a radiant of α=277.5°, δ=+39° for July 9-10.[37]

During 1969, members of the Crimean Yaroslavl and Dnepropetrovsk amateur astronomer societies participated in a visual and telescopic study of the Alpha Lyrids. One of the shower's original discoverers, Martynenko, coordinated the survey with the help of N. I. Bondar, N. M. Kremneva and V. V. Frolov. Meteors from this stream were detected during July 9-19, with the greatest activity coming during July 11-16. Based on 19 radiants determined from the plots of 825 visual and telescopic meteors, the average position was α=280.9°, δ=37.8°. The radiant diameter was generally given as two degrees, and the general direction of the radiant's daily motion was to the southeast. The meteors were described as white and fast, with an average magnitude of 4.1. During the period of observation the authors concluded that the Alpha Lyrids were "one of the most active showers in the range of bright stellar magnitudes, up to 3.5^m, inclusively, and the most active in the faint range, 3^m-6^m."[38]

Previous observations of this shower seem to be nonexistent. No trace of activity is present in any of the 19th Century publications, nor does convincing evidence exist in Cuno Hoffmeister's *Meteorströme* and the records of the American Meteor Society. According to Robert Mackenzie (director of the British Meteor Society) the visual ZHR of this shower reaches 1.5.[39]

---

[36]Grygar, J., Kohoutek, L., Kviz, Z., and Mikusek, J., *BAC*, **11** (1960), p. 85.

[37]Grygar, J., Kohoutek, L., Kviz, Z., and Mikusek, J., *BAC*, **11** (1960), p. 85.

[38]Bondar, N. I., Kremneva, N. M., Martynenko, V. V., and Frolov, V. V., *SSR*, **6** (January 1973), pp. 112-113.

[39]Mackenzie, Robert A., *BMS Radiant Catalogue*. Dover: The British Meteor Society (1981), p. 19.

In an attempt to establish the orbit of this stream, the Author has searched through the lists of photographic and radar meteor orbits. No possible candidates were found among the first group of orbits, but three radar orbits were found. These meteors had been detected during July 15-19, 1969, by Zdenek Sekanina during the 1968-1969 session of the Radio Meteor Project and revealed an average radiant of α=286.8°, δ=35.5°.

Orbit

The three radar meteors found by the Author among the 39,145 orbits obtained by Sekanina were averaged to reveal the following orbit.

ω	Ω	i	q	e	a
231.8	113.7	34.3	0.860	0.652	2.471

July Phoenicids

Observer's Synopsis

The duration of this shower extends from July 9-17. Maximum occurs on July 14 (λ=112°), from a radiant of α=32°, δ=–48°. The maximum hourly rate reaches 30 for radio-echo observations, but is no greater than 1 visually. The radio-echo observations also reveal the radiant to be diffuse.

History

The first observation of this shower is attributed to A. A. Weiss (Adelaide Observatory, South Australia) because of his detection of a radiant at α=31°, δ=–44° during July 12-17, 1957. Weiss had been conducting a radio-echo survey in an attempt to locate active meteor showers visible in the Southern Hemisphere. He said the shower possessed "a marked asymmetry in activity across the stream, with a fairly sharp maximum at λ=112°. The activity is patchy and the radiant tends to be diffuse."[40]

An idea of this stream's activity was quickly obtained by Weiss as he proceeded to reexamine the radio-echo observations begun at Adelaide in 1953. In that first year's records, the July Phoenicids were recognized from observations made on July 9, 10, 13 and 16, and a radiant of α=30°, δ=–43° was determined. Weiss was also confident that observations had been made in 1954 and 1956 since increased activity had been noted around July 13, "which could be due to a radiant which transits near 08 hr. This is sufficiently close to the transit time of from 06.49 to 07.12 found from the radiant equipment to identify this activity also with the Phoenicid radiant." The shower was re-observed by

---

[40]Weiss, A. A., *MNRAS*, **120** (1960), p. 397.

Weiss in 1958, so that he concluded that the activity was annual, with a 15-day duration centered on July 14, at which time the activity reached a maximum radio-echo rate of 30 from a radiant of $\alpha=32°$, $\delta=-48°$.[41]

Although Weiss showed the July Phoenicids to be an exceptional radar shower, naked-eye observations have shown it to be very weak. During the period of 1969-1980, Michael Buhagiar (Perth, Western Australia) observed 20,974 meteors. In 1981, he compiled them into a list which revealed 488 visual radiants. During the 12-year period of study, Buhagiar observed this shower on only two occasions. He gave the duration as July 11-15, while a maximum hourly rate of one came on July 14 from a radiant of $\alpha=34°$, $\delta=-50°$.[42]

According to Jeff Wood, observations by the Western Australia Meteor Section have produced somewhat inconclusive results as to the activity of this stream. Members failed to detect any activity from the region in 1977, and the 1979-1980 observations failed to cover the shower's period of activity with observing gaps of July 2-19 in 1979, and July 7-18 in 1980.[43]

The conclusion that might be reached concerning this meteor shower is that, although strong activity seems to be present in the radio-echo (and probably telescopic) range, visual activity is very low. Even though the Australian observations revealed rates of only one per hour, it should be noted that Robert A. Mackenzie (director of the British Meteor Society) gives the maximum ZHR of this shower as five. He agrees with the July 14 date of maximum, and gives the duration as July 3-18.[44]

Orbit

Insufficient data has been gathered to allow the determination of the velocity and, hence, the orbit of this stream. The Author has therefore computed the following parabolic and elliptical orbits, the latter of which is based on an assumed semimajor axis of 2.0 (a good average for meteor streams).

ω	Ω	i	q	e	a
25.5	292.0	86.1	0.967	1.0	∞
37.0	292.0	79.0	0.946	0.527	2.0

The Author notes a strong similarity between this parabolic orbit and the slightly hyperbolic orbit of comet Gale (1912 II). There is also a close similarity with comet Peltier-Whipple (1932 V), which has an orbital period of 291 years, although the biggest problem here is the ascending node. Jack D. Drummond

---

[41] Weiss, A. A., *MNRAS*, **120** (1960), pp. 395 & 398.

[42] Buhagiar, Michael, *WAMS Bulletin*, No. 160 (1981).

[43] Wood, Jeff, Personal Communication (October 15, 1986).

[44] Mackenzie, Robert A., *BMS Radiant Catalogue*. Dover: The British Meteor Society (1981), p. 19.

(New Mexico State University) predicted that a shower from this latter comet would reach maximum on September 8 from α=57°, δ=–39°.[45]

	ω	Ω	i	q	e	a
1912 II	25.6	297.5	79.8	0.716	1.000	-1587.758
1932 V	38.5	344.8	71.7	1.037	0.976	43.891

Alpha Pisces Australids

Observer's Synopsis

The duration of this southern shower extends from July 16 to August 13. Maximum occurs on July 30 (λ=127°), at which time the radiant is at α=337°, δ=–28°. The maximum ZHR tends to reach 3-5. The meteors are generally slow and white.

History

This meteor shower was plainly observed on several occasions during the 19th Century and into the 20th Century, but, despite the numerous observations, the stream was virtually ignored during the period of 1938-1952.

The discovery seems best attributable to Alexander S. Herschel, who observed a radiant of α=338°, δ=–28° on July 28, 1865. The first apparent confirmation was made by E. F. Sawyer (Cambridge, Massachusetts), who plotted four very slow and bright meteors from a radiant of α=337°, δ=–33° on July 28, 1878. Other observations were made by Cruls (Rio de Janeiro Observatory) during July 25-30, 1881, when several meteors were detected from α=343°, δ=–25°, and by William F. Denning during July 28-31, 1898, when four slow and white meteors were plotted from α=338°, δ=–25°.[46] As can be seen, only Cruls' observation was from the Southern Hemisphere, so that the other three observers isolated the radiant even though it was at a low altitude of 20° or less. During the first four decades of the 20th Century, two major groups of observers—one in Germany and the other in New Zealand—shed new light on this stream's duration and radiant.

In his 1948 book *Meteorströme*, Cuno Hoffmeister separated the 5406 German-observed visual radiants into two separate groups and managed to detect the Alpha Pisces Australids in each collection. The first detection involved 10 radiants observed during the period of 1910-1930. The average radiant was given as α=336°, δ=–28°, while the date of maximum was given as July 29 (λ=125°). From the second group of radiants the shower was again detected.

---

[45]Drummond, Jack D., *Icarus*, **47** (1981), p. 507.
[46]D1899, p. 284.

Primarily involving observations obtained in 1937, when Hoffmeister was observing in South Africa, the average radiant was given as α=338°, δ=–29°, while the date of maximum was given as August 2 (λ=129°). After combining all of the shower's radiants, Hoffmeister concluded that the stream produced notable activity during July 29-August 2 from an average radiant of α=337°, δ=–28°. The date of maximum was given as July 31 (λ=127).[47] Hoffmeister added that the simultaneous rise and fall of activity from this stream and the Delta Aquarids, as well as the large apparent radiant of the latter shower, brought to light the possibility of an association.

Using observations made by New Zealand observers during the 1920's and early 1930's, Ronald A. McIntosh found seven apparently separate areas of activity from the relatively small constellation of Pisces Austrinus which reached maximum at different times between mid-July and mid-August. The most prominent shower was called the "Alpha Pisces Australids" (the same name Denning gave to the 19th Century observations), the existence of which was based on 24 visual radiants. The duration was given as July 26-August 8, during which time the radiant moved from α=337°, δ=–33° to α=350°, δ=–30°. The other six radiants isolated by McIntosh are as follows:[48]

*Pisces Australids: Based on 3 visual radiants, the duration was given as July 28-August 3, while the average radiant was at α=326°, δ=–26°.

*Theta Pisces Australids: Based on 2 visual radiants, the duration was given as August 12-14, while the average radiant was at α=327°, δ=–32°.

*Beta Pisces Australids: Based on 11 visual radiants, the duration was given as July 14-22, while the average radiant moved from α=330.5°, δ=–30° to α=339°, δ=–30°.

*Lambda Pisces Australids: Based on 5 visual radiants, the duration was given as August 5-14, while the average radiant moved from α=334°, δ=–27.5° to α=339°, δ=–26°.

*Epsilon Pisces Australids: Based on 3 visual radiants, the duration was given as August 13-14, while the average radiant was at α=338°, δ=–24°.

*20 Pisces Australids: Based on 2 visual radiants, the duration was given as August 8-9, while the average radiant was at α=340.5°, δ=–27°.

As noted earlier, there is a period of complete neglect of this meteor shower following the 1937 observations of Hoffmeister's group, but observations were

---

[47]H1948, pp. 80-85.

[48]M1935, p. 717.

finally resumed in 1953. In that year, radar equipment at Christchurch, New Zealand observed a "highly probable" Pisces Australid radiant at α=328°, δ=–27° during July 21-26. The shower was considered quite strong.[49]

During 1960, a more elaborate radar survey was conducted by B. L. Kashcheyev and V. N. Lebedinets at the Kharkov Polytechnical Institute (USSR). They detected 32 meteors from this stream during July 16-August 13, and concluded that maximum came on July 26 (λ=123°), at which time the radiant was at α=340°, δ=–26°. From this data, the first orbit was computed which indicated a 45° inclination and a semimajor axis of 4.31 AU.[50]

The long absence of visual observations finally ended in 1965, when Edward F. Turco (Cranston, Rhode Island), a member of the American Meteor Society, observed the Alpha Pisces Australids during three nights, centered on July 29/30. "The meteors I saw were not too bright, though I was surprised with two exceptions, both fireballs." He pointed out that this was the first time he had ever detected this shower.[51]

The fact that Turco only noted the shower in 1965, despite being a regular observer, as well as member and regional director of the American Meteor Society for many years, brings forth an excellent explanation as to the reason for the shower's neglect during 1938-1952. Meteors from this shower are only visible from the United States and Europe under special circumstances that not only involve weather, but also require an unusual number of meteors heading toward the north. Observers in the Southern Hemisphere should find the shower as an annual fixture in their late July skies. It will be remembered that Hoffmeister noted excellent activity from this radiant during his observations in South Africa in 1937, and McIntosh noted numerous radiants in the area based upon observations made by him and fellow New Zealand observers during the 1920's and 1930's.

The first notable Southern Hemisphere visual survey of meteor activity following that of McIntosh, was conducted by Michael Buhagiar (Perth, Western Australia) during 1969-1980. Recording a total of 20,974 meteors, Buhagiar compiled a list of 488 probable visual radiants. Radiant 472 was called the "Alpha Pisces Australids" and was based on 10 visual radiants determined during the 12-year period. The duration was given as July 27-August 10, while a maximum hourly rate of 4 was said to be emanating from α=344°, δ=–30° on July 31. None of the nearby Pisces Austrinus radiants noted by McIntosh were prominent enough to make Buhagiar's list.[52]

---

[49]Ellyett, C. D., and Roth, K. W., *AJP*, **8** (1955), pp. 395-396.

[50]KL1967, p. 188.

[51]Turco, Edward F., *Review of Popular Astronomy*, **62** (October 1968), p. 7.

[52]Buhagiar, Michael, *WAMS Bulletin*, No. 160 (1981).

Nearly simultaneous with Buhagiar's survey, the Royal Astronomical Society's New Zealand Meteor Section was also conducting a decade-long survey for a list of Southern Hemisphere meteor showers. Covering the 1970's, the observations allowed section director John E. Morgan to compile a list of 213 probable visual radiants. The "Alpha Pisces Australid" shower was detected as usual. The duration was given as July 23-August 2, during which time the radiant moved from α=344°, δ=–31° to α=347°, δ=–34°. The maximum ZHR was given as 4. Interestingly, McIntosh's "Beta Pisces Australid" shower was also detected by the New Zealand observers. Given a short duration of July 19-21, the average radiant was at α=337°, δ=–34°, while the maximum ZHR was 3. Another radiant was also found which seems closely related to the McIntosh's August portion of the Alpha Pisces Australids. Called the Xi Gruids, their duration was given as July 24-August 8, during which time the radiant moved from α=343°, δ=–32° to α=350°, δ=–32°.[53]

Since 1977, the Western Australia Meteor Section (WAMS), under the directorship of Jeff Wood, has been extremely active. The Alpha Pisces Australids were observed during July 23-29, 1977. A maximum ZHR of 3.65±2.11 was observed on July 28, at which time the radiant was located at α=343°, δ=–30°. During 1979 meteors were observed from this shower during July 27-August 5. A maximum ZHR of 3.82±0.52 came on July 28 from an average radiant of α=343°, δ=–28°. A full moon occurred on July 27, 1980, thus interfering with complete observations of this shower maximum. However, observers still gave the shower's duration as July 19-August 4, and detected a maximum ZHR of 2.00±0.33 on August 3. The average radiant was then given as α=336°, δ=–32°,[54] and the Author believes this indicates the 1980 observations were a conglomeration of several showers in the region.

Orbit

The orbit of the Alpha Pisces Australid stream, as determined by Kashcheyev and Lebedinets, is as follows:

ω	Ω	i	q	e	a
114	303	45	0.17	0.96	4.31

The Author has located one photographic meteor among the more than 2000 orbits obtained during the Harvard Meteor Project of 1952-1954. It is listed in MP1961 and was designated 3443. It was detected on July 28, 1952, from α=338°, δ=–27°.

ω	Ω	i	q	e	a
131	305	42	0.19	0.98	7.50

[53]Wood, Jeff, Personal Communication (October 15, 1986).

[54]Wood, Jeff, Personal Communication (October 15, 1986).

The Author has also looked through the 39,145 radio meteor orbits obtained by Zdenek Sekanina during the two sessions of the Radio Meteor Project conducted during the 1960's. Five meteors were located, which indicated a duration of July 24-30. The stream seems to cross its ascending node on July 29, with an average radiant of α=341.9°, δ=–26.9°. The orbit follows:

ω	Ω	i	q	e	a
137.8	305.1	39.9	0.186	0.893	1.743

Additional July Showers

Sigma Capricornids

The first observations of this minor stream were made by Cuno Hoffmeister during 1937. Although normally observing at Sonneberg, Germany, Hoffmeister conducted an expedition to South West Africa during May 7, 1937 to February 5, 1938. On July 9 (λ=106.5°) a radiant was detected at α=296°, δ=–15°. Another radiant was detected on July 12 (λ=109.3°), the position of which was α=298°, δ=–14°, while a final observation came on July 14 (λ=111.2°), when the radiant was at α=296°, δ=–13°.[55]

The only additional record of visual observations comes from Western Australia, when observations made during July 5-6, 1980, revealed a radiant of α=298°, δ=–13°. A maximum ZHR of 2.23±0.18 came on July 6. The shower was referred to as the "Alpha Capricornids."[56]

If it was not for the meticulous nature of the observations made by Hoffmeister and the Western Australians, no visual detection of this shower would be on record. No doubt the congestion of meteor showers in this region of the sky during July and August have contributed to its anonymity. Fortunately, important details have been gathered about this stream thanks to two important radar surveys conducted during the 1960's.

C. S. Nilsson (Adelaide Observatory, South Australia) conducted a radio survey during 1960-1961. During July 14-25, 1961, five meteors were detected from an average radiant of α=306.9, δ=–15.4°. Although the average date of activity was given as July 20 (λ=117.1°), it should be pointed out that the radio equipment was not in operation during June 20-July 13, so it is possible that only part of the shower was observed.[57] The orbit was given as

ω	Ω	i	q	e	a
289.8	117.1	3.9	0.37	0.87	2.86

---

[55]H1948, p. 236.

[56]Wood, Jeff, Personal Communication (October 15, 1986).

[57]N1964, pp. 227-229.

A second radar detection was made by Zdenek Sekanina during the 1961-1965 session of the Radio Meteor Project. He found the shower's duration to be June 18-July 30. The nodal passage came on July 9.6 (λ=106.9°), at which time the radiant was at α=297.6°, δ=–18.7°.[58] Sekanina gave the orbit as

ω	Ω	i	q	e	a
290.3	106.9	2.1	0.431	0.758	1.782

Sekanina also made the suggestion that this stream might be associated with the Apollo asteroid 2101 Adonis. The possibility of the association seemed very good as he computed a D-criterion value of only 0.097. The orbit of this asteroid follows:

	ω	Ω	i	q	e	a
Adonis	41.056	351.207	1.370	0.442	0.764	1.873

Finally, although no photographic survey seems to have isolated this stream, the Author has found three definite meteor orbits among the list of 2,529 photographic orbits obtained during the Harvard Meteor Project of 1952-1954. All of the meteors were detected during July 17-21, 1953. The average radiant was α=307.8°, δ=–15.2° and the orbit is

ω	Ω	i	q	e	a
290.9	116.8	4.5	0.379	0.853	2.584

Tau Capricornids

The meteor shower was discovered by Ronald A. McIntosh and was listed in his classic 1935 paper, "An Index to Southern Meteor Showers." McIntosh had isolated ten visual radiants observed by himself and other New Zealand meteor observers which indicated a duration extending from July 10-20. The radiant was noted to move from α=302°, δ=–14° to α=317°, δ=–11°. The shower had been referred to as the "Beta Capricornids II."[59]

One radio-echo survey provides the best evidence for this stream's existence. Zdenek Sekanina analyzed the data accumulated during the 1968-1969 session of the Radio Meteor Project and found the stream's duration to be June 16-July 29. The date of the nodal passage was given as July 12.3 (λ=109.5°), at which time the radiant was at α=310.6°, δ=–14.7°.[60] The orbit was given as follows:

ω	Ω	i	q	e	a
311.2	109.5	4.5	0.272	0.792	1.310

Curiously, Sekanina gave details of another stream observed during the 1968-1969 session, which he called the "Sigma Capricornids." The duration was

---

[58] S1973, pp. 257 & 259.

[59] M1935, p. 716.

[60] S1976, pp. 280 & 295.

given as June 2-July 29. The indicated nodal passage was June 24.3 (λ=92.3°), at which time the radiant was at α=292.4°, δ=−13.6°. The following orbit was given:

ω	Ω	i	q	e	a
309.4	92.3	8.2	0.332	0.707	1.133

The similarity of the two orbits is quite striking (more so than exists between this orbit and that given earlier for the Sigma Capricornids). It is also interesting that these two radio-echo orbits possess many of the same orbital elements as the Northern Iota Aquarids (see Chapter 8), except that, once again, the ascending node and, thus, the date of maximum is considerably different. As the Tau Capricornids are apparently a minor stream, more will be discussed about its relationship with other streams in the next chapter under the section on the Iota Aquarids.

In recent years, members of the Western Australia Meteor Section have provided some valuable data concerning the rates of this shower. During 1977, meteors were observed over the period of July 12-13. A maximum ZHR of 3.15±0.58 occurred on July 13, from α=305°, δ=−10°.[61] Jeff Wood indicates the region is very complex and claims that, in general, a stream the Australian observers refer to as the "Alpha Capricornids" is active during June 16-August 12, with four definite maximums occurring on July 6, 19, 26 and 30. The average radiant is at α=305°, δ=−13°, while the maximum ZHRs range from 5-10 meteors.[62]

A search through the photographic records reveals only one possible member. This meteor was photographed during the Harvard Meteor Project on June 29, 1952. Its radiant was at α=299.3°, δ=−16.1°, while its orbit was given as follows:

ω	Ω	i	q	e	a
314.2	97.6	7.4	0.223	0.852	1.502

No convincing visual evidence exists in D1899 or H1948, so that this stream might be considered a producer of generally faint, possibly telescopic, meteors.

Omicron Draconids

The visual support for this shower is extremely weak, but it seems possible that William F. Denning (Bristol, England) plotted 21 meteors from a radiant of α=284°, δ=+57° during July 16-18, 1876. He described the meteors as very rapid and faint.[63] Thereafter, only three doubly observed meteors of magnitude

---

[61] Wood, Jeff, Personal Communication (October 15, 1986).

[62] Wood, Jeff, Personal Communication (October 24, 1985).

[63] D1899, p. 273.

1-2 appear to be related to Denning's radiant. These were reported by English observers in 1908 and 1914.

The shower's official discovery should be attributed to Allan F. Cook, Bertil-Anders Lindblad, Brian G. Marsden, Richard E. McCrosky and Annette Posen, who isolated three photographic meteors detected during the Harvard Meteor Project of 1952-1954. The duration was given as July 6-24, while the average radiant was α=271°, δ=+59°. Only a parabolic orbit was given.[64]

ω	Ω	i	q	e	a
190	113	43	1.01	1.0	∞

Although the evidence was not considered strong, the authors noted that the orbit they had determined bore a strong resemblance to that of comet Metcalf (1919 V).

ω	Ω	i	q	e	a
185.7	121.4	46.4	1.115	1.000	–5023.833

Official confirmation of this stream's existence came in 1976 as Zdenek Sekanina published his analysis of the data acquired during the 1968-1969 session of the Radio Meteor Project . He found the Omicron Draconids to possess a duration extending over the period of July 14-28. The date of the nodal passage was given as July 17.7 (λ=114.7°), at which time the radiant was at α=284.7°, δ=+60.9°.[65] The following orbit being revealed.

ω	Ω	i	q	e	a
192.2	114.7	46.2	1.006	0.768	4.329

No additional observations seem to have been made, as the extensive records of the American Meteor Society, as well as the more than 5000 visual radiants given in Cuno Hoffmeister's book *Meteorströme*, fail to reveal a trace.

---

[64]Cook, A. F., Lindblad, B.-A., Marsden, B. G., McCrosky, R. E., and Posen, A., *SCA*, **15** (1973), p. 3.

[65]S1976, pp. 281 & 296.

Chapter 8: August Meteor Showers

Iota Aquarids

Observer's Synopsis

This stream consists of two fairly diffuse branches. The Southern Iota Aquarids possess a duration extending from July 1-September 18. The August 6 maximum (λ=133°) possesses an hourly rate of 7-8 and a radiant position of α=337°, δ=–12°. The Northern Iota Aquarids occur during August 11-September 10. Maximum occurs on August 25 (λ=152°), at which time 5-10 meteors per hour can be seen from α=350°, δ=0°. Both streams produce meteors with an average magnitude slightly fainter than 3. The daily motion of the southern stream's radiant is +1.07° in α and +0.18° in δ, while the northern radiant moves +1.03° in α and +0.13° in δ.

History

This may represent one of most confused of the well-known annual meteor streams due to a pair of very diffuse radiants. To make matters worse, visual observations are complicated by the addition of the two Delta Aquarid streams and several streams coming out of Capricornus during July and August.

The first apparent observation of Iota Aquarid activity was made by William F. Denning (Bristol, England), who, during 1877-1888, plotted several meteors from α=338°, δ=–12° during the period of August 3-5.[1] This radiant was falsely identified as belonging to the Delta Aquarid stream, which had been found just a few years earlier. The Author believes this was actually the first observation of the Southern Iota Aquarid stream. Although the number of plotted meteors from this radiant was low, it was capped by the observation of a stationary meteor on August 5, 1888, the radiant of which was at α=336°, δ=–11°.[2] Denning was also the first apparent observer of the Northern Iota Aquarids, as he plotted 10 meteors from α=350°, δ=0° during August 21-23,

---

[1]D1899, p. 284.

[2]D1923A, p. 42.

1879. In this instance, Denning mistakenly grouped this radiant in with other radiants forming his stationary "Beta Piscid" stream of June-October.[3]

Observations of both streams continued to be made into the 20th Century, though the radiants tended to be grouped with meteors of other streams. The first person to actually recognize the Iota Aquarid stream was Alphonso King (England), who plotted five meteors from α=331°, δ=–12° during July 27-30, 1911. The meteors were described as possessing "Longish paths" with no trains.[4] King's identification referred to the Southern Iota Aquarids, but he gave no indication of a possible northern branch.

Ronald A. McIntosh (Auckland, New Zealand) appears to have isolated both branches of the Iota Aquarid stream from the observations made by New Zealand observers during the 1920's and 1930's. In his classic paper, "An Index to Southern Meteor Showers," two showers—called the "Iota Aquarids"—were listed which most likely referred to the southern branch. Shower number 272 was based on 7 visual radiants and indicated a radiant movement from α=330°, δ=–14° to α=339°, δ=–10° during July 31-August 11. Shower 280 was based on 13 visual radiants and indicated the radiant moved from α=332°, δ=–15° to α=338°, δ=–12° during July 25-August 5. For the Northern Iota Aquarid stream there were also two possibly associated showers. Shower number 301, was based on 4 visual radiants. It had been called the "5 Piscids" and during August 13-21 its radiant averaged α=346°, δ=0°. Shower 315 was called the "14 Piscids" and was also based on 4 visual radiants. With a duration of August 13-25, it possessed an average radiant of α=352°, δ=–3°.[5]

Although neither of the Iota Aquarid showers were officially recognized by Cuno Hoffmeister, the Author has found that both Iota Aquarid streams were detected by Hoffmeister and his German colleagues. The southern branch was seen on five occasions during 1932-1937. These observations indicated a duration of August 3-10, with the average radiant being α=338.2°, δ=–12.2°. Three radiants of the Northern Iota Aquarids were detected during 1930-1937. The indicated duration was August 26-30, while the average radiant was α=354.0°, δ=–2.7°.[6]

The first official recognition that the Iota Aquarid stream was composed of two branches should be credited to a 1957 paper by Frances W. Wright, Luigi G. Jacchia and Fred L. Whipple. While examining "all ι-Aquarid meteors found on Harvard plates during the interval June 28 to September 1, for all years of observation up through 1955," they noted a meteor found on August 18, 1952,

---

[3]D1899, p. 285.

[4]K1916, p. 547.

[5]M1935, p. 717-718.

[6]H1948, pp. 213, 217, 237-239 & 249.

which "seemed to indicate a possible northern stream...." They identified two additional photographic meteors from 1952—one on August 21 and the other on September 1. Using only the two earliest meteor orbits, the authors used a least-squares solution to determine the radiant's daily motion as 1.04° in α and –0.08° in δ (the use of only two meteors makes these values highly uncertain—Author). For the Southern Iota Aquarids, they used five double-station and five single-station meteors to determine the radiant's daily movement as +1.04° in α and +0.02° in δ.

Unfortunately, despite the photographic meteors indicating only late August activity for the Northern Iota Aquarids, the authors operated under the assumption that this stream possessed a date of maximum similar to that of the southern branch. Subsequently they identified a visual radiant listed by McIntosh (not discussed earlier) as representing this stream. The radiant was designated shower number 275 and was based on 11 visual radiants. During its duration of July 28-August 2, the radiant was said to have moved from α=331°, δ=–8° to α=334°, δ=–8°.[7] From these photographic and visual details the authors concluded that the mid-point of the shower's activity occurred at a solar longitude of 132.5° (August 4), with the radiant at α=330.87°, δ=–4.95°. Based on the data already given, the maximum of the Northern Iota Aquarids definitely seems to occur in late August (this fact is also supported by the radio-echo data soon to be discussed). Subsequently, the adoption of August 4 as the maximum of the Northern Iota Aquarids (despite the indications of the photographic meteors) caused confusion for several years to come.

The first problem that emerged involved the radar survey of C. S. Nilsson (Adelaide Observatory, South Australia). During 1960-1961, Nilsson had gathered orbital data on 2200 radio meteors, which led to the determination of 71 meteor stream orbits. Using the work of Wright, Jacchia and Whipple as a guide, two possible Iota Aquarid streams were noted which came to maximum at the end of July. Designated as streams 61.7.3 and 61.7.11, they were said to respectively represent the southern and northern branches. Nilsson noted that differences did exist between his work and the 1957 study, and he considered both of his identifications as uncertain.[8]

From what is now known of the Iota Aquarid streams, the Author believes that stream 61.7.3 was more likely associated with the Southern Delta Aquarids. Stream 61.7.11 (observed during July 25-August 3 from α=326.2°, δ=–12.3°) actually represented the southern branch of the Iota Aquarids, instead of the northern branch, while a stream designated 61.8.2 (detected during August 16-24, from α=343.5°, δ=+0.8°) was actually the Northern Iota Aquarid stream.

---

[7]Wright, F. W., Jacchia, L. G., and Whipple, F. L., *AJ*, **62** (September 1957), pp. 228-229.
[8]N1964, pp. 227-229 & 232-233.

The other problem associated with the statement that the Northern Iota Aquarid maximum occurred in early August, involved the lists of meteor showers published in the years that followed. Amateur groups worldwide subsequently adopted the August 4 date and many had not corrected this error at the time this book was published. A possible negative result involves the complete lack of visual observations of this northern shower during the 1960's, and only a handful during the 1970's.

The first accurate delineation of these two streams came about during the two sessions of the Radio Meteor Project, which had been conducted by Zdenek Sekanina at Havana, Illinois, during the 1960's. During the 1961-1965 session the Southern Iota Aquarids were found to possess a duration of July 14-August 27. The nodal passage was identified as occurring on August 9.2, at which time the radiant was at α=335.3°, δ=–9.1°. The Northern Iota Aquarids were described as having a duration extending over the period of August 13-28. The nodal passage came on August 25.1, when the radiant was at α=351.9°, δ=–1.1°.[9] During the 1968-1969 session, the Southern Iota Aquarids were found to have a duration of July 1-September 18. The nodal passage came on August 10.4, when the radiant was at α=343.0°, δ=–3.2°. The Northern Iota Aquarid activity extended over the period of August 11-September 10. The nodal passage came on August 26.0, at which time the radiant was at α=349.5°, δ=+0.3°.[10]

Some of the first significant visual details of these two showers were obtained in 1977 by observers in Florida. Norman W. McLeod III was able to determine the first average magnitude estimates of each stream. The Southern Iota Aquarids generally produced the brighter meteors, with a value of 3.05 being determined from 43 magnitude estimates. The Northern Iota Aquarids were found to have an average magnitude of 3.32, as determined from 34 magnitude estimates. Bill Gates commented that the northern stream reached a peak activity rate of 12 meteors per hour on August 20.[11]

Southern Hemisphere meteor observers have provided some interesting results on the Iota Aquarid streams in recent years. In 1981, Michael Buhagiar (Perth, Western Australia) published details of his 20,974 meteor observations made during 1969-1980. Among the 488 visual radiants determined, there was no convincing evidence supporting the existence of the Southern Iota Aquarids. A possible radiant representing the Northern Iota Aquarids was seen during August 21-24. From a total of 3 observations, it was concluded that a maximum hourly rate of 5 came from α=354°, δ=0° on August 22.[12]

---

[9]S1973, pp. 257-259

[10]S1976, pp. 282-283 & 296-297.

[11]*MN*, No. 38 (Mid-October 1977), p. 8.

[12]Buhagiar, Michael, *WAMS Bulletin*, No. 160 (1981).

Members of the Western Australia Meteor Section (WAMS) have also obtained inconsistent observations of these streams. No convincing observations have been made of the Northern Iota Aquarids, but several interesting details have come forth on the Southern Iota Aquarids. In 1978, 41 meteors of this stream revealed an average magnitude of 3.36, while 15.7% left persistent trains.[13] In 1980, the WAMS detected activity from this shower during August 2-10. A maximum ZHR of 2.02±0.45 came on August 2, from a radiant of α=335°, δ=–16°.[14] Combining all of the WAMS observations, director Jeff Wood concludes that the duration of the Southern Iota Aquarids extends from July 16 to August 19. A maximum ZHR of 7-8 occurs on August 6 from a radiant of α=335°, δ=–15°.[15]

Orbit

Three reliable orbits are given for the Southern Iota Aquarids. It should be noted that the radar equipment in N1964 was not in operation during August 5 to 15, therefore, since only the first half of the shower was monitored, it is not unlikely that the Ω would actually be larger than given here.

	ω	Ω	i	q	e	a
N1964	312.5	125.3	6.9	0.23	0.85	1.563
S1973	307.5	136.2	1.4	0.277	0.836	1.692
S1976	319.1	137.3	4.4	0.249	0.762	1.045

The previous orbits were all determined using radio-echo techniques, but it should be noted that several Southern Iota Aquarid orbits have been computed from photographic meteors. These indicate that the values of ω and Ω are each reversed by about 180°. The orbit from MP1961 was based on 10 meteors, that from JW1961 was based on 6 meteors, and SH1963 was derived from 4 meteors.

	ω	Ω	i	q	e	a
MP1961	129.6	310.3	4.2	0.22	0.92	2.78
JW1961	126.3	303.3	3.6	0.25	0.90	2.50
SH1963	133.5	301.1	1.2	0.23	0.88	2.00

Three reliable orbits were determined for the Northern Iota Aquarids during radio-echo surveys conducted in the 1960's. Once again a note should be added about Nilsson's orbit. Since the radar equipment was not in operation during August 25-September 21, the shower's August 25th maximum was missed. Thus, it seems probable that the value of Ω could be larger than given by Nilsson.

---

[13]Wood, Jeff, *MN*, No. 45 (April 1979), p. 12.

[14]Wood, Jeff, Personal Communication (October 15, 1986).

[15]Wood, Jeff, Personal Communication (October 24, 1985).

	ω	Ω	i	q	e	a
N1964	310.4	146.0	7.9	0.301	0.75	1.205
S1973	313.5	151.5	3.2	0.242	0.823	1.366
S1976	307.4	152.4	5.2	0.302	0.777	1.356

As can be seen, the orbits of the two streams are quite similar. Generalizations would give the Southern Iota Aquarids a smaller orbital inclination and perihelion, while the northern branch would possess a smaller argument of perihelion and semimajor axis.

One final note should be added about these meteor streams. As noted in Chapter 7, the Tau Capricornid orbit bears a striking resemblance to that of the Northern Iota Aquarids. The only significant difference is in the date of the nodal passage. There is also a possibility that the Southern Iota Aquarids could be added to this already striking similarity. What this implies is that the original orbit was perturbed in such a way that its shape and size remained basically unchanged—only the line of nodes and, thus, the longitude of perihelion were affected. During this evolutionary change, three significant events seem to have occurred which are now responsible for the maximums of these three showers. The solar longitudes of these events were 110°, 137° and 152°. Two possible reasons for these three distinct maximums have been considered by the Author. Either they are due to variations in the strength of planetary perturbations, or they appeared when the parent body underwent sudden increases in dust output.

Alpha Capricornids

Observer's Synopsis

The duration of this shower extends from July 15 to September 11. Maximum seems to occur during August 1 (λ=128.6°) from an average radiant of α=306.7°, δ=–8.3°. The maximum ZHR ranges from 6-14, while the meteors are generally described as slow. The shower has the reputation of producing some of the brightest meteors of the major showers, with the average magnitude being estimated as about 2.2.

History

Although the Iota Aquarids might be one of the most confused annual streams, the last three decades have revealed the Alpha Capricornids to not be far behind. This visual shower has been well known since the 19th Century, but as photographic and radio-echo surveys were conducted the shower began to appear more complex. Today astronomers generally seem to agree that two or three distinct maximums occur during the time the Alpha Capricornids are active.

This shower seems to have been discovered in 1871 as N. de Konkoly (Hungary) plotted six meteors from $\alpha=305°$, $\delta=-4°$ during July 28-29. Before the end of the decade two additional observations were made. On July 28, 1878, William F. Denning (England) made a probable observation of this shower when he plotted five meteors from $\alpha=305°$, $\delta=-14°$ and E. Weiss (Hungary) observed the shower during July 25-28, 1879, when he plotted four meteors from $\alpha=305°$, $\delta=-7°$.[16]

The Alpha Capricornids became known as a consistent producer of meteors during the period of late July and early August, and in 1899, Denning commented that, at the time of Perseid activity, this shower was rich with very slow and "often bright" meteors.[17] This latter comment of bright meteors has become the trademark of this stream and Denning provided interesting facts to back this statement in 1920. He pointed out that he and his fellow English observers had determined the real paths of 25 meteors from this shower during the period of July 15-August 28, which possessed a mean radiant of $\alpha=305.2°$, $\delta=-10.4°$. He added that he had personally "seen at least 34 bright meteors from it during the period from July 27 to Aug. 5."[18]

Several visual surveys of the early 20th Century helped to expand the knowledge of this shower. One of the first was Ronald A. McIntosh's 1935 paper "An Index to Southern Meteor Showers." Primarily using observations made by observers in New Zealand during 1927-1934, McIntosh combined 15 visual radiant determinations of the Alpha Capricornids to reveal that the radiant moved from $\alpha=300°$, $\delta=-11°$ to $\alpha=308°$, $\delta=-10°$ during July 22-31. A second radiant was also noted which may have been the first indication of the complex nature of the Alpha Capricornids. It was said to possess a radiant that moved from $\alpha=300°$, $\delta=-9°$ to $\alpha=305°$, $\delta=-8°$ during July 23-31.[19]

It is interesting that the Alpha Capricornids are not well represented among the 5406 visual radiants listed in Cuno Hoffmeister's 1948 book *Meteorströme*. Subsequently it is not listed in his table of annual meteor showers. On the other hand, Hoffmeister's analysis of the active annual showers began with a preliminary list of 238 radiants. One of those, designated number 56, was given an average activity date of July 29 ($\lambda=126°$), at which time the radiant was at $\alpha=314°$, $\delta=-12°$.[20] Although this may not have been produced by the main Alpha Capricornid shower, it could represent yet another early clue to the complex nature of this stream.

---

[16]D1899, p. 278.

[17]D1899, p. 222.

[18]Denning, W. F., *The Observatory*, **43** (July 1920), p. 263.

[19]M1935, p. 716.

[20]H1948, p. 80.

While the German observers failed to detect convincing proof of this shower's existence, members of the American Meteor Society were quite successful in observing activity. The Author notes that during the period of 1929-1953 no less than 21 radiants were observed. The indicated duration of the shower was July 15 to August 5, while the average radiant was at α=303.1°, δ=–12.5°.

The picture of a fairly consistent Alpha Capricornid stream became disrupted in 1956 when Frances W. Wright, Luigi G. Jacchia and Fred L. Whipple (Harvard College Observatory, Massachusetts) examined associated photographic meteors. Using 12 doubly-photographed meteors, the authors first established the overall look of the shower. The duration was given as July 16-August 22, while the date of maximum was determined as August 2 (λ=129.0°). The average radiant was α=308.5°, δ=–9.7°, while the daily motion was determined as +52'±2.4' in α and –2'±2.6' in δ. As can be seen, the motion of δ was quite uncertain, with the potential existing for either a northward or southward movement. The reason for this was a large scatter in the δ values of the collected photographic meteors. The authors stated that the scatter was possibly due to the existence of two or more streams producing the photographic Alpha Capricornid meteors.

The authors combined the double-station meteors with 36 single station meteors to derive a graph showing the scatter of the meteors around the radiant. They found "the mean scatter for the early part of the stream, to July 27, to be 83'. The mean scatter for ten days around August 1 is 93', and from August 7 to August 22, there is a fairly large increase in scatter to an average of 155', quite independent of mass (luminosity)." The authors added that an "irregular frequency distribution of meteors with date" existed. They concluded that two concentrations definitely seemed present. The first was given a duration of July 16-August 1, while the second began on August 1 and ended on August 22.[21]

The Author has examined the lists of multiple-station photographic meteors obtained from the United States and the Soviet Union, and has isolated 29 probable members of the Alpha Capricornid stream. The two streams noted by Wright, Jacchia and Whipple are present, as well as a third stream. Details of these follow:

> *Stream I: This represents the main stream of the Alpha Capricornid complex. It is based on 17 meteors and possesses a duration of July 16 to August 29. The nodal passage occurs on August 1 (λ=128.6°) from an average radiant of α=306.7°, δ=–8.3°. The radiant's daily motion is +0.84°±0.07° in α and +0.21°±0.02° in δ.

---

[21]Wright, F. W., Jacchia, L. G., and Whipple, F. L., *AJ*, 61 (March 1956), pp. 61-69.

*Stream II: This was the secondary stream noticed by Wright, Jacchia and Whipple. It is based on 5 meteors, which indicate a duration of August 8-21. The nodal passage occurs on August 15 (λ=141.9°), at which time the radiant is at α=322.4°, δ=–13.1°.

*Stream III: This third stream is composed of 7 meteors, which indicate a duration of July 15-August 1. The nodal passage occurs on July 25 (λ=122.2°), at which time the radiant is at α=302.7°, δ=–12.5°. It is especially interesting that its semimajor axis of 2.069 AU is 20-25% smaller than that determined for the other two streams.

The Alpha Capricornid stream was detected during the 1960's by both sessions of the Radio Meteor Project. From the 1961-1965 session, the duration was determined as July 30-September 11. The nodal passage came on August 20.2 (λ=146.8°), at which time the radiant was at α=326.4°, δ=–11.9°.[22] For the 1968-1969 session, a similar duration was given as July 25-September 9, but the nodal passage was determined as August 9.6 (λ=136.6°), while the average radiant was at α=314.8°, δ=–7.1°.[23] The orbit of each stream is listed in the "Orbit" section below. It can be seen that the stream detected in 1961-1965 bears a striking resemblance to the orbit of the Author's "Stream II" discussed above, while the stream detected during 1968-1969 is quite similar to "Stream I," which is the primary component of the Alpha Capricornid stream. These also happen to be the two streams detected by Wright, Jacchia and Whipple.

Numerous astronomers have tried to identify the object responsible for the formation of the Alpha Capricornid stream. The Russian astronomer E. N. Kramer was the first to examine the problem, when, in 1953, he concluded that the most likely candidate was comet 1457 II. In 1954, H. J. Bernhard, D. A. Bennett and H. S. Rice suggested a link to comet 1881 V (periodic comet Denning, now known as Denning-Fujikawa). In 1956, Wright, Jacchia and Whipple suggested that comet 1948 XII (periodic comet Honda-Mrkos-Pajdusakova) "may be a parent comet for the later α-Capricornids."[24] In 1973, Sekanina indicated that the comet most likely capable of producing this stream was, again, 1948 XII,[25] but in his 1976 analysis of radio-echo data he suggested the Apollo asteroid Adonis as a candidate.[26] Unfortunately, the question of which object produced the Alpha Capricornid stream remains unsolved. None of

---

[22]S1973, pp. 257 & 259.

[23]S1976, pp. 282 & 296.

[24]Wright, F. W., Jacchia, L. G., and Whipple, F. L., *AJ*, **61** (March 1956), pp. 61-62.

[25]S1973, p. 267.

[26]S1976, p. 307.

the suggested bodies matches the orbit of this stream perfectly and all of these authors rely on the fairly diffuse nature of this stream as an indication that it is old and far removed from the original orbit.

Although most of the information on this shower has come from observer's in the Northern Hemisphere, the radiant does possess a fairly low altitude for most observers, thus reducing the number of meteors which can be seen. As an example, several members of the American Meteor Society in New York, Texas and Florida observed (or tried to observe) activity during July 23-August 1, 1970. Hourly rates seemed to reach a fairly consistent maximum of 2-3 between July 30 and August 1.[27] According to the British Astronomical Association, meteor observers can expect to see a maximum ZHR of 8 on August 2 (λ=129°), from a radiant of α=309°, δ=–10°.[28]

For the most part, it appears that the quantity of meteors visible to observers in the Northern Hemisphere prevents an accurate isolation of the individual branches of the Alpha Capricornid stream. Such does not seem to be the case in the Southern Hemisphere, where Capricornus resides quite near the zenith. During the period 1969-1980, Michael Buhagiar (Perth, Western Australia) observed 20,974 meteors. Among them was included the Alpha Capricornid shower, the hourly rate of which he gave as 14.[29] Hourly rates of this magnitude, allow a more complete examination of the visual complexity of the Alpha Capricornid shower.

The 1979 observations of the Western Australia Meteor Section (WAMS) may possibly indicate just how complex this stream might be. The most significant detail offered by these observations is that more than one peak in activity was noted. In fact, three distinct dates of activity were revealed: July 22, July 28 and August 5 (the latter being affected by increasing moonlight). The multiple radiants noted on the first two dates may or may not be significant, depending on whether they are confirmed in the future. The Author believes it may be more acceptable to simply average these radiants into one July 22 and July 28 radiant.

The first maximum of July 22 revealed three weak Alpha Capricornid radiants. The longest duration shower lasted from July 20-27, and possessed a maximum ZHR of 1.74±0.25 from a radiant of α=307°, δ=–11°. The other two radiants lasted 2-3 days centered on July 22 and appeared to be subcenters of the longer duration group. One had a ZHR of 1.08±0.18 and a radiant of α=304°, δ=–14°, while the second had a ZHR of 1.46±0.21 and a radiant of α=308°, δ=–9°. The second maximum of July 28 possessed two showers with durations

[27]*MN*, No. 3 (October 1970), p. 3.

[28]Handbook of the B.A.A. (1982).

[29]Buhagiar, Michael, *WAMS Bulletin*, No. 160 (1981).

of July 27-28. One shower was located at α=304°, δ=–12° and had a ZHR of 1.56±0.33, while the other shower was at a position of α=306°, δ=–11° and had a ZHR of 3.42±0.51. Finally, with moonlight beginning to interfer, the Australian observers found another radiant reaching maximum on August 5. The duration of this shower extended over August 3-5. The ZHR peaked at 6.20±1.79, while the radiant was at α=309°, δ=–10°.[30] Using all observations of WAMS members, section director Jeff Wood has concluded that the ZHR typically reaches 5-10.[31]

Additional details of the Alpha Capricornid meteors have been obtained by numerous observers in recent years. In the following table it will be noted that the fairly consistent estimates of the average magnitude reveal a value slightly fainter than 2. Details of the number of Alpha Capricornids possessing persistent trains are unfortunately lacking.

Alpha Capricornid Magnitudes and Trains

Year(s)	Ave. Mag.	# Meteors	% Trains	Observer(s)	Source
1960-1976	1.69	355	----	McLeod	*MN*, No. 32
1966-1982	2.41	106	----	West	*MN*, No. 61
1966-1981	2.03	38	5.1	Swann	*MN*, No. 62
1967-1981	2.45	143	----	Hill	*MN*, No. 63
1977	2.6	31	----	Gates	*MN*, No. 38
1978	2.21	61	18.0	WAMS	*MN*, No. 45
1982	2.09	241	----	WAMS	*MN*, No. 62
1982	1.72	14	----	Lunsford	Personal Comm.
1986	2.52	55	----	Roggemans	Personal Comm.
1986	2.38	8	----	Lunsford	Personal Comm.

The observers cited are Norman W. McLeod III (Florida); John West (Texas); David Swann (Texas); Rick Hill (North Carolina); Bill Gates (Florida); Western Australia Meteor Section; Robert Lunsford (California); Paul Roggemans (Belgium).

As with all major meteor showers, the question of the color of the meteors frequently comes up. For the Alpha Capricornids, the primary color seems to be yellow. This seems to be the consensus of the British Astronomical Association, as the shower was described as "rich in yellow fireballs,"[32] and the Western Australia Meteor Section, where color estimates made in 1978 revealed 23.3% of the Alpha Capricornids to be yellow.[33]

---

[30]Wood, Jeff, Personal Communication (October 15, 1986).

[31]Wood, Jeff, Personal Communication (October 24, 1985).

[32]Handbook of the B.A.A. (1982).

[33]Wood, Jeff, *MN*, No. 45 (April 1979), p. 12.

Orbit

The three orbits computed by the Author are listed below. Stream "I" is the main branch of the Alpha Capricornids and was determined from 17 photographic meteor orbits. Stream "II" is identical to the secondary stream noted by Wright, Jacchia and Whipple and was determined from 5 meteors. Stream "III" was determined from 7 meteors.

	ω	Ω	i	q	e	a
I	266.5	128.6	7.8	0.597	0.785	2.777
II	268.1	141.9	1.1	0.582	0.780	2.645
III	272.2	122.2	5.4	0.573	0.723	2.069

The Alpha Capricornids were detected in both sessions of the Radio Meteor Project conducted during the 1960's. It should be noted that both sessions poorly covered the period of July 19-25, so that the values of Ω given below could potentially be smaller.

	ω	Ω	i	q	e	a
S1973	267.2	146.8	0.9	0.630	0.659	1.850
S1976	267.9	136.6	6.1	0.620	0.677	1.920

Kappa Cygnids

Observer's Synopsis

With a duration spanning July 26 to September 1, this shower has been observed almost annually since about the mid-19th century. The maximum is complex, but seems to mainly occur around August 19 (λ=146°), with an average rate of about 6 meteors per hour. There is evidence to suggest widely varying activity rates and occasional increases in fireball production. The radiant is at α=290°, δ=+54° when the shower reaches maximum. The meteors are of moderate speed and are usually described as bluish-white in color. Their average brightness is about 3. The daily motion seems to be about +0.5° in α and +0.2° in δ.

History

The first apparent radiant determination for this meteor shower was made by N. de Konkoly (Hungary), who, during August 11-12, 1874, observed seven meteors from α=291°, δ=+50°.[34] The discovery was made while Konkoly observed the Perseid maximum and, subsequently, no additional observations were made of this new radiant.

William F. Denning first observed the Kappa Cygnids during 1877. As had been the case with Konkoly, Denning saw the first meteors while observing

[34]Denning, W. F., *Memoirs of RAS*, **53** (1899), p. 276.

the Perseid maximum, but he continued to observe for several days thereafter. Overall, meteors were seen during August 10-16, with the radiant being determined as α=292°, δ=+48°.[35] Denning again detected activity from this shower during 1885, 1886 and 1887, but he considered 1893 as the year activity was most notable.

Writing in *The Observatory* in September 1893, Denning wrote how he was "struck with the frequency and brightness of meteors from a contemporary radiant on the N.W. limits of Cygnus near the star κ. Altogether I observed 28 paths directed from the point 292°, +53°, and must have missed others." The activity was noted during August 5-16. He described the meteors as "rather swift, with short courses, and in a majority of cases the nucleus, before final disruption, burst out very suddenly and left a short streak, marking the spot where it occurred." Denning added that other British observers independently noted the unusual activity from Cygnus—most notably Henry Corder (Bridgwater), J. Evershed (Kenley), and R. A. Batt (Leyton Road)—with several bright fireballs being noted from the region during August 13-17. One very notable finding made by Denning involved the eastward movement of the radiant, with his observations of August 5, 6, and 8 revealing a position of α=290°, δ=+53°, his observations of August 13 and 14 revealing a position of α=292°, δ=+53°, and his August 16 observations indicating a position of α=296°, δ=+53°.[36]

Since most of the August meteor observations made during the 19th Century were typically centered on the date of the Perseid maximum, the actual date of the greatest Kappa Cygnid activity was not clearly understood until early in the 20th Century. Observers then began to notice that the shower was strongest around a week following the Perseid maximum. For example, during 1922, the English observers J. P. M. Prentice and A. Grace Cook independently plotted more meteors from the Kappa Cygnid shower than for any other August shower, excluding the Perseids. The former observer plotted 12 meteors during August 15-20, from an average radiant of α=291°, δ=+52°. Cook plotted 18 meteors during August 15-17, 20 and 26, from a radiant of α=291°, δ=+50°. She called it the "Theta Cygnids."[37] A probable third independent observation came from across the Atlantic as W. H. Christie (Victoria, British Columbia) plotted 6 meteors from α=299.5°, δ=+50.8° on the night of August 21/22.[38]

Other interesting observations of this shower came in the years that followed. In 1925, the Russian Society of Amateurs plotted 23 meteors during

---

[35]D1899, p. 276.

[36]Denning, W. F., *The Observatory*, **16** (September 1893), pp. 317-318.

[37]Denning, W. F., *The Observatory*, **45** (October 1922), pp. 332-333.

[38]Olivier, Charles P., *FOR*, No. 4 (1929), p. 22.

August 16-20, and revealed an average radiant of α=291.4°, δ=+54.0°.[39] Two excellent early observations of this stream were made by American Meteor Society (AMS) members on August 14 of 1936 and 1951. In the first year, Balfour Whitney (Oklahoma) plotted 7 meteors from α=285°, δ=+57°,[40] while Richard Widner (Oregon) plotted 5 in the latter year from α=289°, δ=+50.5°.[41]

Curiously, despite the frequent appearance of this shower in various observational records, it has just as frequently been overlooked. In the previous two paragraphs the observations by Christie, Whitney and Widner mark the only Kappa Cygnid observations among over 6000 radiants on record with the AMS. Similarly, in Cuno Hoffmeister's book *Meteorströme*, not only is this shower overlooked as an annual producer of meteors, but among the 5406 visual radiants obtained by German observers during 1908-1938, only one observation barely qualifies as a Kappa Cygnid radiant. On August 21, 1930 (λ=147.2°), a radiant was detected at α=289°, δ=+61° and was given a weight of 6 on a scale of 1 to 10.[42]

Although additional visual observations were occasionally reported during the 1940's and 1950's, a newer more substantial view of the Kappa Cygnids was finally achieved in 1954 when Fred L. Whipple published the first list of photographic meteor orbits. He determined the orbits of 144 meteors from double-station photographs, and 5 were identified as Kappa Cygnids. These meteors indicated a duration of August 9-22, while the average radiant was α=291.5°, δ=+53.0°. Whipple commented that the orbits indicated this stream has a "short-period comet orbit with high inclination, period 7-8 years, [aphelion] of 7-8 a.u. The long duration...suggest remnants of a large comet."[43]

The next important event in meteor studies involved radio-echo observations and the Kappa Cygnids were finally recognized by this technique in the early 1960's. Zdenek Sekanina detected this stream during the 1961-1965 session of the Radio Meteor Project. Activity was detected during August 23-28 from an average radiant of α=298.9°, δ=+62.4°. Although the date of the nodal passage was given as August 26.1 (λ=152.5°),[44] it should be noted that the radar equipment was not in operation during August 18-22, so that an earlier nodal passage and subsequent average date of maximum activity is quite probable. The stream's occasional disappearance, noted earlier among visual observations, is also evident in Sekanina's 1968-1969 session of the Radio Meteor Project.

---

[39]Denning, W. F., *The Observatory*, 49 (August 1926), p. 257.

[40]Olivier, Charles P., *FOR*, No. 41 (1938), p. 7.

[41]Olivier, Charles P., *FOR*, No. 99 (1954), p. 261.

[42]H1948, pp. 221 & 223.

[43]W1954, pp. 202-203 & 211.

[44]S1973, pp. 257 & 260.

Despite the radio equipment being operated during August 11-15, 17-18 and 25-30, the Kappa Cygnids were either too weak to be delineated by the computer or totally nonexistent.

Whether the Kappa Cygnids are actually absent in some years cannot be answered since the stream has never been the focus of intensive observations. What is apparent is that the hourly rates of the shower do seem to vary. The first determination of the Kappa Cygnid activity levels came from an observation by Denning on August 22, 1879, when 52 of this shower's meteors were seen during 5 hours of clear sky (the Author computes a ZHR of 13 for this observation).[45] With increased visual observations, the variations in activity have been especially noted in recent years. In 1974 the Hungarian Meteor Team obtained a peak ZHR of 23.6±5.1.[46] In 1982 observers of the Nippon Meteor Society estimated a peak ZHR of 14.1,[47] while members of the Dutch Meteor Society obtained a maximum ZHR of only 2-3 in 1984.[48]

Aside from occasional hourly rates, visual details of the Kappa Cygnid meteors have rarely been obtained. Some of the few examples of magnitude estimates are listed below. No recent estimates of the percentage of meteors leaving persistent trains have been made.

Kappa Cygnid Magnitudes and Trains

Year(s)	Ave. Mag.	# Meteors	% Trains	Observer(s)	Source
1977	3.34	67	—	Lunsford	Personal Comm.
1978	3.00	13	—	Lunsford	Personal Comm.
1986	3.00	58	—	Roggemans	Personal Comm.

The observers cited are Robert Lunsford (California); Paul Roggemans (Belgium).

The Author has sifted through the published lists of photographic meteor orbits and has isolated 9 meteors that are probable members of the Kappa Cygnid stream. They indicate a duration extending from July 26 to September 1. The date of the nodal passage is August 17 (λ=143.4°), at which time the radiant is at α=287.6°, δ=+53.8°. The indicated daily motion of the radiant is +0.50°±0.06° in α and +0.15°±0.12° in δ.

Orbit

The only radar study to detect this stream was Zdenek Sekanina's 1961-1965 session of the Radio Meteor Project. As noted earlier, the failure of the

---

[45]Denning, W. F., *Nature*, **20** (September 11, 1879), p. 458.

[46]*MN*, No. 35 (Mid-March 1977), p. 6.

[47]*Meteoros*, **14** (May 1984), p. 31.

[48]Veltman, Rudolf, *Radiant*, **7** (July-August 1985), p. 75.

equipment to operate during August 18-22 probably indicates that Ω should be smaller. The orbit was given as

	ω	Ω	i	q	e	a
S1973	203.1	152.5	42.9	0.979	0.621	2.583

Nine photographic meteor orbits were gathered by the Author from W1954, MP1961, B1963, BK1967, CF1973 and C1977. The average orbit was

ω	Ω	i	q	e	a
202.7	143.4	38.0	0.977	0.793	4.720

Upsilon Pegasids

Observer's Synopsis

The activity of this stream seems to occur between July 25 and August 19, with maximum occurring on August 8, from α=349°, δ=+18°. Activity levels may be irregular, but generally remain between 2-5 per hour. The meteors are swift and tend to be primarily yellow. The average magnitude is about 3.5, with only about 6% of the meteors being brighter than magnitude 2. Only 13% of the Upsilon Pegasids show persistent trains.

History

Harold R. Povenmire (Florida) discovered this shower on August 8, 1975, while observing the Perseids. Three meteors were noted in a very short time which originated from within the Great Square of Pegasus, but clouds prevented more detailed observations. The next evening, Povenmire again detected the shower and was able to determine the radiant as α=350°, δ=+19°.

Moonlight interfered with observations during 1976, but moonless nights prevailed during early August 1977. An analysis by *Sky & Telescope* said observers detected hourly rates as high as 8 to 13 on the evening of August 12 and 13, and Povenmire was quoted as saying that meteors were detected up to a week before and after these dates.[49] On the other hand, brief details published in *Meteor News* seemed to cast doubt on the high rates and the results of 7 observers were summarized as indicating a maximum of only 1.71 per hour.[50]

Moonlight was again no problem in 1978, but the confirming reports of activity were also accompanied by reports with a negative tone. From Los Padres National Forest in California, Mark Davis and Donald Gutierrez observed 20 meteors which originated within 20° of the Great Square of Pegasus during the morning of August 12. On the other hand, Richard Nolthenius (Alpine,

[49]*Sky & Telescope*, **54** (November 1977), p. 441.
[50]*MN*, No. 38 (Mid-October 1977), p. 8.

California) made the comment, "Only six Pegasids were seen during 2 1/4 hours of observing on the morning of the 13th. Due to the high rate of sporadic meteors this year, my impression was that the number radiating from the Great Square was not much more than would be expected by chance."[51] Povenmire said that about 700 persons around the world contributed observations, with some of the detailed observations including color estimates of primarily yellow to white, and a radiant diameter of about three degrees.[52]

As these examples demonstrate, there were widely varying estimates of the strength of Upsilon Pegasid activity. Subsequently, these conflicting observations brought controversy and veteran meteor observer Norman W. McLeod III (Miami, Florida) questioned the shower's very existence in the January 1979 issue of *Meteor News*.

McLeod listed at least 12 major and minor meteor showers which were active during early August and in his evaluation of Upsilon Pegasid activity he said, "Several of the named showers can account for the bulk of such meteors." He went on to discuss his own observations, which generally revealed "only about 3 meteors/night from the Pegasid radiant" in early August. He added that his best night was August 4/5 when 7 meteors were observed and that during the Perseid maximum "only 1 Pegasid/night was seen."[53]

The Author believes the Upsilon Pegasids were victim to a series of unfortunate events which led many people to believe they were a potentially major shower which peaked at the time of the Perseid maximum. This was far from the truth. First of all, *Sky & Telescope* unfortunately did not discuss the observations to their fullest, as the Upsilon Pegasid discussion occurred at the end of a multi-page article dealing with the Perseids. The fact was that the high rates quoted came from inexperienced observers who, as McLeod pointed out above, simply confused meteors from several active showers with the Upsilon Pegasid activity. What the magazine should have noted was that *experienced* observers tended to reveal hourly rates of about 0-2 during the time of the Perseid maximum and it is this fact that leads to the next problem. Despite Povenmire's discovery of notable activity during August 8 and 9, 1975, observers in 1977-1978 rarely observed any earlier than August 11. Since it is now apparent that the Upsilon Pegasids rapidly weaken after their August 8 maximum, no one would have expected these observers to see more than a few of these meteors each night the Perseids were active.

Although a few meteor observers have continued to make this shower a controversial subject, many others have helped to refine the current knowledge

---

[51]*Sky & Telescope*, **56** (November 1978), p. 473.

[52]Povenmire, Harold, *The Upsilon Pegasid Meteor Shower*. (1986), unpublished.

[53]McLeod, Norman W., III, *MN*, No. 44 (January 1979), p. 4.

of the Upsilon Pegasid stream. In the years following 1978, observers became more aware of the earlier maximum for the Upsilon Pegasids and the picture soon developed of a shower which produces maximum rates of 2 to 3 per hour.

The most significant visual study of this shower was made during July 30-August 16, 1983 by the Hungarian Meteor and Fireball Observation Network. In the publication *Meteor*, Tepliczky István collated the observations and revealed some interesting statistics concerning this shower's members. First of all, the meteor colors were tabulated into the following percentages: blue, 7%; blue-white, 14%; white, 28%; yellow-white, 5%; yellow, 32%; orange, 14%. Secondly, a good record of the train activity of the shower was kept. Of 155 Upsilon Pegasids observed, 20, or 13%, left persistent trains. Finally, ZHRs were computed for each day, with a maximum of 5 being recorded on three occasions—August 1/2, 6/7 and 16.[54] Maximum was plainly indicated as occurring during August 1-7, with ZHRs remaining between 4 and 5, and the average ZHR being 4.6.

Aside from the estimates of hourly rates, one of the next most important observations of any meteor shower is the determination of the average brightness of its meteors. The average magnitude of the Upsilon Pegasids tends to be fainter than for the other showers active during the first half of August and this indicates a proportionately higher percentage of faint meteors than normal. Some examples of the average magnitudes are given below. Data on the percentage of meteors leaving persistent trains has not been frequently given, but, as noted earlier, it seems no higher than 13%.

Upsilon Pegasid Magnitudes and Trains

Year(s)	Ave. Mag.	# Meteors	% Trains	Observer(s)	Source
1977-1985	3.62	93	—	McLeod	Unpublished[55]
1979	3.11	18	—	Lunsford	Personal Comm.
1980	3.73	11	—	Lunsford	Personal Comm.
1981	3.10	12	—	Lunsford	Personal Comm.
1981	2.90	21	—	Parviainen	Unpublished[56]
1986	3.59	17	—	Lunsford	Personal Comm.

The observers cited are Norman W. McLeod, III (Florida); Robert Lunsford (California); Pekka Parviainen (Finland).

As with all meteor showers, observations become more significant when photographs are involved, and Povenmire's annual request for photographs of

---

[54]István, pp. 21-22.

[55]Personal Communication—Norman McLeod, III (Florida) to Harold Povenmire (Florida).

[56]Personal communication—R. Persson (Sweden) to Harold Povenmire (Florida).

the region have revealed several good specimens. Most notable has been the detection of two meteors that appeared nearly stationary, thus giving an indication of the general correctness of the original radiant estimate. In 1980, Steve Hunt (Kansas City, Missouri) recorded a meteor "coming nearly head-on from the radiant," while, in 1983, Leighton Venn (Bartlesville, Oklahoma) photographed a nearly stationary Upsilon Pegasid which, when farthest from the radiant, was only one degree away.[57]

One of the best examples of a photographic Upsilon Pegasid was not photographed by a regular watcher of this shower, but, instead, it was detected by five cameras of the European Meteor Network on August 19, 1982. Network director, Dr. Zdenek Ceplecha (Ondrejov Observatory, Czechoslovakia), made a careful study of this fireball and proved that it was an Upsilon Pegasid.[58] The meteor was designated EN190882 and, when brightest, was about magnitude –13.8, or brighter than the full moon! Ceplecha's subsequent calculation of a precise orbit gave the first indication of what the orbital plane of this stream might look like (see "Orbit" section below).

Despite the evidence that has accumulated since this shower's discovery, continued controversy as to the reality of this stream inspired the Author to conduct an extensive search through publications of the last 100 years to uncover past appearances. Overall, seven radiants are present which indicate a duration extending at least from August 3 to August 11. The two earliest observations were made by P. Denza and Alexander S. Herschel. The former observer plotted several meteors from α=350°, δ=+24° on August 11, 1869, while Herschel's observations of 1860-1881 revealed a radiant at α=345°, δ=+15° on August 11.[59] Other observations around the apparent date of maximum are as follows:

Upsilon Pegasid Radiants

Date (UT)	RA(°)	DEC(°)	Meteors	Source
1913, August 3	347	16	13	H1948
1924, August 4	347.5	22	4	*FOR*, No. 4
1937, August 4.1	352	18	?	H1948
1964, August 8.7	342	22	4	*FOR*, No. 155
1921, August 9	352	18	?	H1948

Finally, in addition to searches for visual radiants, the Author has also searched photographic records for double-station meteors. These meteors not only offer reliable radiants, but, in many cases, they even offer orbital data. What may be the very first double-station Upsilon Pegasid appears to have been

[57]Povenmire, Harold, *The Upsilon Pegasid Meteor Shower*. (1986), unpublished.
[58]Ceplecha, Zdenek, *SEAN Bulletin*, 7 (Sept. 30, 1982), pp. 13-14.
[59]D1899, p. 287.

detected on August 3, 1902, during the Yale Photographic Meteor Survey, and emanated from α=346.4°, δ=+17.8°.[60] The inability to compute a velocity prevented the calculation of an orbit and the radiant is not considered absolutely certain. The next probable photographic Upsilon Pegasid appears among the 2529 double-station meteor orbits computed by Richard McCrosky and Annette Posen from data gathered during the Harvard Meteor Project. Designated 3607, the meteor was detected on August 18.3, 1952, from α=349°, δ=+22°. Its magnitude was 0.1, while the semimajor axis was computed as 13.47 AU.[61] On August 13.8, 1958, a photographic survey conducted at Dushanbe (USSR) revealed a meteor from α=344.5°, δ=+18.4°. The subsequent orbital calculation by Dr. P. B. Babadzhanov revealed a slightly hyperbolic orbit with an eccentricity of 1.04.[62]

Orbit

A bright fireball photographed by the European Meteor Network on August 19, 1982, proved to be a member of the Upsilon Pegasid stream. Ceplecha computed the following orbit:

ω	Ω	i	q	e	a
306.9	145.34	84.7	0.198	0.999	198.0

Adding the two additional photographic meteor orbits briefly discussed earlier, the Author has computed the following weighted mean orbit. The orbit is still parabolic since the eccentricities of all three meteor orbits range from 0.98 to 1.04. Since two of the three meteors possess eccentricities less than 1, it seems possible that this stream is in a long-period orbit.

ω	Ω	i	q	e	a
303.9	143.3	79.5	0.218	1.0	∞

Perseids

Observer's Synopsis

This meteor shower is generally visible between July 23 and August 22. Maximum occurs during August 12/13 (λ=139°), with the radiant located at α=48°, δ=+57°. The hourly rate typically reaches 80, although some years have been as low as 4 and as high as 200. The meteors tend to be very fast, possess an average magnitude of 2.3 and about 45% leave persistent trains. The radiant advances by a rate of 1.40°/day in α and 0.25°/day in δ.

---

[60] Olivier, Charles P., *FOR*, No. 38 (1937), p. 7.

[61] MP1961, p. 64.

[62] Simakina, E. G., *SSR*, **2** (1968), p. 131.

History

This is the most famous of all meteor showers. It never fails to provide an impressive display and, due to its summertime appearance, it tends to provide the majority of meteors seen by non-astronomy enthusiasts.

The earliest record of its activity appears in the Chinese annals, where it is said that in 36 AD "more than 100 meteors flew thither in the morning."[63] Numerous references appear in Chinese, Japanese and Korean records throughout the 8th, 9th, 10th and 11th centuries, but only sporadic references are found between the 12th and 19th centuries, inclusive. Nevertheless, August has long had a reputation for an abundance of meteors. The Perseids have been referred to as the "tears of St. Lawrence", since meteors seemed to be in abundance during the festival of that saint on August 10th,[64] but credit for the discovery of the shower's annual appearance is given to Quételet (Brussels), who, in 1835, reported that there was a shower occurring in August that emanated from the constellation Perseus.[65]

The first observer to provide an hourly count for this shower was Eduard Heis (Münster), who found a maximum rate of 160 meteors per hour in 1839. Observations by Heis and other observers around the world continued almost annually thereafter, with maximum rates typically falling between 37 and 88 per hour through 1858. Interestingly, the rates jumped to between 78 and 102 in 1861, according to estimates by four different observers, and, in 1863, three observers reported rates of 109 to 215 per hour.[66] Although rates were still somewhat high in 1864, generally "normal" rates persisted throughout the remainder of the 19th-century.

Computations of the orbit of the Perseids between 1864 and 1866 by Giovanni Virginio Schiaparelli (1835-1910) revealed a very strong resemblance to periodic comet Swift-Tuttle (1862 III).[67] This was the first time a meteor shower had been positively identified with a comet and it seems safe to speculate that the high Perseid rates of 1861-1863 were directly due to the appearance of Swift-Tuttle, which has a period of about 120 years. Multiple returns of the comet would be responsible for the distribution of the meteors throughout the orbit, but meteors should be denser in the region closest to the comet, so that meteor activity should increase when the comet is near perihelion (as has been demonstrated by the June Boötids, Draconids and Leonids).

---

[63] Imoto, S., and Hasegawa, I., *SCA*, **2** (1958), pp. 135 & 138.

[64] McKinley, D. W. R., *Meteor Science and Engineering*. New York: McGraw-Hill Book Company, Inc., 1961, p. 6.

[65] Denning, W. F., *Nature*, **20** (September 11, 1879), p. 457.

[66] Denning, W. F., *Nature*, **20** (September 11, 1879), p. 457.

[67] Schiaparelli, G. V., *Sternschnuppen*.

As the 20th-century began, the maximum annual hourly rates of the Perseids seemed to be declining. Although rates were above Denning's derived average rate of 50 per hour during five years between 1901 and 1910, the observed rate in 1911 was only 4 and for 1912 it was 12. Denning wondered whether the shower was declining,[68] but hourly rates seemed to return to "normal" in the years that followed. Quite unexpectedly the shower suddenly exploded in 1920, when rates were estimated to be as high as 200 per hour. This was extremely unusual as it came at a time when the parent comet was nearing aphelion! Although a few weaker-than-normal years occurred during the 1920's, the Perseids regained their consistency thereafter, and, except for abnormally high rates of 160 and 189 during 1931 and 1945, respectively, nothing unusual was observed up through 1960.

During 1973, Brian G. Marsden predicted Comet Swift-Tuttle would arrive at perihelion on September 16.9, 1981 (±1.0 years).[69] This immediately generated excitement among meteor observers as the potential for enhanced activity unfolded. This excitement seems to have been fully justified, as the average rate of 65 per hour during 1966-1975 suddenly jumped to over 90 per hour during 1976-1983—with the high being 187 in the latter year.[70] Although meteor observers seemed content with their observations of the enhanced activity from Swift-Tuttle, comet observers were less enthusiastic as the comet was never recovered. The possibility exists that this comet might not actually arrive at perihelion until 1992, as considered by Marsden after noting that a comet seen by Kegler during 1737 might represent an earlier return of Swift-Tuttle. If this were actually the case, then the Author suggests that the recent enhanced activity of the late 1970's and early 1980's might have been caused by whatever generated the early activity noted by Heis in 1839, so that the best is yet to come.

Since the 1983 peak, hourly rates for the Perseids seem to have been declining. With a full moon occurring just a day before maximum in 1984, the Dutch Meteor Society still reported unexpectedly high rates of 60 meteors per hour.[71] In 1985, reported rates generally fell between 40 and 60 meteors per hour in *dark* skies,[72] and results were generally the same in 1986.[73] Thus, the rates of the Perseids may have now dropped back to normal, with comet Swift-Tuttle possibly having passed through perihelion unobserved.

---

[68]Mackenzie, Robert A., *Solar System Debris*. Dover: The British Meteor Society (1980), pp. 16-17.

[69]Marsden, B. G., *AJ*, **78** (September 1973), p. 658.

[70]Mackenzie, Robert A., *Meteoros*, **14** (May 1984), p. 26.

[71]Mackenzie, Robert A., *Meteoros*, **15** (February 1985), p. 18

[72]di Cicco, Dennis, *Sky & Telescope*, **70** (November 1985), pp. 504-505.

[73]di Cicco, Dennis, *Sky & Telescope*, **72** (November 1986), p. 546.

From the 1860s onward, studies of the Perseids began to include more than just hourly rates. Numerous observers began to plot the paths of meteors onto star charts to derive the points from which the meteors seemed to be radiating. The most prolific observer of this stream was William F. Denning, who, between 1869 and 1898, observed 2409 Perseids[74] and became the first person to derive a daily ephemeris of the radiant's movement. In 1901, he published his most precise radiant ephemeris as follows:[75]

Perseid Radiant Ephemeris

Date	RA(°)	DEC(°)
July 27	27.1	53.2
July 29	29.3	53.8
July 31	31.6	54.4
Aug. 2	33.9	55.0
Aug. 4	36.4	55.5
Aug. 6	38.9	56.0
Aug. 8	41.5	56.5
Aug. 10	44.3	56.9
Aug. 12	47.1	57.3
Aug. 14	50.0	57.7
Aug. 16	52.9	58.0

[A recent plotting of 102 precise photographic meteor orbits by the author supports the general accuracy of the above ephemeris with the daily motion of the radiant being computed as α=+1.40°, δ=+0.25°]

In addition to this main radiant near Eta Persei, there have been indications that several secondary showers are also active. Minor activity near the main Perseid radiant has been noted on several occasions up to the present time and may have been noted as long ago as 1879, when Denning pointed out that he had "detected the existence of two other simultaneous showers from Chi and Gamma Persei."[76] This latter shower is one of the most active of the secondary radiants and seems to have been frequently observed during the twentieth century—especially with telescopic aid. The following observations represent some of the details.

> *In 1921, Ernst Öpik (Dorpat) observed the Perseids telescopically on August 10 and 12. On the latter date, he noted that 9 meteors emanated from an oval area 5.7°x2.2° across centered at α=40.0° δ=+55.6°.[77]

---

[74]Denning, W. F., *Memoirs of RAS*, **53** (1899), p. 210.

[75]Denning, W. F., *MNRAS*, **62** (1901), p. 169.

[76]Denning, W. F., *Nature*, **20** (September 11, 1879), p. 458.

[77]Öpik, E. J., *Astr. Nach.*, **217** (1922), p. 41.

*On August 11, 1921, C. P. Adamson (Wimborne, Dorset) seemed to have detected both the normal radiant and this southern component visually by claiming the Perseids emanated from an elongated area extending from α=43°, δ=+57° to α=49°, δ=+58°.[78]

*On August 10, 1931, C. B. Ford and B. C. Darling found a telescopic radiant at α=40.9°, δ=+54.4°.[79]

*On August 8, 1932, Öpik found a radiant at α=39°, δ=+54°.[80]

*During August 12-13, 1934, Ford found a telescopic radiant at α=43.1°, δ=+55.2°.[81]

One of the most recent examples of the complexity of the Perseid meteor shower was revealed in three studies of the radiant conducted during 1969 to 1971, by observers in the Crimea. In addition to the main radiant near Eta Persei, they confirmed the existence of the major radiants near Chi and Gamma Persei, as well as minor radiants near Alpha and Beta Persei. These meteor showers are generally short-lived and possess radiants that move nearly parallel to the main radiant. The following are summaries of the most consistent of the secondary Perseid radiants.[82, 83]

*The Gamma Perseids *mainly* occur during August 11 to 16 from an average radiant of α=41°, δ=+55°. The radiant diameter averages about 2°. Rates rise and decline with those of main radiant.

*The Chi Perseids occur during August 7 to 16 from an average radiant of α=35°, δ=+56°. The radiant diameter is about 2°. Maximum seems to occur between the 9th and 11th.

*The Alpha Perseids occur during August 7 to 24 from an average radiant of α=51°, δ=+50°. The radiant diameter averages about 1.5°. Maximum seems to occur somewhere between the 12th and 17th.

*The Beta Perseids occur during August 12 to 18 from an average radiant of α=47°, δ=+40°. The radiant diameter averages about 1°. Rates are irregular. Weakest branch of the Perseid cluster.

---

78 *Nature*, **107** (August 18, 1921), p. 793.

79 Olivier, Charles P., *PA*, **39** (November 1931), p. 547.

80 Öpik, Ernst, *Harvard College Observatory Circular*, No. 388 (1934), p. 36.

81 Olivier, Charles P., *PA*, **42** (October 1934), p. 469.

82 Martynenko, V. V., and Smirnov, N. V., *SSR*, **7** (1973), pp. 104-109.

83 Martynenko, V. V., Vagner, L. Ya., and Levina, A. S., *SSR*, **7** (1973), pp. 163-165.

These secondary centers of activity have been predominantly visual displays; however, time was taken to seek out some of these other radiants during the Jodrell Bank radio-echo survey of the 1950's. Only the Alpha Perseids were noted with confidence. Detected in both 1951 and 1953, the radiant was very diffuse and 8° in diameter centered at α=54°, δ=+48°. It was detected between August 8 and 11, and the highest radio-echo rate reached 37 per hour (the main Perseid radiant reached radio-echo rates of 50 per hour during the same years).[84]

Other studies conducted by amateur and professional astronomers during the last 30 to 40 years have involved specific details of shower members. One especially interesting statistic that has been brought forward was the trend that the Perseids seem to be brighter before the date of maximum than afterward. In 1953, A. Hruska (Czechoslovakia) found the average magnitude to be about 2.5 during August 8 to 12; however, on August 12/13 it had dropped to 2.8 and by August 14/15 it had fallen to 3.4.[85] In 1956, Zdenek Ceplecha also showed a similar, though less pronounced decline in brightness. During August 4 to 10, the average Perseid was near magnitude 2.68, while during August 10 to 15 it was 2.94. The extremes came on August 6/7 (magnitude 2.31) and August 13/14 (magnitude 3.18).[86] Just as Hruska and Ceplecha's studies show conflicting patterns representing the decline in the Perseid magnitude distribution during August, two very recent studies seem to support both views.

During 1983, members of the Spanish astronomical group Agrupacion Astronomica Albireo, under the direction of Eduardo Martinez Moya, obtained an excellent series of Perseid magnitude observations, which seemed to support Hruska's study. Between August 1 and 13, 1983, the average daily magnitude varied from 1.75 to 2.04. Thereafter, it dropped to 2.19 by the 14th, 2.52 by the 15th, 2.77 by the 17th, 2.92 by the 19th and 3.45 by the 20th. Robert Mackenzie (director of the British Meteor Society) claims the magnitude distribution of the Perseids "gives an indication of the particle mass variation in the cross-section of the stream encountered by the Earth."[87] This variation seems to support Hruska's study.

Another excellent series of magnitude estimates were made by Paul Roggemans (Brussels, Belgium) during July 27 to August 16, 1986. Observing in darker skies than the Spanish group, Roggemans detected 1315 Perseids and gave the average magnitude of the shower as 3.10.[88] Roggemans' estimates

---

[84]Bullough, K., *Jodrell Bank Annals*, **1** (June 1954), pp. 80-85.

[85]Hruska, A., *BAC*, **5** (1954), p. 14.

[86]Ceplecha, Z., *BAC*, **9** (1958), p. 236.

[87]Mackenzie, Robert A., *Meteoros*, **14** (May 1984), p. 26.

[88]Roggemans, Paul, Personal Communication (September 24, 1986).

were very consistent throughout the shower's duration with variations being typically less than 10% on any given day; however, there were two exceptions. The first came on August 5/6 and 6/7, when the average magnitude dropped to a low of 3.54. The second drop occurred on August 9/10 and 10/11 when the average magnitude reached 3.71. This set of observations seems to support Ceplecha's study.

All of the above magnitude studies (and many more not discussed here) seem to have one thing in common—they point to an irregular mass distribution within the Perseid stream. Filamentary structure seems the best explanation. During some years, the filaments are encountered in rapid succession by Earth's passage through the Perseid stream, thus accounting for the consistent magnitude estimates followed by a steady decline. In other years, the filaments are spread out across the stream's width, thus causing the consistent average magnitude estimates to be disrupted by periods of activity from primarily brighter or fainter meteors.

Another statistic that has been brought forward during the last 30 to 40 years has been the percentage of Perseids that possess persistent trains. This is a major factor long noted in the separation of Perseids from other active showers occurring during the first half of August. Miroslav Plavec used the records made at the Skalnate Pleso Observatory (Czechoslovakia) to produce one of the most ambitious studies of train phenomena to date. He studied 8,028 meteors observed between 1933 and 1947, and found the following percentages: 45% possessed trains in 1933, 60% in 1936, 35% in 1945 and 53.5% in 1947. The variations could not be correlated to sunspot numbers.[89] Taking an average of meteor train activity noted in various publications between 1931 and 1985, the author has found the average value to be 45% for nearly 60,000 meteors.

Orbit

More orbits have been computed for the Perseids than for any other meteor stream, with the first coming during the early 1860's. During the last few decades photographic and radio-echo techniques have enabled the first precise orbital determinations. The Author has accumulated 102 precise photographic meteor orbits from the various United States, Soviet Union and Czechoslovakian lists. The following orbit was revealed.

ω	Ω	i	q	e	a
149.2	139.5	113.2	0.942	0.902	9.641

Two major radar surveys revealed the Perseids during the 1960's. B. L. Kashcheyev and V. N. Lebedinets found the Perseids during a 1960 survey at the Kharkov Polytechnical Institute (BL1967), while Zdenek Sekanina

[89]*Sky and Telescope*, **12** (January 1953), pp. 74-75.

determined an orbit from data gathered during the 1969 session of the Radio Meteor Project (S1976).

	ω	Ω	i	q	e	a
BL1967	150.9	136.5	113.1	0.95	0.91	11.11
S1976	152.5	139.7	110.2	0.960	0.881	8.040

The orbit of periodic comet Swift-Tuttle (1862 III) is as follows:

ω	Ω	i	q	e	a
152.8	138.7	113.6	0.960	0.963	24.326

Alpha Ursa Majorids

Observer's Synopsis

This meteor stream seems to produce a weak meteor shower during August 9 to 30. The maximum occurs during August 13 to 14, from an average radiant of α=165°, δ=+63°, and tends to produce meteors at the rate of 4 per hour. There is evidence that this shower may be strong telescopically and with radar, but few details are available.

History

This stream is relatively new in the observational records. It was discovered by W. A. Feibelman while observing the Perseids in 1961. He claimed that between "3:00 and 5:00 a.m. E.D.T. on August 14/15, four meteors per hour were seen to radiate from the direction of α Ursa Majoris toward the zenith." Rates of one per hour persisted during the next two nights. The meteor paths were typically 20° to 30° long and the velocity, compared to the Perseids, was considered low. Feibelman added, "There were no terminal bursts, and the light intensity tapered off gradually."[90]

Apparent confirmation came from the Mt. Bezovec meteor expedition, led by Z. Kviz and J. Kvizová. During August 6 to 17, 1961, three groups of eight observers were using the independent counting method to "determine the meteor perception probability."[91] Although radiants were not computed from the data, the direction of motion could be established and an increase in meteor activity was noted from the region of α Ursa Majoris on the evenings of August 13/14 and 14/15.

Attempts by the Author to track down visual activity from this stream prior to 1961 have been unsuccessful. No trace of this shower appears in the extensive radiant lists of the American Meteor Society, nor in lists published in Britain,

[90]Feibelman, W. A., *Astrophysical Journal*, **136** (1962), pp. 315-316.

[91]Kviz, Z., and Kvizová, J., *BAC*, **15** (1964), p. 31.

Germany and Russia. On the other hand, one photographic meteor was found. On August 17, 1952, cameras of the Harvard Meteor Project photographed a –0.1-magnitude meteor that originated from a radiant of α=169°, δ=+65°. The orbit is listed below.

Whether the meteor shower was in existence prior to 1961, seems uncertain at present; however, one source seems to indicate that the shower persisted after 1961. The Author has examined the 39,145 radio meteors obtained by Zdenek Sekanina during the two sessions of the Radio Meteor Project of the 1960's, and has isolated 27 radio meteor orbits. Although the existence of such a shower was missed by the initial computer search carried out by Sekanina, it is not the first to have been missed and its significance does not readily appear until both lists are compared.

The data presented is indeed scanty and, if not for the radio-echo meteor data, the shower would not have been included in this book. Nevertheless, it is interesting to note that during the five years of sensitive radio-echo data, meteors were detected each year, so that the stream may at least be considered a possible producer of an annual telescopic meteor shower. The visual information obtained by Feibelman is also interesting, but may simply represent an outburst of activity.

Finally, there is one other point that must be added which could explain the shower's visual appearance in 1961. The orbit of this meteor stream possesses nearly the same orbital elements as comet Alcock (1959 IV). Although the orbit of this comet is not considered a meteor producer in a list compiled by Jack Drummond in 1981,[92] the fact that the meteors are predominantly small, and more susceptible to perturbations than larger meteors, could explain the stream's smaller perihelion distance and subsequent encounter with Earth. The unusual outburst of bright meteors that appeared in 1961, may have been directly due to a cloud of material following comet 1959 IV.

Orbit

Thus far, the photographic records of meteors have revealed only one rough orbit. That orbit, which comes from MP1961, is as follows:

ω	Ω	i	q	e	a
123	144	44	0.80	0.88	6.63

A more precise orbit was determined by the Author from 27 radio meteors extracted from the 39,145 radio meteor orbits obtained by Sekanina during the Radio Meteor Project. The average orbit is as follows:

ω	Ω	i	q	e	a
125.6	145.0	47.7	0.851	0.635	2.332

[92]Drummond, Jack D., *Icarus*, 47 (1981), pp. 500-517.

Finally, for the sake of comparison, the orbit of comet Alcock (1959 IV) is also given. The agreement between the angular orbital elements is excellent.

	ω	Ω	i	q	e	a
1959 IV	124.7	159.2	48.3	1.150	1.000	----

Additional August Showers

August Eridanids

This somewhat diffuse radiant was noted by the Author while examining the 39,145 radio meteor orbits obtained by Zdenek Sekanina during the Radio Meteor Project of the 1960's. Six radio meteors were identified which suggest a duration of August 2-27. Maximum seems to occur around August 11/12 from an average radiant of α=49.6°, δ=–4.9°. What makes this stream orbit especially interesting is its similarity to the orbit of periodic comet Pons-Gambart (1827 II). The orbit of this stream based on six radio meteor orbits is as follows:

ω	Ω	i	q	e	a
17.1	319.9	140.2	0.985	0.644	4.093

The orbit of Pons-Gambart is as follows:

	ω	Ω	i	q	e	a
1827 II	19.2	319.3	136.5	0.807	0.946	14.944

Interestingly, no visual evidence exists to directly support this radiant; however, there is strong evidence of activity 8°-10° north among the records of the American Meteor Society, as the following illustrates:

August Eridanids Radiants

Date (UT)	RA(°)	DEC(°)	Meteors	Observer
1954, Aug. 7.4	45.0	+4.8	5	Worley
1954, Aug. 8.4	48.0	+1.5	5	Worley
1954, Aug. 9.4	48	+2	7	Worley
1958, Aug. 11.1	48	+3	4	Metzger
1926, Aug. 11.7	57.3	+0.3	3	Alden
1929, Aug. 13.0	57	–2	3	McIntosh
1941, Aug. 17.2	59.5	+3.5	5	AAFI

The observers cited are Charles E. Worley (California); Eli Metzger (Israel); Dr. H. L. Alden (South Africa); Ronald A. McIntosh (New Zealand); Amateur Astronomers of the Franklin Institute (Pennsylvania).

Among the list of 5406 German radiants compiled by Cuno Hoffmeister in his 1948 book *Meteorströme*, are three additional radiants which possess positions similar to those compiled from the American Meteor Society. They were detected during August 8-19, 1937, while Hoffmeister was observing from South-West Africa. By combining these radiants with those of the AMS, the Author has computed the following parabolic orbit.

ω	Ω	i	q	e	a
351.4°	317.4°	151.3°	1.002	1.0	∞

What seems to be indicated by this data, is that the August Eridanids are very old. The radio-echo stream and the visual stream are probably related and indicate an extensive mass separation has occurred.

Comet Pons-Gambart was only seen for a month in 1827. Its orbital period of 57.5 years is subsequently considered to possess an error of ±10 years and no recovery has ever been made. Interestingly, in 1979, Hasegawa computed a parabolic orbit for a comet seen in 1110. He concluded that it is possibly identical to comet Pons-Gambart. Thus, there is evidence that this lost periodic comet has been in a similar orbit for nearly 1000 years.

	ω	Ω	i	q	e	a
1110	358°	320°	136.5°	0.83	1.0	∞

Gamma Leonids

This daylight meteor shower was detected by two radio-echo surveys conducted during the 1960's and is significant as it may represent a recurrence of January's Delta Cancrid stream.

C. S. Nilsson (Adelaide Observatory, South Australia) was the first to observe this stream, when the radio program he was directing detected four meteors during August 18-22, 1961. The radiant was determined as α=152.7°, δ=+21.0°,[93] while the following orbit was computed.

ω	Ω	i	q	e	a
90.6	148.0	7.5	0.595	0.75	2.381

The second radio-echo detection of this stream was made by Zdenek Sekanina during the 1969 session of the Radio Meteor Project. The daylight shower was observed during August 14-September 12. The date of the nodal passage was given as August 25.4 (λ=151.8°), at which time the radiant was at α=155.5°, δ=+20.1°.[94] The following orbit was computed.

ω	Ω	i	q	e	a
87.3	151.8	7.0	0.571	0.710	1.969

Sekanina considered that a slight chance existed that this stream was associated with the Delta Cancrids of January. The D-criterion was given as 0.285, so that it was a borderline association.

---

[93]N1964, pp. 227-229 & 233.

[94]S1976, pp. 283 & 297.

Chapter 9: September Meteor Showers

Gamma Aquarids

Observer's Synopsis

This shower is active during September 1-14, with maximum occurring on September 7 (λ=164°) from α=333°, δ=–5°. Its highest rates of activity only reach 1-4 meteors per hour.

History

This meteor shower was first observed in England during 1921, when A. Grace Cook and J. P. M. Prentice independently detected activity in early September. The former observer plotted five meteors from α=334.5°, δ=–2° during September 1, 3, and 8, while Prentice plotted four meteors from α=335°, δ=–2° during September 6-8.[1] During September 7-8, 1923, Prentice reobserved the shower and managed to plot five meteors from a radiant of α=334°, δ=–3°.[2]

Independent observations are always important in the quick confirmation of any scientific discovery, thus, the 1921 observations seem to offer strong evidence for the existence of an active radiant near Gamma Aquarii. But while the 1921 observations were both made by observers in the same country, independent observations during the period of 1927-1934 were made by groups of observers located in New Zealand and Germany.

The New Zealand observations were analyzed by Ronald A. McIntosh (Auckland, New Zealand) in his 1935 paper "An Index to Southern Meteor Showers." The indicated duration of the Gamma Aquarids was September 1-14 while the average radiant was α=335°, δ=–2°.[3] The German observations were included among the 5406 visual radiants listed in Cuno Hoffmeister's 1948 book *Meteorströme*. Despite the observations covering the period of 1908 to 1938, Gamma Aquarid radiants were only detected during 1929, 1932, and 1934. The

---

[1] Denning, W. F., *The Observatory*, **44** (October 1921), p. 318.

[2] Denning, W. F., *The Observatory*, **46** (November 1923), p. 347.

[3] M1935, p. 717.

1929 radiant was detected on September 8 from α=333°, δ=+5°. The 1932 radiant was detected on September 5 from α=333°, δ=–10°, and the 1934 radiant was detected on September 4 from α=330°, δ=–11°.[4]

Although the German observations of September 5, 1932 and September 4, 1934, appear to be slightly further south than indicated by the earlier English observations, calculations by the Author reveal the Gamma Aquarid radiant to possess a northeasterly motion, thus, it must be expected that the radiant of September 4-5 would be southwest of the September 8 position. A possible confirmation of the early position of this radiant might be present in a 1964 paper by Charles P. Olivier entitled "Catalogue of Fireball Radiants," in which a radiant with a mean date of September 5 was given an average position of α=333°, δ=–8°. Based on five observed fireballs, the indicated duration extended for three days before and after the mean date.[5]

Only one radio-echo survey has ever revealed the Gamma Aquarid stream. During the 1961-1965 session of the Radio Meteor Project, activity was noted during September 10-11. Dr. Zdenek Sekanina gave the date of the nodal passage as September 10.8 (λ=167.7°), at which time the radiant was at α=335.6°, δ=–1.7°.[6] The stream's absence from other radio-echo surveys can usually be explained as due to the equipment not being in operation during early September; however, the second session of the Radio Meteor Project did not reveal this radiant, despite its being in operation during September 9-12, 1969. Sekanina describes this session as being more sensitive than that of 1961-1965, yet the Author finds absolutely no evidence of activity. Thus, the possibility may exist that this stream produces a periodic display rather than an annual one.

Although this stream was not noted in any of the analyses of photographic meteor orbits during the 1950's, 1960's and 1970's, the Author has identified two photographic meteors from a list of 2529 orbits computed by Richard E. McCrosky and Annette Posen that appear to definitely belong to the Gamma Aquarid stream. They were both detected on September 11, 1952, by the Harvard Meteor Project and indicate a radiant of α=335°, δ=–3°. The magnitudes of the meteors were given as –0.4 and +1.6, while their beginning heights were 98.7 and 87.0 kilometers, respectively.[7]

During the 1970's two positive detections of this stream came from observers in the Southern Hemisphere. The meteor section of the Royal Astronomical Society of New Zealand succeeded in observing this shower on several occasions during the decade. Section director John E. Morgan compiled a

---

[4]H1948, pp. 212, 217 & 221.
[5]Olivier, Charles P., *FOR*, No. 146 (1964), p. 14.
[6]S1973, pp. 257 & 260.
[7]MP1961, p. 66.

list of 213 active radiants which included the Gamma Aquarids. Meteor rates of four per hour were noted on September 10, from α=335°, δ=–3°. During 1980, the Western Australia Meteor Section observed meteors from this shower during September 10-13. A maximum ZHR of 1.12±0.24 was observed on September 10 from α=335°, δ=–3°.[8]

Orbit

The orbit determined by Zdenek Sekanina using data obtained during the 1961-1965 session of the Radio Meteor Project is

ω	Ω	i	q	e	a
250.6	167.7	4.4	0.725	0.718	2.572

Two photographic meteor orbits have been isolated by the Author from MP1961. Both meteors appeared on September 11, 1952, and revealed the following average orbit.

ω	Ω	i	q	e	a
248.0	168.0	3.5	0.775	0.685	2.460

Alpha Aurigids

Observer's Synopsis

This shower's duration seems to persist from August 25 to September 6. Maximum occurs from α=85°, δ=+41° on September 1 (λ=158°). The annual maximum ZHR may be as high as 9, but outbursts of over 30 occurred in 1935 and 1986.

History

The Alpha Aurigids were discovered by Cuno Hoffmeister and A. Teichgraeber (Sonneberg, Germany) on the night of August 31/September 1, 1935 (λ=158°). The maximum hourly rate was reported as 30, while the radiant was determined as α=84.2°, δ=+42.0°. The meteors possessed an average magnitude of 2.62, while 74% of those brighter than magnitude 3.5 left trains. V. Guth immediately noted the similarity between this radiant and the predicted radiant of α=90.2°, δ=+39.3° for comet Kiess (1911 II).[9]

In his 1948 book, *Meteorströme*, Hoffmeister noted that the link with comet Kiess made the circumstances of the 1935 observations curious. He pointed out that the comet's orbit was nearly parabolic, making the shower's sudden appearance 24 years after its perihelion passage difficult to explain.

---

[8]Wood, Jeff, Personal Communication (October 15, 1986).
[9]H1948, p. 119.

Hoffmeister examined his own annual observations made near the end of August and in early September, and noted probable detections of this shower in 1911, 1929, and 1930. In the former year, 5 of the 55 meteors he had plotted on September 2 (λ=159.2°) converged at α=84°, δ=+43°. During 1929, he found radiants of α=85°, δ=+38° on September 1 (λ=158.1°), α=87°, δ=+38° on September 3 (λ=159.8°), and α=89°, δ=+39° on September 4 (λ=161.0°). In 1930, one radiant was found at α=82°, δ=+38° on August 31 (λ=156.4°). Hoffmeister concluded that, although activity seems to have been present since the comet's perihelion passage, there is no evidence that the Alpha Aurigids are a permanent shower. He added that the strong 1935 shower was probably due to an isolated meteor group in the comet's orbit.[10] Hoffmeister's conclusion seemed well founded, as additional observations failed to appear in the records of American, European, or Russian observers in the four decades following 1935; however, three significant observations have been made in recent years.

During 1979 and 1980 members of the Western Australia Meteor Section (WAMS) succeeded in observing the Alpha Aurigids. In the former year, observations were made over the period of August 25-September 2. Maximum activity came on September 2, when the ZHR reached 8.52±1.87 from a radiant of α=87°, δ=+42°. In 1980, observations were made during August 31-September 6. Maximum came on the 6th, when the ZHR reached 9.11±0.96, from α=82°, δ=+38°.[11]

The latest detection of this shower came on September 1, 1986, when I. Tepliczky (Hungary) observed 24 meteors from α=94°, δ=+36.4° between 00:47 and 02:11 (UT). Around 01:25 (λ=158.34°), the ZHR reached 39.6±8.1. The meteors ranged in brightness from magnitude –4 to +4, with an average of +0.5.[12]

Orbit

The Author has determined the following average orbit from the six visual radiants listed in Hoffmeister's *Meteorströme* and the two radiants given by the WAMS.

ω	Ω	i	q	e	a
126.7	158.7	149.0	0.807	1.0	∞

The orbit of comet Kiess (1911 II) was calculated by Brian G. Marsden in 1978, and is

ω	Ω	i	q	e	a
110.4	158.0	148.4	0.684	0.996	184.621

---

[10]H1948, p. 120.

[11]Wood, Jeff, Personal Communication (October 15, 1986).

[12]*MN*, No. 76 (January 1987), p. 7.

Piscids

Observer's Synopsis

This weak and diffuse annual shower is active during the period of August 12 to October 7. A maximum ZHR of about 5 occurs on September 11 (λ=168°) from an average radiant of α=0°, δ=+4°. A probable northern branch might reach maximum around October 12 (λ=198°) from α=26°, δ=+10°.

History

Usually considered the "highlight" of September, this meteor shower, like all showers visible during the month, has been poorly observed due to an overall lack of observations. It seems apparent that the radiant is quite diffuse, though additional observations might help to resolve the stream's possibly complex structure.

Only two apparent observations of this shower seem to exist in the 19th Century and they were both made by William F. Denning (Bristol, England). The first observation came on September 14-15, 1879, when he plotted seven "very slow" meteors from α=1°, δ=–5°. The next observation involved the plotting of five meteors from α=4°, δ=–2° during September 3-5, 1885.[13] Although both radiants were called "Piscids," neither Denning or any other astronomer recognized this as a possible annual shower. The Author notes that both radiants barely qualify as representing this stream.

Cuno Hoffmeister was the first to officially recognize the Piscid stream. Analyzing the observations made by German observers between 1908 and 1938, he identified numerous probable observations of this shower. He concluded that the duration extended from August 16 to October 8, while maximum came on September 12 (λ=169°) from α=0°, δ=+4°. Hoffmeister described the shower as distinct, but weak. He added that two radiants were possibly present—both moving parallel to one another, but one located north of the ecliptic and the other south.[14]

Although Hoffmeister gave no exact details about his possible northern and southern Piscid showers, a table was given which demonstrated the general motion of the radiant. The Author notes some interesting patterns in this table. First, between August 16 and September 10, the declination changed only slightly, with the average being about +1.2° (based on 49 radiants). On the other hand, the declination for the period September 16 to October 8 averaged +8.1° (based on 38 radiants). The number of observed radiants remained fairly consistent during September 1, 8, 12, and 18, which may indicate a fairly flat

[13]D1899, p. 225.

[14]H1948, pp. 91 & 139-140.

maximum, rather than a distinct peak. On September 29, the number of observed radiants reached its highest point. Blending this declination and activity analysis, the Author believes Hoffmeister's data points to a southern stream reaching a flat maximum during the first half of September, while a northern branch is active around September 29. Whether the September 29 peak was due to a maximum of the northern branch or simply a combination of both branches cannot be determined.

The addition of photography in the field of meteor studies had began during the 1890's, but the Harvard Meteor Project of 1952-1954 was the most ambitious program. From this survey came the orbits of over 2500 meteors, which supplied a very large database for future studies of meteor showers. During 1971, Bertil-Anders Lindblad began a computerized study of these meteor orbits in an attempt to isolate active streams.

Lindblad detected two streams which he called the "Piscids"—each receiving a provisional stream number of "31" and "92." Stream 92 was given a long duration extending from August 31 to November 2. The radiant was given as $\alpha=10°$, $\delta=+6°$. But while this stream possessed similar characteristics to Hoffmeister's Piscid stream, the orbit revealed several strong dissimilarities, notably a 20° difference in the argument of perihelion, a 21° difference in the ascending node, and a 0.12 AU difference in the perihelion distance. Stream 31 seemed a more likely candidate, although it did occur one month later than Hoffmeister's stream and possessed a fairly short duration of September 25-October 19. The indicated date of the nodal passage was October 13 ($\lambda=199.1°$), at which time the radiant was at $\alpha=26°$, $\delta=+14°$.[15]

During 1973, Lindblad, as well as Allan F. Cook, Brian G. Marsden, Richard E. McCrosky and Annette Posen reevaluated several of the streams isolated by Lindblad's 1971 study. One of the most important discoveries was that over half of the meteors representing stream 92 were actually the Andromedids (see November). The authors concluded that the remaining meteors of stream 92 and the meteors of stream 31 formed the respective southern and northern branches of the Piscids.[16] Although this seemed to confirm Hoffmeister's 1948 suggestion of two Piscid streams, the Author still has doubts about the validity of stream 92.

The 1971 study listed 33 members of stream 92, but this was whittled down to 14 meteors in 1973 after the Andromedids were subtracted. The Author believes 4 of the remaining "Southern Piscids" are actually members of the Southern Taurid stream of November. The remaining 10 meteors were considered a family by the authors of the 1973 study simply because their

---

[15]L1971B, pp. 16-17 & 21-22.

[16]C1973, p. 4.

longitudes of perihelion were very nearly the same. In actuality these meteors change in perihelion distance from 0.25 to 0.66 AU and in argument of perihelion by 50°. From a study of the Perseids, Delta Aquarids, Orionids, and Geminids, the Author believes meteors should be assigned to particular streams if all orbital elements, except that of the ascending node, are a close match—especially when dealing with showers of long-duration. Applying this theory, the Author notes that 8 of the 10 meteors assigned to the "Southern Piscids" form two separate streams (one of which is the October Cetids)—neither of which seem to represent the observations of the Piscids noted by Hoffmeister. In fact, no convincing photographic Piscid appears in the 1952-1954 Harvard Meteor Project, partially due to the cameras being shut down during September 2-9.

The Piscids were well represented during the two sessions of the Radio Meteor Project conducted by Zdenek Sekanina during the 1960's. During the 1961-1965 survey Piscid meteors were detected during August 14-October 4. The date of the nodal passage was calculated as September 10.7 (λ=167.6°), at which time the radiant was at α=359.8°, δ=+3.4°. The link with Hoffmeister's Piscids was considered very good (D-criterion of 0.110).[17] The 1968-1969 session revealed Piscids over the period of August 12 to October 7. The nodal passage was said to have occurred on September 16.1 (λ=172.8°), at which time the radiant was at α=8.5°, δ=6.9°.[18]

Sekanina's radio-echo data seem to indicate that a long-duration shower does reach maximum in the Pisces region in early September; however, even though there was no trace of a secondary stream, the orbits obtained during the 1961-1965 study and that of 1968-1969 were somewhat different. The Author believes these orbital differences, as well as the differences in the dates of nodal passage, offer supporting evidence that two branches of the Piscid stream do exist, but that their production of very diffuse and weak showers make them difficult to separate when comparing orbital elements. It is probable that only a close monitoring of activity levels will properly separate these two streams.

Recent observations of this shower have been made by members of the Western Australia Meteor Section. According to section director Jeff Wood, the September full moon of 1979 was followed by observed activity from the Delta Piscium region during September 22-23. A maximum ZHR of 1.32±0.22 came on September 23 from an average radiant of α=10°, δ=+11°. During 1980, Piscids were detected during September 10-12. A maximum ZHR of 1.24±0.26 was observed on September 11 from α=4°, δ=+9°. In addition, the 1980 observations also revealed showers which were referred to as the "Southern

---

[17]S1973, pp. 257, 260, & 267.

[18]S1976, pp. 285 & 299.

Piscids" and "Northern Piscids." The former shower was observed during September 11-27. A maximum ZHR of 2.08±0.20 came on September 19 from an average radiant of α=4°, δ=0°. The "Northern Piscids" were detected during October 5-16. A maximum ZHR of 2.94±0.25 came on October 12 from α=26°, δ=+8°.[19] Although the latter shower seems to definitely represent an observation of the northern branch of the Piscid stream, the Author believes the "Southern Piscids" might not actually be associated with the southern branch.

Orbit

The orbit of this stream was first determined by Hoffmeister. Based on visual radiants, it is

ω	Ω	i	q	e	a
296	169	3.5	0.40	0.72	1.43

What later became known as the "Northern Piscids," stream number 31 of Lindblad's 1971 photographic meteor survey possessed the following orbit.

ω	Ω	i	q	e	a
290.8	199.1	3.4	0.399	0.797	2.062

Two orbits were determined by Sekanina from data gathered during the two sessions of the Radio Meteor Project. The first orbit represents the material obtained during the 1961-1965 survey, while the second orbit was computed using 1968-1969 data.

ω	Ω	i	q	e	a
298.5	167.6	3.8	0.344	0.816	1.868
306.3	172.8	3.5	0.311	0.769	1.349

The Author has collected seven photographic meteor orbits from MP1961 and B1963. The average stream orbit follows:

ω	Ω	i	q	e	a
306.6	174.7	6.5	0.271	0.853	1.844

If two separate branches are present within the Piscid stream, their presence might best be demonstrated by comparing the values of ω and q above. Admittedly, these orbits do not confirm Hoffmeister's belief that one radiant lay north and the other south of the ecliptic.

Additional September Showers

Eta Draconids

No trace of this meteor shower seems to exist in the major lists of visual radiants, including D1899, H1948 and the records of the American Meteor

[19]Wood, Jeff, Personal Communication (October 15, 1986).

Society; however, both sessions of the Radio Meteor Project, conducted by Zdenek Sekanina during the 1960's, revealed consistent evidence of this stream's existence.

The first session covered the years 1961-1965. Activity was noted during August 28 to September 23. Sekanina said the nodal passage came on September 12.1 (λ=168.9°), from α=248.8°, δ=+63.4°.[20] The second session covered the years 1968-1969, but activity was only detected during September 9-12. The nodal passage was determined to have occurred on September 12.2 (λ=169.0°), from α=246.3°, δ=+63.8°.[21] The established Eta Draconid orbits from these two surveys are

	ω	Ω	i	q	e	a
S1973	167.9	168.9	33.3	0.996	0.526	2.102
S1976	165.7	169.0	31.9	0.993	0.465	1.856

Whether this stream produces a strictly telescopic shower is not currently known. The radiant is not well placed in the sky, as it crosses the zenith during daylight hours. The best time for observations is during the early evening hours following sunset—a period usually ignored by observers due to generally low meteor activity.

Gamma Piscids

The discovery of this stream should be attributed to Zdenek Sekanina, who established its existence from data gathered during the two sessions of the Radio Meteor Project. The 1961-1965 session revealed only nine definite meteors from the Gamma Piscids during the period of September 10-October 5. The date of the nodal passage was determined as September 24.0 (λ=180.6°), while the average radiant was α=349.6°, δ=+2.9°.[22] The 1968-1969 session picked up 35 members of this stream during August 26-October 22. The date of the nodal passage was calculated as September 22.2 (λ=178.8°), at which time the radiant was at α=342.3°, δ=+7.7°.[23] The orbital elements were given as

	ω	Ω	i	q	e	a
S1973	253.7	180.6	3.9	0.705	0.702	2.366
S1976	247.8	178.8	7.6	0.741	0.716	2.606

The Author has found no trace of activity from this stream during the 19th century and the first half of the 20th century among the writings of William F. Denning, Alphonso King, Ronald A. McIntosh, Cuno Hoffmeister or Ernst Öpik. On the other hand, the Author has located four photographic meteors

---

[20]S1973, pp. 257 & 260.

[21]S1976, pp. 285 & 299.

[22]S1973, pp. 257 & 260.

[23]S1976, pp. 285 & 299.

among the 2529 orbits published by Richard E. McCrosky and Annette Posen in 1961.[24] These not only add support to the very existence of this stream, but obviously indicate that visual observations should be possible.

The photographic meteors were detected during the 1952-1954 Harvard Meteor Project and indicate a duration of September 19-October 2. The apparent date of the nodal passage is September 26 (λ=183.0°), while the average radiant is α=351°, δ=+10°. The average orbit is

ω	Ω	i	q	e	a
255.3	183.0	8.0	0.695	0.723	2.509

Interestingly, a fairly recent visual detection of this shower was made during the 1970's. A decade-long visual survey of meteors was conducted by Michael Buhagiar (Perth, Western Australia) which involved the plotting of 20,974 meteors and the determination of 488 radiants. The Gamma Piscids were detected on five occasions, revealing a duration of September 21-28. Maximum was said to have generally occurred on September 25 from α=350°, δ=+5°. The highest hourly rate was typically given as 4.[25]

---

[24]MP1961, pp. 67-69.

[25]Buhagiar, Michael, *WAMS Bulletin*, No. 160 (1981).

Chapter 10: October Meteor Showers

October Arietids

Observer's Synopsis

The duration of this meteor shower extends from September 7 to October 27. Maximum occurs around October 8 (λ=194°) from α=32°, δ=+8°, with an hourly rate of 3-5. The radiant's daily motion is +0.90° in α and +0.35° in δ.

History

This shower is part of a cluster of radiants that is active in the Taurus-Aries region during the period of September through November. The complex nature of this region was first noted in 1928, when William F. Denning published the details of thirteen radiants. One of these radiants, the "Xi Arietids" was said to be active during September and October from an average radiant of α=30.9°, δ=+9.6°, while another radiant—the "Sigma Arietids"—was said to be active during October from α=41.9°, δ=+13.7°.[1] At the time Denning was still a strong believer in the stationary radiant theory, but the currently known movement of the October Arietids would seem to link these two radiants.

F. W. Wright and Fred L. Whipple were the next people to isolate the October Arietids when, during a 1950 investigation of the photographic Taurid meteors, they found 8 photographic meteors which came from an average radiant of α=+41.3°, δ=+10.3°. The mean date of activity was October 20.22.[2] This study involved 102 meteors photographed during the period of October 15-December 2, in the years 1896-1948. The authors expressed their opinion that the October Arietids might form a continuous stream with the southern Taurids, though they suggested further studies would be needed to clarify the situation.

The use of radar equipment came of age during the 1950's, but, despite numerous observations of the Taurids, the equipment was not yet sensitive

---

[1]Denning, W. F., *JBAA*, **38** (1928), p. 302.

[2]Wright, F. W., and Whipple, F. L., *Harvard College Observatory Reprints*, Series 2, No. 35 (1950), pp. 1-44.

enough to separate diffuse radiants close together. This changed in the early 1960's when two radio-echo studies detected the October Arietids: B. L. Kashcheyev and V. N. Lebedinets (Kharkov Polytechnical Institute) detected 18 meteors during October 10-21, 1960, from an average radiant of α=40°, δ=+15° and C. S. Nilsson (Adelaide Observatory, South Australia) detected 30 meteors during October 20-31, 1961, from an average radiant of α=44.8°, δ=+12.4°.

Nilsson may have also detected the October Arietids during the last days of September. He contemplated the matter after noting a similarity between a stream he designated 61.9.1 and the "Arietids" (designated 61.10.1). The stream was detected during September 22-29, from an average radiant of α=18.1°, δ=+4.9°. It was based on 19 meteors and is listed below with the October branch. Nilsson found the orbit of this stream to closely resemble the orbit of the June daytime shower called the Zeta Perseids. He added, "It is just possible that the two streams 9.1 and 10.1 are both part of a wide band of meteors, but consideration of the characteristics of the October Arietids and November Taurids does not encourage this view."[3] The Author points out that the radio equipment at Adelaide was not operational during September 30 to October 19, 1961.

The two sessions of the Radio Meteor Project conducted during the 1960's at Havana, Illinois, also succeeded in detecting the October Arietids. Zdenek Sekanina said the 1961-1965 session revealed activity extending from September 11 to October 23. The date of the nodal passage was given as October 1.4 (λ=187.8°), at which time the radiant was at α=23.9, δ=+8.8.[4] From data compiled during the 1968-1969 session, Sekanina gave the duration as September 7 to October 22. The date of the nodal passage was given as October 11.5 (λ=197.8°), while the average radiant was at α=32.3°, δ=+10.2°.[5]

The October Arietids are a neglected shower among visual observers and, if not for the photographic and radar programs of the last 30 years, virtually nothing would be known of this stream. Most of this neglect is probably due to the Southern Taurid stream which moves through a region of the sky a few degrees to the southwest. This latter shower is also a larger producer of meteors, with rates typically near 10 per hour.

One of the most significant visual studies occurred while seven Russian astronomers observed the Orionids during October 17-21, 1973. Coordinated by G. N. Sizonov, the observations took place from the meteor station of Armavir middle school #18. Combined observations of the October Arietids revealed an average magnitude of 3.37, based on 43 magnitude estimates, and a maximum ZHR of 3.4. Sizonov showed the highest ZHR to have occurred on October 20

[3]N1964, p. 242.
[4]S1973, pp. 257 & 260.
[5]S1976, p. 288.

at which time the average distance between particles within the stream was 484 kilometers, and the spatial density of particles per cubic kilometer was 8.90 x 10^{-9}.[6]

From analyzing the photographic meteors of both the October Arietids and the Southern Taurids, the Author has noted that the radiants of each stream remain fairly close together. At the time they both appear in mid-September, the October Arietids are 7° east of the Southern Taurids, while they are about 4° northeast at the end of October. Both streams also possess diffuse radiants, so that it is nearly impossible to visually separate one from the other. This analysis also revealed the October Arietid daily motion to be +0.90° in α and +0.35° in δ.

Orbit

Using 48 photographic meteors obtained from W1954, MP1961, B1963 and BK1967, the following orbit was obtained by the Author:

ω	Ω	i	q	e	a
125.2	15.4	6.4	0.280	0.846	1.818

This stream was also detected during the two sessions of the Radio Meteor Project. Zdenek Sekanina determined the following orbits:

	ω	Ω	i	q	e	a
S1973	126.9	7.8	1.4	0.273	0.841	1.723
S1976	122.5	17.8	2.9	0.333	0.768	1.435

During 1960, C. S. Nilsson detected the October Arietids in late October and, possibly, an earlier occurrence of the same stream in September. The October Arietid orbit, designated 61.10.1, was based on 30 meteors, while the similar September stream orbit, designated 61.9.1, was based on 19 meteors.

	ω	Ω	i	q	e	a
61.9.1	125.7	2.3	5.4	0.34	0.83	1.64
61.10.1	118.3	33.0	5.9	0.28	0.83	2.00

Finally, the orbit determined by Kashcheyev and Lebedinets in BL1967 is

	ω	Ω	i	q	e	a
BL1967	131	22	1.2	0.24	0.86	1.74

Delta Aurigids

Observer's Synopsis

Meteors from this stream may be observed during the interval of September 22 to October 23. Maximum occurs between October 6 and 15 from an average radiant of α=84.8°, δ=+51.9°. The radiant's daily motion is +1.2° in α and +0.1° in δ.

---

[6]Sizonov, G. N., *SSR*, **9** (1975), pp. 52-54.

History

This meteor shower was discovered in 1979, while Jack D. Drummond, Robert K. Hill and Herbert A. Beebe (New Mexico State University) were determining the orbits of 13 meteors doubly photographed during 1976 and 1977 at the NASA-NMSU Meteor Observatory in southern New Mexico. Two of the 13 meteors had nearly identical orbits and had been photographed on October 13 and 18, 1977. It was concluded that the possible date of maximum might have occurred during October 15-16, 1977, from an average radiant of α=95.7°, δ=+52.5°.[7]

The first observations of this shower occurred during October 1980, as Norman W. McLeod, III (Florida) and Drummond (New Mexico) carried out independent visual surveys. McLeod's observations began early as he noted 1 meteor in 4.6 hours on October 6, 1 in 0.8 hours on the 10th and 0 in 2.7 hours on the 12th. Both men observed on the 13th, with McLeod noting 2 in 1 hour and Drummond detecting 3 in 2 hours. Drummond's observations of the 14th and 15th revealed 2 in 2 hours and 5 in 2 hours, respectively—the latter being the highest determined hourly rate of 2.5 per hour. Observations continued from the 16th to the 22nd, with the Delta Aurigids being either nonexistent or possessing hourly rates of 1 or less.[8]

During 1981, Drummond isolated 5 probable members of this stream that had been photographed by American and Russian surveys conducted between 1950 and 1962. Combining these 5 with the 2 meteors from the 1977 NASA-NMSU survey, Drummond concluded that the shower's duration might be from September 29 to October 18. The mean date of activity was October 8, with the radiant then being α=87.8°, δ=+50.2°. The shower's daily motion was given as +1.0° in α and +0.1° in δ. Lastly, the stream's mean orbit, based on the 7 photographic meteors, is a retrograde orbit with a period of 115 years.[9]

From both the photographic studies and the visual survey of 1980, Drummond pointed out that only the second half of the indicated shower duration had been covered and that the determination of the shower's true maximum "must await further observation from the first half."[10]

The Author has recently searched through the 39,145 radio meteor orbits obtained during the two sessions of the Radio Meteor Survey and isolated 24 probable Delta Aurigids. The orbital details will be further discussed in the "Orbit" section below, but several interesting features can be discussed here.

---

[7]Drummond, Jack D., Hill, Robert K. and Beebe, Herbert, A., *The Astronomical Journal*, **85** (April 1980), pp. 496-497.

[8]Drummond, Jack D., *Icarus*, **51** (1982), p. 657.

[9]Drummond, Jack D., *Icarus*, **51** (1982), pp. 656-657.

[10]Drummond, Jack D., *Icarus*, **51** (1982), p. 658.

First, while comparing the radio meteor orbits with the photographic orbits, it becomes apparent that the radio meteors seem to reach peak concentration about one week before the photographic meteors. The radio meteors also possess a much smaller semimajor axis of 2.3 AU, compared to the photographic meteors' semimajor axis of 18.7 AU. At first glance, it might seem that there is a significant indication of mass distribution within this stream, but when all of the data was combined and reexamined it appears that the following very distinct filaments may be present in this stream:

*Filament "A" reaches maximum on September 30 (λ=186.5°), from an average radiant of α=87.8°, δ=+54.1°. It possesses the largest perihelion distance of the four possible filaments and was composed of 11 meteors—one being photographic.

*Filament "B" reaches maximum on October 7 (λ=193.2°), from an average radiant of α=87.5°, δ=+50.2°. It possesses the largest semimajor axis of the four filaments (4.5 AU) and is the probable main core of this stream. It was based on 13 meteors, with 6 being photographic.

*Filament "C" is the weakest filament, but seems to reach maximum on October 13, from an average radiant of α=91.1°, δ=+47.8°. It possesses the highest inclination (133°) and the smallest perihelion distance (0.75 AU) of the four filaments. It was based on 4 meteors, with 2 being photographic.

*Filament "D" reaches maximum on October 2 (λ=188.2°), from an average radiant of α=74.3°, δ=+55.0°. It possesses the smallest inclination (116°) and the smallest semimajor axis (2.2 AU) of the four filaments. It was based on 5 radio meteor orbits.

Searches for previous sightings of the Delta Aurigid shower have not proved extremely successful, though one very promising observation occurred during October 14-17, 1876, when William F. Denning plotted 6 meteors from an average radiant of α=90°, δ=+58°.[11] The meteors were described as rapid, with streaks. This radiant is the closest 19th century match for the Delta Aurigids, but is only borderline according to the D-criterion.

More positive observations of the Delta Aurigid stream are found in Cuno Hoffmeister's *Meteorströme*. Three excellent radiants are given for October 1, 1910 (α=86°, δ=+57°), October 9, 1920 (α=84°, δ=+48°), and October 7, 1935

---

[11]D1899, p. 246.

(α=83°, δ=+48°). The first radiant seems a good example of a filament "A" shower, while the latter two radiants probably belong to filament "B". Seven additional radiants also appear, but are strictly borderline, according to the D-criterion. These seven radiants might belong to filaments "A", "B" and "C". Filament "D" continues to be elusive in the records of visual observations and may be composed mostly of small, telescopic particles—if it exists at all.

There seems to be no appearance of this shower in the records of the AMS, or in visual observations compiled in D1912B, K1916, D1923A, D1923B, Ö1934 and RM1974.

Orbit

Combining 10 photographic meteors obtained from W1954, MP1961, B1963, C1964, BK1967 and D1980, with 24 radio meteor orbits obtained from the orbital data obtained during the two sessions of the Radio Meteor Project, the Author found this stream to possess the following average orbit.

ω	Ω	i	q	e	a
227.1	190.9	125.9	0.859	0.720	3.068

Some differences were noted between the photographic and radar data as the following two orbits indicate:

	ω	Ω	i	q	e	a
Photo	229.5	195.3	131.1	0.823	0.956	18.705
Radar	226.4	188.6	123.9	0.878	0.617	2.292

Finally, despite the slight differences between the radar and photographic orbits, the Author combined all of the orbital data and found the following four distinct orbits which *may* represent filaments within the Delta Aurigid stream.

	ω	Ω	i	q	e	a
A	210.2	186.5	123.4	0.951	0.621	2.509
B	229.4	193.2	129.5	0.848	0.810	4.463
C	243.5	199.3	132.6	0.753	0.793	3.638
D	245.2	188.2	116.4	0.773	0.648	2.196

Stream "A" was based on 11 meteors, "B" on 13 meteors, "C" on 4 meteors and "D" on 5 meteors. As can be seen, these four branches cross the ecliptic between solar longitudes of 186.5° and 199.3°, or, roughly between September 30 and October 13. Streams "A" and "B" apparently produce the strongest activity and have maximums between September 30 and October 7.

Only one known comet seems to have an orbit similar to that of the Delta Aurigids—comet Bradfield (1972 III). The link should not be considered definite, but its orbit is given here as follows:

	ω	Ω	i	q	e	a
1972 III	257.7	159.6	123.7	0.927	0.998	494.778

Draconids ("Giacobinids")

Observer's Synopsis

This shower is associated with periodic comet Giacobini-Zinner. The duration may extend from October 6 to 10, though the point of maximum is very sharply defined within a 4-hour interval on October 9 (λ=195°). The radiant at maximum is α=262°, δ=+54°, but the annual maximum hourly rates are not consistent. The radiant rarely produces any recognizable shower except during years especially close to the parent comet's perihelion passage. The best examples of these rare exceptional showers occurred during 1933 and 1946, when the hourly rates were near 6000. The meteors are slow and tend to be relatively faint. They are generally yellow.

History

The discovery of this meteor shower resulted from predictions by several astronomers that the periodic comet Giacobini-Zinner might produce a radiant in early October.

The first to make such a prediction was the Reverend M. Davidson, who, in 1915, examined the periodic comets observed since 1892 to isolate any that might be capable of producing meteor showers. One of the comets which met the established criteria was Giacobini-Zinner. Davidson found that the comet's orbit would be fairly close to Earth on October 10, 1915, and predicted that, if the debris from this comet had spread outward by about two million miles, a shower might be active from α=267°, δ=+50°.[12] This was about two years after the parent comet had passed perihelion. During the first half of October, William F. Denning recorded "a number of meteors" from α=267°, δ=+49°.[13]

Davidson revised his prediction in 1920 (primarily due to a discovered error in his earlier prediction), saying that the distance between the orbits of the comet and Earth amounted to 5 1/2 million miles. He subsequently suggested that maximum would most likely occur on October 9 from α=251.5°, δ=+55.9°.[14] Later that same year, Denning became the first observer to make a definite observation of this new shower, as he observed 5 meteors from α=268°, δ=+53° during October 6-9. These meteors were described as slow. Giacobini-Zinner had passed perihelion during the Spring of 1920.

The comet was next expected at perihelion at the end of 1926, and predictions for a meteor shower in October of that year were made by both Davidson and A. C. D. Crommelin. Both men gave October 10 as the expected

---

[12]Davidson, M., *JBAA*, **25** (April, 1915), p. 292.

[13]Denning, W. F., *MNRAS*, **87** (November 1926), p. 104.

[14]Davidson, M., *MNRAS*, **80** (1920), p. 739.

day of maximum, and they gave similar radiants of α=261°, δ=+53.5° and α=265°, δ=+54°, respectively. The orbits of the comet and Earth were found to intersect in 1926, though the expected shower would occur 70 days before the comet passed that point of its orbit. Around October 10.4, observers in England were made aware that some unusual activity was present.

What gained the attention of many observers on October 9, 1926, was the appearance of a fireball. The event was noted by hundreds in the British Isles and 35 reports allowed the radiant to be determined as α=262°, δ=+55°. The meteor moved slowly and lit up the sky. A persistent train lasted about 30 minutes "during which time it underwent curious changes of form and exhibited drift amongst the stars."[15]

Several other radiants were determined in 1926, which acted to confirm the predictions of Davidson and Crommelin. J. P. M. Prentice (Stowmarket, England) observed for 3 hours centered on October 9.9, and detected 16 meteors from α=263°, δ=+54°. He described the meteors as slow and estimated the radiant diameter as 6°. He claimed that the hourly rate may have been near 17 had his observations been continuous. Observations made by Alphonso King (Ashby) and Denning (Bristol) gave a radiant of α=255°, δ=+56°.[16]

The years following 1926 were closely monitored by a few observers trying to catch another glimpse of the Draconid shower. During 1927-1932, Prentice observed extensively around October 9-10, but no activity was detected; however, the skies opened up on October 9, 1933.

Comet Giacobini-Zinner passed perihelion on July 15, 1933, and, on the date of the predicted maximum, Earth crossed the area of the comet's descending node just 80 days after the comet. Astronomers were not prepared for what was in store for them, but, as evening twilight fell over Europe, observers noted the beginnings of something unusual. Within just a couple of hours the number of Draconids skyrocketed, and, at 20:00 UT, one of the best displays of the 20th century was in progress. Some of the more impressive statistics follow:

> *From Ireland, the Reverend W. Ellison (Armagh Observatory) reported that meteors fell as frequently as snowflakes and he gave the radiant as α=266°, δ=+55°. W. Milligan (near Omagh) saw thousands, including 100 during one 5-second interval. The radiant was α=264.5°, δ=+54.5°.[17]
>
> *At Birchircara, Malta, R. Forbes-Bentley observed over 22,500 meteors in just a few hours and estimated the

---

[15]Denning, W. F., *MNRAS*, **87** (November 1926), pp. 104-105.

[16]Denning, W. F., *MNRAS*, **87** (November 1926), pp. 104-105.

[17]King, A., *The Observatory*, **56** (November 1933), pp. 348-349.

peak rate hit 480 per minute at 20:15 UT. The meteors were described as mostly faint with only 5% reaching first magnitude. The radiant was estimated as α=262.5°, δ=+55°.[18]

*In Russia, N. S. Sytinskaja (Leningrad) collected numerous observations between 18:00 and 22:00. At maximum, rates reached 100 per minute at Leningrad, 300 per minute at Pulkovo and 200 per minute at Odessa.[19]

*From Spain, P. M. Ryves (Zaragoza) estimated a maximum rate of 100 per minute and an average radiant of α=266°, δ=+53.5°. He said the meteors were generally faint with the "great majority of the meteors being 3rd to 5th mag."[20]

As can be gathered from these reports (and many others not included here), the shower's maximum rate reached 100 per minute, or about 6000 per hour, around 20:15 UT on October 9. The meteors were slow, generally faint and were usually yellow.

Following the 1933 appearance, the Draconids again fell to nonexistence. The comet's next perihelion date was February 17, 1940, and there were numerous predictions of a possible strong return in October 1939; however, Earth crossed the comet's orbit 136 days ahead of the comet which meant bad news for meteor enthusiasts, as no storm—or even a small shower—appeared. During 1940 to 1945, activity continued to be absent, but astronomers were already making predictions for the very favorable 1946 return. In that year, the comet was expected at perihelion on September 18, so that Earth would cross the comet's orbit just 15 days after the comet!

The Draconids were best placed for observers in the Western Hemisphere during October 9/10, with excellent meteor counts not only coming from all across the United States, but Canada and even Venezuela. European observers did detect the Draconids, but the radiant was very low over the horizon and, though spectacular, it was at least one-fourth the strength of that seen in America. Observations were also made in Czechoslovakia for slightly more than one hour prior to morning twilight. Some of the more interesting observations are as follows:

*At the University of Oklahoma Observatory, Professor Balfour S. Whitney led a team of 10 observers

---

[18]King, A., *The Observatory*, **56** (November 1933), pp. 348-349.

[19]King, A., *The Observatory*, **56** (November 1933), p. 349.

[20]King, A., *The Observatory*, **56** (December 1933), p. 380.

during the period of 1:23 and 4:34 UT. Taking counts every 10 minutes, they determined that maximum came around 3:50 UT (October 10) when estimated hourly rates were near 3000.[21]

*From Southern California, numerous counts were tabulated at Griffith Observatory. From that observatory, Roland Michaelis and Karl Bouvier observed rates of 55 per minute between 3:45 and 4:02 UT. E. L. Forsyth (Fallbrook) detected 63 per minute at 3:50 and George W. Bunton (Sunland) observed 180 per minute around this same time.[22]

*In England, Prentice evaluated the British observations and noted a maximum rate of 965 per hour. The radiant was then very low over the horizon and it was determined that the ZHR was actually 2250.[23]

*At Skalnate Pleso Observatory (Czechoslovakia) observers fought clouds, moonlight and morning twilight to observe the Draconids. Maximum occurred at 3:53 UT with a corrected rate of 6800 per hour. Two secondary maxima occurred at 3:23 and 3:40. Analysis revealed the half-time of the shower to have been 0.65 hours.[24]

What may have been the highlight of this event was the appearance of a large blue-white fireball over Southern California at 3:38 UT. Forsyth said it left a yellow train which lasted over three minutes. As the train drifted and became diffuse, it took on the shape of a horseshoe.[25]

The 1946 event marked an important first for meteor astronomy—the detection of a meteor shower by radar. In the United States alone, 21 radar systems were operated at frequencies of 100, 600, 1200, 3000 and 10000 Mc/sec. From these instruments only the radar operating at 100 Mc/sec detected meteor echoes. The majority of all meteor activity occurred between 3:00 and 4:30 UT on October 10.[26] Other radar equipment operating in London and the Soviet Union operated at frequencies between 3.5 Mc/sec and 212 Mc/sec and confirmed that maximum occurred between 3:00 and 4:00. Most interesting was

---

[21]Olivier, Charles P., *FOR*, No. 67 (1947), pp. 29-30.

[22]Cleminshaw, C. H., *PASP*, **58** (December 1946), pp. 362-363.

[23]Prentice, J. P. M., *The Observatory*, **67** (1947), p. 3 and *JBAA*, **57** (1947), p. 86.

[24]Kresak, L., *MN*, No. 43 (Mid-October 1978), pp. 4-5.

[25]Cleminshaw, C. H., *PASP*, **58** (December 1946), p. 362.

[26]Stewart, J. Q., Ference, M., Slattery, J. J., and Zahl, H. A., *Sky & Telescope*, **6** (1947), pp. 3-5.

a record obtained by J. A. Pierce, who used a 3.5 Mc/sec pulsed ionospheric sounder and found that meteors were so numerous that a temporary ionosphere was formed at a height of 90 km.[27] The meteoric ionosphere lasted three hours and was confirmed elsewhere.

Following 1946, both visual and radio-echo techniques were utilized in searches for this shower during 1947-1951. Visual observers detected no meteors possibly associated with the Draconids, while radio-echo observations at Jodrell Bank detected "no activity during the Giacobinid epoch in excess of the background sporadic rate (that is, not greater than 4 or 5 per hour)."[28]

A possible shower was predicted for October 9, 1952. Various calculations revealed Earth would cross the comet's orbit 193 days ahead of the comet. In addition, the closest distance between the orbits of Earth and Giacobini-Zinner was very similar to that of 1933, or about 0.0057 AU; however, on this occasion, the comet's orbit would actually pass inside of Earth's orbit. Visual observations by British observers revealed only the barest hint of activity shortly after sunset on October 9/10, but, just a few hours earlier, daylight observations had been made using the radio-echo apparatus at Jodrell Bank in England.

The Jodrell Bank team first noted the Draconid rate rising above that of the sporadic background at 14:20 UT on October 9. Meteor echoes were counted during 10-minute intervals: 3 were noted at 15:00, there were 6 at 15:10, 10 at 15:20, 11 at 15:30 and 17 appeared at 15:40. The highest rates occurred at 15:50, when a 10-minute rate of 29 was reported—indicating an hourly rate of 174. The following decline in activity was very rapid, and one-half hour after maximum the 10-minute rate had declined to only 3. The last definite sign of activity occurred at 16:40, when the rate was 2. The Jodrell Bank observers concluded that maximum occurred at a solar longitude of 196.25° from α=262°, δ=+54°.[29]

The 1959 return of Giacobini-Zinner was very favorable and, since Earth arrived at the comet's orbit just 21.6 days before the comet passed through the region some believed a meteor storm would occur; however, the comet's perihelion distance had been pulled closer to the sun so that the closest distance between the orbits of Earth and the comet was 0.058 AU. Subsequently, no shower was observed. The comet's 1966 return also failed to produce meteors due to unfavorable geometric conditions.

Giacobini-Zinner passed only 0.58 AU from Jupiter during 1969, which acted to increase its perihelion distance to 0.99 AU—Draconid showers were again possible. Searches for activity began during October 1971, with Earth

---

[27]Pierce, J. A., *Phys. Review*, **71** (1947), pp. 88-92.

[28]Davies, J. G., and Lovell, A. C. B., *MNRAS*, **115** (1955), p. 23.

[29]Davies, J. G., and Lovell, A. C. B., *MNRAS*, **115** (1955), pp. 25-26.

crossing the comet's orbit 309 days before the comet. No notable activity was observed during October 7 to 10, by members of the American Meteor Society, as hourly rates remained around one.[30]

The 1972 Draconids were looked for with much anticipation. Not only was Earth going to cross the comet's orbit 58.5 days after the comet, but the two orbits were separated by only 0.00074 AU! Unfortunately, despite these promising statistics, the shower was quite a disappointment. Observers in the United States obtained the highest visual rates when 10-15 per hour were detected on October 8/9. Maximum had been predicted for 17:00 UT on October 8, which made Japan the best location for observations. Unfortunately, the Japanese observers were met with cloudy skies. Despite this hindrance, the Hiraiso Branch of the Radio Research Laboratory operated a 27.1 MHz radar. A peak of 84 returns in 10 minutes was noticed at 16:10 UT on October 8, followed by a secondary peak of 69 returns in 10 minutes at 21:00 UT.[31] Further predictions for Draconid showers in 1978-1979 and 1985-1986, were not met with noticeable displays.

Using the orbit of periodic comet Giacobini-Zinner as representing that followed by the Draconids, Ken Fox (Queen Mary College, England) projected the orbit of this stream backward and forward for 1000 years. He found the distances between Earth and meteor stream orbits to have been too great for showers to occur in the years 950 or 2950.[32]

Orbit

During the two sessions of the Radio Meteor Project, which was conducted during the 1960's, Zdenek Sekanina isolated enough meteors for the following two orbits to be determined:

	ω	Ω	i	q	e	a
S1970	184.0	192.2	51.5	0.997	0.623	2.647
S1976	187.6	194.5	49.2	0.994	0.553	2.221

Epsilon Geminids

Observer's Synopsis

The duration of this meteor shower extends from October 10 to 27, with a maximum of 1-2 meteors per hour occurring on October 18 (λ=204°), from α=103°, δ=+25°. The radiant's daily motion is +0.7° in α and –0.1° in δ.

---

[30]*MN*, No. 8 (Mid-October 1971), p. 11.

[31]Marsden, Brian G., IAU *Circular*, No. 2451, October 16, 1972.

[32]F1986, pp. 523-525.

History

The first instance of this shower being recognized as a fairly consistent producer of meteor activity was in 1899, when William F. Denning listed a shower called the Delta Geminids in his great "General Catalogue of Radiant Points of Meteoric Showers and of Fireballs and Shooting Stars observed at more than one Station." Although the shower was listed as being active from September through January, it is obvious after examining Denning's list that several streams were responsible for the activity—most notably, a shower was definitely active shortly after mid-October.

According to Denning's article, the first sighting of this shower may have been on October 19, 1868, when T. W. Backhouse (Sunderland) plotted 6 meteors from α=100°, δ=+18°. Denning had observed the shower on two occasions, the first being October 15-20, 1879 (α=106°, δ=+23°), and the second being October 14-21, 1887 (α=105°, δ=+22°). Other probable radiants were also listed, including one observed by D. Booth (Leeds) during October 15-20, 1887 (α=99°, δ=+22°),[33] which apparently confirmed Denning's observation of the same year and might indicate a stronger than normal return.

Since 1899, observations of this shower have occasionally appeared in the literature, but these were purely accidental observations, as no specific searches were carried out. During October 18, 20 and 24, 1922, J. P. M. Prentice plotted 13 very rapid meteors from an average radiant of α=103.5°, δ=+18.5°.[34] On October 25, 1932, Ernst Öpik's Arizona Expedition for the Study of Meteors obtained an excellent radiant determination of α=108°, δ=+25°.[35]

Cuno Hoffmeister (Germany) was the next person to recognize an active mid-October shower in Gemini. In his classic book *Meteorströme*, Hoffmeister listed a shower with an average radiant of α=101°, δ=+26°. This shower had been noted on 12 occasions between 1909 and 1933, and the probable date of maximum was October 19 (λ=206°).[36] This book was published in 1948, but the Epsilon Geminids again fell into obscurity.

The shower was next "discovered" in 1958, when Richard E. McCrosky and Annette Posen (Harvard College Observatory, Massachusetts) identified six photographic meteor orbits from data gathered during the Harvard Meteor Project of 1952-1954. The indicated duration of activity was October 17 to 27, and the date of maximum was believed to be October 17. The average radiant position was α=102°, δ=+26°.[37]

---

[33]D1899, p. 250.

[34]Denning, W. F., *The Observatory*, **45** (December 1922), p. 402.

[35]Ö1934, p. 36.

[36]H1948, p. 82.

[37]MP1961, p. 26.

The shower's next milestone came in 1960, when it was first detected by radio-echo techniques. B. L. Kashcheyev and V. N. Lebedinets (Kharkov Polytechnical Institute, USSR) isolated thirteen Epsilon Geminids during October 10-27. The date of the nodal passage was indicated as October 16 (λ=203°), at which time the average radiant was α=104°, δ=+25°. Interestingly, this has remained the only radio-echo survey to reveal any trace of this stream.

The first visual study of this shower was conducted by V. Znojil (Public Observatory, Brno, Czechoslovakia) in 1965. It began with telescopic observations from Boleradice and Bohuslavice (Czechoslovakia). Observations were made during October 20-25, with Epsilon Geminids being detected each day. Maximum may have occurred on October 22 (though the low activity obtained by the telescopic observations made the exact date difficult to determine), at which time the average radiant was at α=103°, δ=+25°. The radiant's daily motion was given as +0.7° in α and +0.2° in δ. Znojil combined these observations with previous studies conducted by Hoffmeister (1948), McCrosky and Posen (1958 & 1961), Southworth and Hawkins (1963), and Kashcheyev and Lebedinets (1967) and concluded that the shower reached maximum at a solar longitude of 204.4°. The average radiant was determined as α=101.8°, δ=+26.2° and Znojil's revised estimate of the radiant's daily motion was +0.7° in α and –0.1° in δ.[38]

Several amateur astronomers have observed this shower in recent years. During October 15-28, 1979, Mark Adams observed 19 Epsilon Geminids in 18 hours 36 minutes and determined the average magnitude as 3.32. He estimated that the maximum rate reached 1-3 per hour.[39] Robert Lunsford (California) observed 11 Epsilon Geminids in 24 hours during October 15-21, 1980, and noted that the average magnitude was 2.73.[40] Jeff Wood, director of the meteor section of the National Association of Planetary Observers (Western Australia), gives the shower's duration as October 14-27. He concluded that a maximum rate of 1-2 meteors per hour was detected on October 20, from α=104°, δ=+26°.[41]

What makes the Epsilon Geminids such an interesting stream is that it bears a strong resemblance to the Orionids, except that the argument of perihelion and ascending node are reversed by about 180°. During 1963, Richard B. Southworth and Gerald S. Hawkins pointed out that the orbit of the Epsilon Geminids was similar to the Orionid orbit, but they showed that the major differences between the orbits were in inclination (9°) and longitude of

[38]Znojil, V., *BAC*, **19** (1968), pp. 306-315.

[39]*MN*, No. 49 (April 1980), p. 6.

[40]Lunsford, Robert, Personal Communication (October 11, 1986).

[41]Wood, Jeff, Personal Communication (October 24, 1985).

perihelion (34°).[42] In 1968, Znojil and J. Papousek, showed that while the radiant dispersion of the Epsilon Geminids was "of the same order of magnitude as that of the radiants of the individual branches of the Orionids," the fact that the Epsilon Geminids undergo greater planetary perturbations than the Orionids, makes this stream younger. The authors concluded that a relationship between these two streams would be difficult to explain.[43]

The Author has noted some variances in this stream's activity. As noted earlier, the 1960 Russian study remains the only radio-echo survey to detect the Epsilon Geminids, with N1964, S1973, GE1975, and S1976 failing to show a trace. In fact, the activity of this stream has consistently shown signs of being irregular or periodic in nature. The seven known photographic meteors were detected in 1950, 1952 and 1953, with the middle year possessing five of the seven. As pointed out earlier, both Denning and Booth independently observed the shower in 1887, and obtained similar radiants and durations. Finally, Hoffmeister's 12 visual radiants were observed during 1909-1933, but 4 were seen during a three-day interval in 1931.

Orbit

Seven photographic Epsilon Geminids were noted by the Author in W1954 and MP1961. The stream was also detected in 1960, by B. L. Kashcheyev and V. N. Lebedinets while using radar equipment. The respective orbits are

	ω	Ω	i	q	e	a
Photo	234.3	205.7	172.8	0.791	0.978	35.955
BL1967	223	203	175	0.88	0.75	3.58

Jack D. Drummond wrote in *Icarus* (45, pp. 548-550) that comet Ikeya (1964 VIII) might be responsible for the meteors of this stream.

	ω	Ω	i	q	e	a
Ikeya	290.8	269.3	171.9	0.82	0.99	53.51

Using an orbit similar to the photographic one above (a=26.77 AU), Ken Fox (Queen Mary College, England) projected the orbit of this stream backward and forward for 1000 years.[44] This stream might have produced a meteor shower in 950 AD, which would have reached maximum at the end of September. By 2950 AD, a shower might be present, though maximum will likely occur at the beginning of November. The following orbits were given.

	ω	Ω	i	q	e	a
950	214.0	185.6	173.3	0.80	0.97	26.55
2950	246.1	218.9	173.3	0.76	0.97	25.25

[42]Southworth, R. B., and Hawkins, G. S., *SCA*, **7** (1963), p. 281.

[43]Znojil, V., *BAC*, **19** (1968), pp. 314-315.

[44]F1986, pp. 523-525.

Orionids

Observer's Synopsis

The duration of this meteor shower extends from October 15 to 29, with maximum occurring on October 21 (λ=207.8°) from a radiant of α=95°, δ=+16°. The maximum ZHR is usually about 20 and the meteors are described as fast. The radiant's daily motion is generally given as α=+1.23°, δ=+0.13°. A secondary center of activity, consisting of meteors averaging about one magnitude fainter than those from the main radiant, is located at α=95°, δ=+19° around the time of maximum.

History

The discovery of the Orionid meteor shower, seems to be best attributable to Edward C. Herrick. In 1839, he made the ambiguous statement that activity seemed to be present during October 8 to 15.[45] A similar statement was made in 1840, when he commented that the "precise date of the greatest meteoric frequency in October is still less definitely known, but it will in all probability be found to occur between the 8th and 25th of the month."[46]

The first precise observation of this shower were made by Alexander Stewart Herschel on October 18, 1864. Fourteen meteors revealed a radiant at α=90°, δ=+16°. Herschel next observed the shower on October 20, 1865, with a radiant of α=90°, δ=+15° being revealed from 19 meteors.[47] Thereafter, interest in this stream increased very rapidly—with the Orionids becoming one of best observed annual showers.

The Orionids were frequently observed during the latter years of the 19th century. The radiant was generally more diffuse than some of the other annual showers and this created a strong debate during the first quarter of the 20th century. William F. Denning, as has been mentioned on several earlier occasions, was a strong supporter of the stationary radiant hypothesis and the Orionids were considered one of his best examples. Visual studies had generally failed to detect any motion in the radiant's position during the 10-15 days the shower was under observation each year and Denning strongly believed that *two* prominent radiants were present—the Orionids at α=91°, δ=+15° and the Geminids at α=98°, δ=+14°.[48] During 1913, the pages of the *Monthly Notices of the Royal Astronomical Society* became a forum for the pens of Denning and Charles P. Olivier as the two debated the stationary character of the Orionids;

---

[45]Herrick, Edward C., *Silliman's Journal*, **35** (1839), p. 366.
[46]Herrick, Edward C., *Silliman's Journal*, **39** (1840), p. 334.
[47]Denning, W. F., *MRAS*, **53** (1899), pp. 246-247.
[48]Denning, W. F., *The Observatory*, **41** (January 1918), p. 60.

however, little was resolved and the matter had to await 1923, when the debate was resumed in the pages of *The Observatory*.

Based on observations made by himself and three colleagues from Leander McCormick Observatory, Olivier demonstrated how the radiant clearly moved eastward from day to day—the position for October 18, 1922 being α=91.0°, δ=+15.0°, while the October 26 position was at α=99.2°, δ=+13°.[49] The five positions given by Olivier clearly indicated an advance in right ascension, but the declination actually showed no clear movement. Olivier added that recent observations by members of the American Meteor Society for the dates of October 12 to 31, inclusive, had also indicated motion in the right ascension, but, again, the movement in the declination was not determined.

The very next issue of *The Observatory* contained three letters by supporters of the stationary radiant theory—Denning, J. P. M. Prentice and A. Grace Cook. Denning stated that the "American observers appear to have failed to discover the minor streams, and until they can do this the real character of the chief radiant must continue to elude them, for it will be easy to make it a shifting position, and especially by the plentiful meteors radiating from 99° +13°."[50] Prentice and Cook essentially argued in support of Denning, citing extensive observations made by themselves and other British observers.

In June 1923, Olivier struck back and, as his ammunition, he stressed the importance of a photographic radiant obtained by Professor King at Harvard on October 20, 1922, and a series of excellent radiant determinations made by R. M. Dole between October 17 and 30, 1922. Dole essentially confirmed Olivier's belief that the radiant moved and provided the following list of radiant positions:

Orionid Radiant Ephemeris

Date	RA (°)	Dec (°)
Oct. 17.9	90.7	+15.0
Oct. 18.8	91.5	+14.8
Oct. 19.8	93.4	+14.8
Oct. 21.8	93.0	+15.1
Oct. 23.8	95.9	+16.8
Oct. 24.9	99.1	+16.6
Oct. 25.8	99.6	+16.6
Oct. 26.8	100.5	+16.8

Olivier said King's photographic radiant fell "exactly where a moving radiant would give it on Oct. 20..." with the position being determined as α=94.1°, δ=+15.8°.[51]

---

[49]Olivier, Charles P., *The Observatory*, **46** (January 1923), pp. 17-18.

[50]Denning, W. F., *The Observatory*, **46** (February 1923), p. 47.

[51]Olivier, Charles P., *The Observatory*, **46** (June 1923), pp. 188-189.

Olivier's belief in a moving radiant gradually won out during the next few years, with several well-known amateur and professional astronomers adding support. Notably, in 1928, Ronald A. McIntosh conducted a very extensive set of observations during the period October 14 to 24, inclusive.[52] Again, the movement of the right ascension was clearly demonstrated, as was the difficulty in determining the motion of the declination. Estimates made over the last 50 years have shown that the declination actually moves slightly northward as each day passes. The 1986 *Handbook of the British Astronomical Association* lists the radiant's daily motion as α=+1.23°, δ=+0.13°.

Referring back to McIntosh's 1928 observations, it should be stressed that the 54 Orionids plotted, revealed more details of the stream than just radiant determinations. McIntosh noted that 25 Orionids, or 46%, left trains, and that the average duration of these trains was 1.3 seconds. His estimates of color revealed 24% to be red and 9% to be blue. Although his observations were frequently hampered by clouds, mist and, finally, the moon, McIntosh did observe on two very clear nights—the 15th and 20th. These nights possessed the highest observed rates of Orionids, with hourly totals being 8 and 6, respectively. *[Although magnitude estimates were made for each meteor, the published material was not broken done sufficiently enough for a precise average to be obtained; however, an educated guess by the Author, based on McIntosh's rough tabulation, produces an average magnitude near 2.5.]*

The first estimates of the activity of this shower came in 1892, when observations placed the hourly rate at 15. During the next 30 years observers became aware that the activity of the Orionids was not consistent, with estimates ranging from a low of 7 in 1900 to a high of 35 in 1922. Unfortunately, there were no organized observations of this shower between 1903 and 1922, so it is not known whether the appearance of comet Halley brought any enhanced activity. It is, however, interesting to note the change in the ZHR since 1930. Between that year and 1953 the average ZHR was about 15, with individual rates never exceeding 20. Between 1960 and 1974, the average ZHR was about 24, with rates of 30 to 40 occurring on 4 occasions. Finally, between 1975 and 1985, the average rate dropped to 18, with the highest rates being only 24.[53] The comet passed closest to the sun in February 1986.

One very unusual feature the Orionids tend to display is an unpredictable maximum. In 1981, observers reported very low rates of less than 10 meteors per hour during the period of October 18 to 21 (maximum predicted for October 21), but high rates of near 20 per hour were noted on the morning of October

---

[52]McIntosh, R. A., *MNRAS*, **90** (November 1929), pp. 161-162.

[53]Mackenzie, Robert A., *Solar System Debris*. Dover: The British Meteor Society (1980), p. 20.

23.[54] Interestingly, a study published in Czechoslovakia during 1982, revealed the Orionids to generally possess a double maximum. The finding was based on observations made during 1944 to 1950, and the maxima occurred at solar longitude 207.8° (primary) and 209.8° (secondary).[55] Shortly thereafter, several visual studies indicated the presence of a "plateau effect" or a long period of maximum devoid of any sharp decline of activity, instead of a double peak. Most notably, the 1984 observations of the Western Australia Meteor Section, show a nearly flat maximum lasting from October 21 to 24,[56] while Norman W. McLeod, III, has frequently noted it to stretch up to 6 days.[57] What appears to be the best explanation of the Orionids' irregular occurring date of maximum, was made by A. Hajduk (Astronomical Institute of the Slovak Academy of Sciences, Bratislava, Czechoslovakia) in 1970.

Hajduk examined the reported activity of the Orionids for the period of October 14 to 28 during the years 1900 to 1967. He particularly noted that the "stream density varies along the orbit," and "there is no fixed periodically recurring position of maximum or secondary maxima."[58] Hajduk concluded that the density changes were not random and that the displacement of activity "can be explained by the presence of some filaments along the stream orbit."[59]

A strong confirmation of Hajduk's filamentary structure was made during 1975, when radar equipment at Ondrejov and Dushanbe was simultaneously utilized during the period of October 17 to 29. The data was analyzed by P. B. Babadzhanov, R. P. Chebotarev (Dushanbe, USSR) and Hajduk. They found radio-echo rates to slowly increase, but, suddenly, at a time generally attributed as the maximum of the Orionids (λ=208°), rates drastically declined. Just as curious was the finding that rates had doubled during the next 24 hours, and then were followed by the normal decreasing rates for every day thereafter. The authors concluded that as Earth entered the Orionid stream "we first intersected a halo with a slight variation of density, then a gap corresponding to λ=208, and this was followed by a steep increase at λ=209."[60] It was claimed that this structure confirmed the presence of filaments.

Hajduk continued the simultaneous observations of the Orionid shower during 1978—this time combining his results at Ondrejov with those obtained by G. Cevolani at Budrio, Italy. On this occasion the period of October 17 to 24

---

[54]*MN*, No. 56 (January 1982), p. 4.

[55]*MN*, No. 60 (January 1983), p. 7.

[56]*MN*, No. 70 (July 1985), p. 7.

[57]*MN*, No. 70 (July 1985), p. 4.

[58]Hajduk, A., *BAC*, *21* (1970), p. 39.

[59]Hajduk, p. 40.

[60]Babadzhanov, P. B., Chebotarev, R. P., and Hajduk, A., *BAC*, 30 (1979), p. 227.

was covered by both stations, while only Budrio continued until October 29. Unlike 1975, the activity curves of both stations indicated a more consistent increase towards maximum. This maximum occurred during October 21 (λ between 207.8° and 208.4°), with a relatively flat peak being noted. Rates remained above one-half of maximum during October 19-22 (λ between 206.5° and 209.5°). The Budrio data also seems to have detected a secondary maximum on October 27 (λ=214.6°), which was explained as due to an encounter with a filament. The only hint of a filamentary structure near maximum was noted on October 20 (λ=207.5°), when a decline in activity was noted at Ondrejov. This decline was not detected at Budrio, causing Hajduk and Cevolani to conclude that, since Ondrejov was capable of detecting fainter and, thus, smaller meteors, a region possessing fewer small particles than normal may have been encountered by Earth.[61]

An interesting feature of the Orionids is the apparent complexity of the radiant. In 1939, J. P. M. Prentice noted two active radiants: the primary radiant, at a solar longitude of 208°, was described as being about 7° wide, centered at α=94.4°, while the declination spread from δ=14.9° to 15.5°. A secondary radiant was noted at α=97.8°, δ=+18.2°.[62] Prentice's visual study is still rated as one of the most complete ever published on the Orionid shower, although a convincing confirmation of the secondary branch had to await a 1965-1966 telescopic survey conducted in Czechoslovakia.

V. Znojil (Public Observatory, Brno) outlined the details of the survey and results during 1968. Telescopes were utilized which possessed 80mm lenses and magnifications of 10. This resulted in a field of view of 7° 22' and a limiting magnitude of 10.8. During 1965, two stations were separated by 23.91 km, while the separation was increased to 63.60 km during 1966.

Znojil found two distinct radiants were present, which he referred to as the northern and southern Orionids. Their average radiants were given as α=95.4°, δ=+17.8° and α=95.6°, δ=+15.9°, respectively. The latter radiant agrees very well with the position usually attributed to the Orionids, while the northern branch is very close to that noted by Prentice 30 years earlier. Znojil's further analysis revealed that the northern branch consistently produced fainter meteors than the southern, with the average difference amounting to 1.02 magnitude. Both radiants possessed a spread in right ascension amounting to 5°, while their declinations covered only 2° of the sky.[63]

Many of the above features just discussed reveal an interesting structure for the Orionids. They demonstrate what may be expected to occur within an old

---

[61]Hajduk, A., and Cevolani, G., *BAC*, **32** (1981), pp. 309-310.

[62]Prentice, J. P. M., *JBAA*, **49** (1939), p. 148.

[63]Znojil, V., *BAC*, **19** (1968), pp. 311-312.

meteor stream. Although observations of hourly rates are one way to detect filaments, future detections might be made in the study of average magnitudes and the percentages of meteors leaving persistent trains. Some recent observations follow.

Orionid Magnitudes and Trains

Year(s)	Ave. Mag.	# Meteors	% Trains	Observer(s)	Source
1973	3.65	210	—	Sizonov	*SSR*, **9**
1976	3.31	—	31.2	McLeod	*MN*, No. 36
1976	3.21	—	20.9	Martinez	*MN*, No. 36
1976	2.89	—	16.8	Matous	*MN*, No. 36
1978	2.33	27	7.4	WAMS	*MN*, No. 45
1979	3.36	391	13.6	WAMS	*MN*, No. 48
1979	3.46	302	39.4	McLeod	*MN*, No. 48
1980	3.48	238	—	McLeod	*MN*, No. 52
1980	2.85	146	—	Lunsford	Personal Comm.
1982	3.05	107	15.0	Lunsford	Personal Comm.
1983	2.69	—	27.5	WAMS	MET, **14**, No. 3
1984	3.10	—	22.8	DMS	MET, **15**, No. 2
1984	3.14	1944	—	WAMS	*MN*, No. 70

Observers in the previous table are as follows: G. N. Sizonov, lead a team of seven observers at Armavir (USSR); Norman W. McLeod, III (Florida); Felix Martinez (Florida); Bert Matous (Missouri); Western Australia Meteor Section; Robert Lunsford (California); Dutch Meteor Society.

Colors have been a point of controversy in meteor astronomy: Are observed colors of meteors real or subjective? During 1978, observers of the Western Australia Meteor Section, headed by Jeff Wood, detected 27 Orionids, of which 0% were yellow and 6.7% were blue-green.[64] However, during 1979, the same group noted that 14.4% were yellow, 3.4% blue, 1.7% orange and 0.8% green. In 1983, Mark T. Adams (Palm Bay, Florida) summed up this situation best when he commented that there was "no clear trend in the observed Orionid colors."[65]

Studies of the evolution of the Orionid shower possess a strong interest due to the stream's link to Halley's comet. This link was indirectly made in 1911, when Charles P. Olivier mentioned the similarity between the orbit of the Orionids and that of the Eta Aquarids of May.[66] Since 1868, this latter stream

---

[64]*MN*, No. 45 (April 1979), p. 12.

[65]*MN*, No. 61 (April 1983), p. 1.

[66]Olivier, Charles P., *Transactions of the American Philosophical Society*, **22** (1911).

had been known to be related to Halley's comet (see the Eta Aquarids in chapter 5); however, this link between Halley's comet and the Orionids is not considered definite, as pointed out by J. G. Porter in 1948.

Porter considered the 0.15 AU separation between the comet's orbit and Earth's orbit enough of a deterrent to make a connection with the Orionids impossible.[67] Although others agreed with Porter's hypothesis, the similarity in the characteristics of the Orionid and Eta Aquarid meteors and activity rates was considered uncanny. Thus, despite the orbital distance of Halley's comet at the time of the Orionid maximum it is generally accepted that the two must be related.

In 1983, B. A. McIntosh (Herzberg Institute of Astrophysics, Ottawa, Canada) and Hajduk (Astronomical Institute of the Slovak Academy of Sciences, Bratislava, Czechoslovakia) published details of a new proposed model of the meteor stream produced by Halley's comet. Using the 1981 study published by Donald K. Yeomans and Tao Kiang, which examined the orbit of Halley's comet back to 1404 BC,[68] McIntosh and Hajduk theorized that "the meteoroids simply exist in orbits where the comet was many revolutions ago."[69] Further perturbations have acted to mold the stream into a shell-like shape containing numerous debris belts possessing stable orbits. These belts are considered as the explanation as to why both the Orionids and Eta Aquarids experience variations in activity from one year to the next.

Orbit

Using 59 photographic meteors obtained from W1954, MP1961, C1964, BK1967, C1977 and D1980, the Author obtained the following orbit:

ω	Ω	i	q	e	a
80.0	28.0	163.7	0.586	1.015	∞

During the two sessions of the Radio Meteor Project conducted at Havana, Illinois, during the 1960's, Zdenek Sekanina determined the following orbits for the Orionid stream:

	ω	Ω	i	q	e	a
S1970	87.4	27.3	164.5	0.561	0.846	3.631
S1976	87.0	27.1	164.4	0.562	0.854	3.850

Finally, the orbit of Halley's Comet is given as follows:

	ω	Ω	i	q	e	a
391 BC	86.8	28.6	163.6	0.588	0.967	17.961
1986	111.8	58.1	162.2	0.587	0.967	17.941

---

[67]Lovell, A. C. B., *Meteor Astronomy*, Oxford: Oxford University Press, 1954. pp. 296.
[68]Yeomans, D. K., and Kiang, T., *MNRAS*, **197** (1981), pp. 633-646.
[69]McIntosh, B. A., and Hajduk, A., *MNRAS*, **205** (1983), p. 931.

The comet was not seen in 391 BC, but, according to calculations by Donald K. Yeomans (Jet Propulsion Laboratory, Pasadena, California) and Tao Kiang (Dunsink Observatory, Republic of Ireland), this was the probable orbit of the comet, according to their elaborate study of the comet's motion.[70] As can be seen it very closely matches the present orbit of the Orionid stream—especially when compared to the photographic orbit.

Using an orbit similar to the photographic one above (but with a semimajor axis of 15.10 AU), Ken Fox (Queen Mary College, England) projected the orbit of this stream backward and forward for 1000 years.[71] The following two orbits were obtained:

	ω	Ω	i	q	e	a
950	81.1	24.4	163.8	0.58	0.96	15.59
2950	80.5	28.5	164.0	0.57	0.96	15.17

The orbit of the Orionids thus seems fairly stable. In 950, the date of maximum was four days earlier from that of the present, while the radiant position was α=90.8°, δ=+15.8°. In 2950, maximum will occur only one-half day later than at present and the radiant will be at α=98.1°, δ=+15.3°. Thus, even though the orbit of Halley's Comet has undergone slight changes during the last 2000 years, the ellipse of material that produces the Orionids is fairly stable.

Sextantids

Observer's Synopsis

This daytime meteor shower possesses a duration extending from September 24 to October 9. Maximum occurs during September 29 to October 4 (λ=185°-190°), from an average radiant of α=153°, δ=–2°. The shower's maximum activity seems weak, but reached nearly 30 meteors per hour in 1957. It may be periodic, with an encounter with Earth occurring every 4-5 years.

History

This daylight meteor shower was discovered by A. A. Weiss in 1957, during a radio survey conducted at Adelaide (South Australia). The radiant position was not firmly established at that time, due to only one aerial being in use, but a theoretical estimate of α=155°±8°, δ=0°±10° was made. Due to the uncertainty in the radiant position, the shower was referred to as the "Sextantids-Leonids." What was firmly established was the shower's duration of September 26 to October 4. In addition, a peak activity of nearly 30 meteor echoes per hour was

[70]Yeomans, Donald K., and Kiang, Tao, *MNRAS*, **197** (1981), p. 643.
[71]F1986, pp. 523-525.

noted during September 29 to October 3.[72] Weiss further noted that no trace of activity had been detected during previous radio-echo surveys.

The next detection of the Sextantid stream was made in 1961. C. S. Nilsson operated the radio equipment at Adelaide during September 21-29, and detected 9 members of this stream during September 24-29. The average radiant was given as α=151.7°, δ=–0.1°. Interestingly, Nilsson noted a similarity between the Sextantid orbit and the orbit of the Geminids of December. He claimed that "statistically, the difference between the Sextantid and Geminid orbits is not significant, and the former could well represent the daytime return of a branch of the latter stream after perihelion passage, if the stream is wide enough."[73] Indeed, as Nilsson pointed out, the closest approach of the Sextantids to Earth in December is 0.34 AU, while the indicated width of the Geminid stream is 0.11 AU.

The width of the Sextantid and Geminid streams presented the largest problem in their being directly related, but other problems also existed. Nilsson noted that the Geminids occurred annually with consistent rates "indicating that the meteoric matter is extended uniformly along the orbit."[74] On the other hand, the Sextantids had only been detected in 1957 and 1961. Surveys in other years should have detected the stream, but since they did not, Nilsson suggested the shower might be periodic.

The next sighting of the Sextantids came in 1969, during the second session of the Radio Meteor Project. The radar equipment was not in operation during September 27 to October 5, but 9 members of this stream were detected during October 7-9. Zdenek Sekanina gave the average radiant as α=156.5°, δ=–8.3°.[75] No possible relationship to the Geminid stream was discussed.

Orbit

The orbit of this stream has been computed by both Nilsson and Sekanina.

	ω	Ω	i	q	e	a
N1964	213.2	3.6	21.8	0.146	0.87	1.124
S1976	212.3	15.1	31.1	0.172	0.816	0.936

The orbits are very similar, though, as is evident by the ascending node, the respective studies did not cover the period of late September and early October adequately. For Nilsson's study the radio equipment was shut down for three weeks following September 29, 1961. For Sekanina's survey the equipment was not in use during September 27 to October 5, 1969.

---

[72]Weiss, A. A., *MNRAS*, **120** (1960), p. 399.
[73]Nilsson, C. S., *AJP*, **17** (1964), p. 159.
[74]Nilsson, p. 159.
[75]S1976, p. 287.

Using the orbit from N1964, Ken Fox (Queen Mary College, England) projected the orbit of this stream backward and forward for 1000 years.[76] No shower was probably detectable in 950 AD, as Fox showed the present stream orbit would have been too far from Earth's orbit; however, the following orbit was obtained for 2950 AD:

	ω	Ω	i	q	e	a
2950	227.3	349.5	16.9	0.15	0.88	1.25

The orbit of the Sextantids thus seems to be changing fairly rapidly. In 2950, maximum will occur about 15 days earlier than at present and the radiant will be at α=150.7°, δ=+0.2°.

Additional October Showers

Eta Cetids

This stream might be similar in nature to the Aurigids of February as they seem to possess a very weak, almost nonexistent shower, with occasional fireballs thrown in. The duration of activity stretches from September 20 to November 2, while the maximum occurs during the first week of October from a radiant of α=15°, δ=–13°.

The first recognition of this area as a producer of fireballs came in 1964, when Charles P. Olivier published the "Catalog of Fireball Radiants" as an American Meteor Society publication offered through Flower and Cook observatories. Designated radiant number 5148, it was based on 6-8 fireballs. The mean date of activity was given as October 4, with a radiant of α=10°, δ=–18°. Activity was also present three days before and after this mean.[77]

One of the most spectacular fireballs from this radiant reached magnitude –20. It was photographed by 9 of the 16 cameras of the Prairie Network (designated 40503) on October 9.30, 1969, from a radiant of α=18.0°, δ=–17.7°. Another very bright stream member was detected by observers in Florida on October 8.3, 1972. Reaching a magnitude of –14, its radiant was determined as α=8°, δ=–11°.

The Author's investigation of the 39,145 radio meteor orbits obtained by Zdenek Sekanina during the two sessions of the Radio Meteor Survey, has uncovered 16 orbits. The following orbit is the average of these radio meteor orbits and 6 photographic meteors obtained from MP1961 and B1963.

ω	Ω	i	q	e	a
75.6	14.8	10.0	0.70	0.67	2.10

---

[76]F1986, pp. 523-525.

[77]Olivier, Charles P., *FOR*, No. 146 (1964), p. 14.

October Cetids

The first sighting of this meteor shower seems to be linked to observations by William F. Denning during 1916 and 1917. During the former date, 4 meteors were seen over the period of October 20 to 25, from a radiant at α=28°, δ=+4°. The meteors were described as rapid.[78] On October 16, 1917, a 1st-magnitude stationary meteor was observed from α=28°, δ=+3°.[79]

Other observations are hard to come by in the records of visual radiants; however, four radio-echo surveys may have detected this stream during the 1960's. Using the radio equipment at Adelaide Observatory (South Australia) in 1961, C. S. Nilsson detected stream members during October 23-30. Although the data revealed a nodal passage on October 27 (λ=213.9°) and an average radiant of α=39.3°, δ=+0.2°, the equipment did not operate during September 30-October 19, so that the early part of the shower's activity was missed. During the 1961-1965 session of the Radio Meteor Project, Zdenek Sekanina isolated a stream which he called the "Beta Cetids." The stream possessed a duration extending from September 8 to October 1. Its nodal passage was given as September 25.7 (λ=182.2°), while the average radiant was α=16.3°, δ=–11.7°.[80] The second session of the Radio Meteor Project detected a possibly associated stream during September 22-October 22, 1969. The apparent date of the nodal passage was October 5.2 (λ=191.6°), at which time the radiant was at α=12.5°, δ=+4.2°. This stream was called the "Delta Piscids."[81] Finally, G. Gartrell and W. G. Elford operated radio equipment at Adelaide Observatory during October 15-19, 1969. Although the indicated nodal passage came on October 16 (λ=203°) from α=28°, δ=+3°, the equipment had been shut down since mid-June, so that the early part of the activity was definitely missed. The orbits revealed by each of these surveys are as follows:

	ω	Ω	i	q	e	a
N1964	94.0	33.9	13.9	0.52	0.86	3.70
S1973	102.4	2.2	14.6	0.493	0.725	1.792
GE1975	96	23	8.2	0.47	0.90	4.17
S1976	93.6	11.6	0.7	0.566	0.693	1.843

As previously noted, the data published in N1964 and GE1975 did not cover the early half of this stream's activity, thus giving the indicated nodal passages little meaning. In addition, the orbits from these two surveys were each based on only three meteors, so that the reliability should not be considered very high.

---

[78]D1923B, p. 47.
[79]D1923A, p. 38.
[80]S1973, pp. 257 & 260.
[81]S1976, pp. 287 & 300.

The orbits obtained in each of Sekanina's surveys should be considered more reliable, as they were based on about 10 and 45 meteors, respectively. Most interesting is the inclination change between the 1961-1965 and 1968-1969 surveys, but this may be a result of the different selection methods being used: the early study used a D-criterion of 0.20, while the second survey used a less stringent value of 0.25. After obtaining the 39,145 radio meteor orbits computed by Sekanina during the 1960's, the Author has found 57 meteors that are probable members of this stream. Çombined with 13 photographic meteors found in MP1961 and C1977, the subsequent orbit was obtained:

ω	Ω	i	q	e	a
95.4	15.2	7.3	0.529	0.745	2.075

By utilizing a D-criterion value of about 0.15, the Author finds that this stream splits into northern and southern branches. The fairly compact northern branch was composed of 46 radio and photographic meteors, while the more diffuse southern stream was composed of 16 meteors.

	ω	Ω	i	q	e	a
N.	95.4	16.2	4.2	0.537	0.740	2.065
S.	96.0	13.6	13.6	0.528	0.744	2.063

The northern branch is definitely the strongest portion of this stream, but neither branch is especially strong visually and the Author believes telescopic aid will probably be necessary for future observations. What makes this stream especially interesting is its orbital similarity to the lost Apollo asteroid Hermes (1937 UB).

	ω	Ω	i	q	e	a
Hermes	90.7	35.3	6.2	0.617	0.624	1.639

This asteroid's ascending node occurs around October 29, which coincides with the end of the October Cetid shower.

October Cygnids

The first detection of this radiant should be attributed to Cuno Hoffmeister (Germany), who observed a radiant at α=305°, δ=+57° on October 9, 1931 (λ=195.6°).[82] Although additional visual observations seem quite rare, this radiant has been detected by both radar and photography.

The Author has found six photographic meteors in W1954 and MP1961, which indicate a duration extending from September 26 to October 10. The nodal passage seems to occur on October 6, at which time the average radiant is at α=311.3°, δ=+54.7°. The average orbit is

ω	Ω	i	q	e	a
207.1	192.7	29.9	0.953	0.734	3.576

---

[82]H1948, p. 252.

Two meteor streams were detected during the 1961-1965 session of the Radio Meteor Project. The "Delta Cygnids" were detected during October 4-10. The date of the nodal passage was given as October 8.9 (λ=195.2°), at which time the radiant was at α=299.7°, δ=+50.7°. The "Alpha Cygnids" were based on 12 meteors detected during September 22-October 11. The date of the nodal passage was given as October 4.4 (λ=190.8°), at which the radiant was at α=316.3°, δ=+52.3°.[83] Their orbits are

	ω	Ω	i	q	e	a
δ	198.6	195.2	25.0	0.976	0.647	2.764
α	216.0	190.8	25.5	0.930	0.538	2.014

Admittedly, the orbits of the two radio-echo streams are somewhat different, but it is evident that at least one radiant is producing activity from this region. The shower's duration seems to extend from September 22 to October 11, and maximum occurs between October 4 and 9. The average position of the radiant is α=311°, δ=+52°.

[83]S1973, pp. 257 & 260.

Chapter 11: November Meteor Showers

Andromedids

Observer's Synopsis

Best known for incredible displays of several thousand meteors per hour on November 27 of 1872 and 1885, today's Andromedids are only weakly represented by displays of about 5 meteors per hour around November 14. The radiant is currently located at $\alpha=26°$, $\delta=+37°$. The associated comet was Biela, which split and was observed as two comets in 1846 and 1852. The comet most likely produced the 1872 and 1885 displays as a result of a total breakup. The shower's present duration extends from September 25 to December 6, although this is primarily due to the existence of numerous filaments caused by the considerable change the comet's orbit has gone through.

History

The history of the Andromedids is directly linked to the history of the remarkable comet Biela. The comet was discovered on three occasions before its periodic nature became known: first by Montaigne (Limoges, France) on March 8, 1772, second by Jean Louis Pons (Marseilles, France) on November 10, 1805, and finally by Wilhelm von Biela (Josephstadt, Germany) on February 27, 1826. The apparitions of 1772 and 1805 involved short observation periods of only 29 and 36 days, respectively, but during 1826, the comet was observed for 72 days, which enabled Biela to mathematically link all three apparitions and declare the discovery of a new periodic comet. The comet was successfully recovered by John Herschel on September 24, 1832.[1]

Comet Biela was missed at the unfavorable return of 1839, but was recovered by Francesco de Vico (Rome, Italy) on November 26, 1845. Although a few observations were made in the next month, interest in the comet increased following Matthew Fontaine Maury's (Washington, DC) January 13, 1846

---

[1]Kronk, Gary W., *Comets: A Descriptive Catalog*. New Jersey: Enslow Publishers, Inc. (1984), p. 223.

observation of two distinct nuclei. Observers reported the nuclei to slowly move away from one another and by the end of March they were separated by 14 arc minutes; however, later investigations revealed the increasing separation was due to the comet's steady approach to Earth and, in truth, the nuclei had remained about 1.6 million miles apart during the entire apparition.

Periodic comet Biela was next observed during 1852. Father A. Secchi (Rome, Italy) recovered the main comet on August 26, but it was September 15 before the first observations of the secondary comet were made. The somewhat unfavorable approach of the comet caused it to enter the Sun's glare at the end of September, and no observations were made following the 29th. As it turned out, 1852 marked the last time observations were made of comet Biela. It was poorly placed for observation in 1859, and several months of extensive, but unsuccessful searches during the very favorable return of 1865-1866 caused astronomers to theorize that the comet had completely broken up.[2]

As the story on comet Biela slowly unfolded, the astronomical world was also becoming aware of a new meteor shower. On the evening of December 6, 1798, Heinrich W. Brandes (Göttingen, Germany) witnessed a large display of shooting stars. He said, "I first noticed them soon after the close of evening twilight, and having no other business, I kept count of the number which appeared in the small segment of the heavens which I could with convenience survey from my seat." His counting revealed rates of about 100 per hour for four straight hours, after which activity dropped off drastically. Brandes noted that occasional glances to other parts of the sky revealed similar quantities of meteors and he estimated "many thousand shooting stars must have been visible above my horizon."[3] The display ended nearly two hours prior to midnight. The only unfortunate aspect of Brandes' observation was the failure to note the radiation point of the meteors. Such was also true on December 7, 1830, when Abbe Raillard (France) recorded the appearance of "many" meteors, but failed to give further details. Fortunately, these two earlier displays were apparently confirmed on December 7, 1838, when observers on the east coast of the United States provided details of a very strong display.

From New Haven, Connecticut, Edward C. Herrick, C. P. Bush, A. B. Haile, J. D. Whitney and B. Silliman, Jr., observed during December 6-15, 1838 and noted meteors falling at a rate of 28 to 62 per hour on the evening of December 7. It was noted that many "large and splendid fireballs...attended with trains" were visible on both December 6 and 7. Herrick uncovered additional observations from Connecticut, New York and Georgia, and concluded that the

---

[2]Kronk, Gary W., *Comets: A Descriptive Catalog*. New Jersey: Enslow Publishers, Inc. (1984), p. 223.

[3]Herrick, Edward C., *Silliman's Journal*, **35** (1839), p. 361.

meteors seemed to radiate "not far from Cassiopeia; or perhaps, more nearly, from the vicinity of the cluster in the sword of Perseus" at an overall rate of between 125 and 175 per hour.[4] The Andromedids were next observed on December 6, 1847, when Eduard Heis (Germany) observed and plotted several meteors from α=21°, δ=+54° [he had originally estimated the radiant as α=25°, δ=+40°, but later revised it—Author].[5]

The 1860's were especially important for the field of meteoritics as Giovanni Virginio Schiaparelli's recognition of comet Swift-Tuttle's production of the Perseids inspired other astronomers to seek additional comet-meteor associations. Early in 1867, Professor Edmond Weiss (Austria), Heinrich Louis d'Arrest (Germany), and Professor Johann Gottfried Galle (Berlin, Germany) independently noted that meteor activity observed in early December of 1798 and 1838 moved in the same orbit as comet Biela. With this link being established, Biela became one of the first comets to be recognized as a meteor shower producer. Weiss and Galle contemplated a return of the Andromedids in 1872, while d'Arrest predicted a reappearance of activity on December 6, 1878.

Weiss continued to investigate the link between Biela and the Andromedids during 1868. Taking the known orbits of comet Biela for 1772, 1826 and 1852, he noted the comet's ascending node was gradually decreasing, thus causing the theoretical dates of maximum to be December 10, December 4 and November 28, respectively. The resulting radiants would have been α=18.7°, δ=+58.1° in 1772, α=22.8°, δ=+47.7° in 1826, and α=23.4°, δ=+43.0° in 1852. Weiss summarized his paper by predicting that activity might be observed from this radiant on November 28 of 1872 or 1879.[6] Meanwhile, although not being immediately revealed until several years later, Giuseppe Zezioli (Italy) observed fairly weak activity from the Andromeda-Cassiopeia border on November 30, 1867. Overall, seven meteors were plotted and a radiant of α=17°, δ=+48° was revealed.[7] This was the first activity noted from this region since 1838; however, although it offered support to Weiss' recognition of the shower's maximum moving into November, this activity was in no way comparable to the activity of 1798 and 1838.

Comet Biela was next predicted to arrive at perihelion in 1872, but the few searches made revealed no trace of either component; however, shortly after sunset on November 27, the pulverized remains of Biela began striking Earth's atmosphere. Father P. F. Denza (Moncalieri, Italy) and three others observed about 33,400 meteors during 6 hours 30 minutes. Around 8 p.m. (November

---

[4]Herrick, Edward C., *Silliman's Journal*, **35** (1839), pp. 361-365.

[5]Herschel, A. S., *MNRAS*, **32** (1872), p. 355.

[6]Weiss, Edmond, *AN*, **72** (1868), p. 81.

[7]D1899, p. 231.

27.79 UT) he said the display "seemed a real rain of fire," when meteors fell at a rate of 400 every minute and a half.[8] J. F. Anderson (Pau, France) obtained excellent meteor counts which revealed rates of 30 per minute shortly after 6:30 p.m., which slowly increased to 36 per minute around 7:45 p.m. (November 27.78 UT) and declined to about 14 per minute by 10:30 p.m.[9]

One of the most complete of the 1872 observations came from Stonyhurst Observatory. Being aware of Weiss' prediction of possible enhanced activity, S. J. Perry began observing after darkness had settled. As soon as he had detected notable activity he directed two assistants to aid in watching the sky. The result was a fairly accurate determination of the radiant position as α=26.6°, δ=+43.8°. It was estimated that maximum had occurred around 8:10 p.m. (November 27.84 UT) when meteors were falling at rates too numerous to count. During the 13 minutes prior to 9:00 p.m. one observer counted 512 meteors, for a rate of about 40 per minute. Perry estimated that a total sky rate would then have been about 100 per minute. The character of the shower's meteors was well established by the Stonyhurst observers as they indicated 90% of the meteors were very faint. Perry said a typical bright Andromedid had the appearance of "a white star with a greenish-blue trail." A peculiar feature of the shower involved the simultaneous appearance of meteors which moved parallel to one another. For example, Perry pointed out that at 9:16 p.m. "five burst out close to γ Andromedae and travelled eastward together."[10]

Although Western Europe was definitely the best place to be for the maximum of the Andromedids in 1872, observations were also made elsewhere in the world. Most notable were the observations by Hubert A. Newton and others in the eastern portion of the United States. Activity was first noted from the Gamma Andromedae region on November 24 when three-fourths of the 40-50 meteors seen each hour radiated from that star. On the 25th hourly rates were 20-25, with about half radiating from near Gamma Andromedae. Overcast skies were present on the 26th, but the storm was well observed on the 27th. Newton said a party of 2-6 observers counted 1000 meteors between 6:38 and 7:34 p.m. (about November 28.0 UT), with the quantity dropping to 750 in the next hour and twenty-five minutes. The meteors were described as slower than the Leonids and generally faint. Newton and A. C. Twining placed the radiant "in the line from the Pleiades to γ Andromedae and 3° beyond that star" (the Author computed this as α=26°, δ=+44°). Newton and Twining described the radiant as longer in α than in δ, with the length not being less than 8°.[11]

---

[8]*Nature*, 7 (December 19, 1872), p. 122.

[9]Anderson, J. F., *Nature*, 7 (December 19, 1872), p. 123.

[10]Perry, S. J., *Nature*, 7 (December 5, 1872), p. 84.

[11]Newton, H. A., *Nature*, 7 (December 19, 1872), p. 122.

In the years immediately following 1872, activity was totally absent from the region of Andromeda. D'Arrest's predicted 1878 appearance of the Andromedids never took place, nor did Weiss' predicted 1879 return. Shortly thereafter, several astronomers predicted the shower would reoccur on November 27, 1885, and a last-minute reminder was published by Crawford (Dun Echt Observatory) a couple of weeks prior to this date.

As the sun set on November 27, 1885, observers immediately became aware of exceptional activity in the sky. James Smieton (Broughty Ferry, Scotland) first began observations at 5:30 p.m. and noted meteors falling at a rate of 25 per minute. By 6:00 p.m. (November 27.75 UT) rates had gradually increased to 100 per minute. Something curious occurred at 6:20 p.m., when "a marked decrease in the intensity of the shower was noted." Thereafter, Smieton noted a steady increase to a peak of 70 per minute around 6:38 p.m., after which the shower steadily declined. The radiant was determined as α=21°, δ=+44°. He described the activity as consisting mainly of "shooting stars," but a large number of meteors "had brilliant phosphorescent trains, which continued to glow for several seconds after the meteors themselves had vanished. Occasionally one of the trains would break up into fragments, and in one instance a curious spiral form was assumed."[12]

William F. Denning (Bristol, England) actually noted activity from the Andromedid region 24 hours earlier, when rates averaged 100 per hour. But on the evening of the 27th, he declared "meteors were falling so thickly as the night advanced that it became almost impossible to enumerate them." He said observers with especially clear skies had rates of about one meteor every second or 3600 every hour.[13]

Additional details of the 1885 Andromedid activity were revealed in the early portion of an 18-page paper written by Newton and published in the *American Journal of Science* in June 1886. It appears that while some observers experiencing clear skies could not accurately count the meteors visible each minute, others gave quite consistent estimates. At Marseilles Observatory (France), E. J. M. Stephan, Alphonse Louis Nicholas Borrelly and Jerome Eugene Coggia independently made several counts near the shower's maximum and said the single observer rate reached 233 per minute. Observers at Palermo obtained a similar estimate of 213 per minute during one 5-minute interval. Using these observations, as well as others made in Beirut and Moncalieri, Newton determined the maximum hourly rate as 75,000 under very clear skies.[14] His indicated time of maximum corresponds to November 27.76 UT.

---

[12]Smieton, James, *Nature*, **33** (December 3, 1885), p. 104.

[13]Denning, W. F., *Nature*, **33** (December 3, 1885), p. 101.

[14]Newton, H. A., *AJS (Series 3)*, **31** (June 1886), pp. 409-412.

Although Newton's study produced an excellent view of the 1885 Andromedids, he went on to look at the physical characteristics of the stream, as well as how it had evolved. From the 1885 observations, he concluded the stream had an overall thickness of 200,000 miles, while "the really dense portion of the stream was less than 100,000 in breadth." Newton commented on the perturbations the comet had experienced from Jupiter during 1794, 1831 and 1841-1842, and theorized the debris encountered by Earth in 1872 and 1885 must have left the comet after the last encounter with Jupiter, otherwise the perturbations "would have scattered the group, and we should have had a much less brilliant star-shower in 1872 and 1885." He also confirmed Weiss' 1868 discovery of the decreasing ascending node of comet Biela by showing how the actual observations of the shower in 1798, 1838, 1847, 1867, 1872, and 1885 indicated the solar longitude of the shower's maximum had also declined from 256.2° to 245.8°—meaning the date of maximum had decreased by nearly 11 days.[15]

During the years immediately following 1885, the Andromedids were again nowhere to be seen, but at the next predicted passage of Biela in 1892 (the comet was of course not located) observers in the United States detected a strong meteor shower. Although it was not of the caliber of the 1872 and 1885 displays, the Andromedids of November 24, 1892 did produce rates of several hundred per hour. In particular, Daniel Kirkwood (California) observed 150 during one 30-minute interval and reported "an intelligent and trustworthy young gentleman counted 350 meteors in half an hour" later in the evening.[16] Another example comes from C. D. Perrine (Alameda, California), who observed 1013 meteors during one interval of an hour and eighteen minutes.[17]

The Andromedids next reached maximum on November 24, 1899,[18] and on November 21, 1904,[19] with hourly rates of 100 and 20, respectively. These rates, when combined with the longer observed durations of November 23-24 in 1899 and November 16-22 in 1904, indicated the stream was rapidly dispersing. The shower was virtually nonexistent in the years immediately following 1904, but, in 1940, *two* apparent peaks of activity were noted: an outburst of 30 faint meteors per hour occurred on November 15, according to R. M. Dole (Cape Elizabeth, Maine), while 5 per hour were detected by J. P. M. Prentice (England) during November 27-December 4.[20] These two peaks inspired

[15]Newton, H. A., *AJS (Series 3)*, **31** (June 1886), pp. 414-425.

[16]Kirkwood, Daniel, *PASP*, **4** (November 26, 1892), pp. 252-253.

[17]Perrine, C. D., *PASP*, **4** (November 26, 1892), p. 255.

[18]Denning, W. F., *MNRAS*, **60** (1900), p. 374.

[19]Denning, W. F., *MNRAS*, **65** (1905), p. 851.

[20]Prentice, J. P. M., *JBAA*, **51** (1941), p. 92.

Prentice to theorize that the Andromedids had divided up into several components.

Although visual activity seemed nonexistent following 1940, some remnants of the Andromedid stream were detected among the over 2000 meteors photographed during the Harvard Meteor Project of the early 1950's. The first official recognition of the photographic Andromedid meteors came in a paper by Gerald S. Hawkins, Richard B. Southworth and Francis Stienon published in 1959. Isolating all November meteors detected during 1950-1956 with α ranging from 0° to 50° and δ ranging from 0° to 50°, they compiled a list of 47 "possible Andromedids." Noting the period of Comet Biela, the authors said associated meteors should have an atmospheric velocity of 20 km/sec and, after sorting out all meteors with velocities between 19 and 21 km/sec, a total of 23 meteors remained. Plotting the meteors by date of appearance, the authors noted a maximum photographic hourly rate of 1.0 was reached on November 14, and they theorized this represented a visual rate of 5 per hour. The duration of activity was given as November 2-22. The authors also examined the duration of the 1872 activity and deduced a thickness of 400,000 miles. They concluded that the failure of activity to appear in 1878 indicated the debris had then spread over less than 4% of the comet's orbit.[21]

A reexamination of the orbit of periodic comet Biela was conducted by Brian G. Marsden and Zdenek Sekanina during 1971.[22] Subsequently, Lubor Kresak computed the new encounter conditions between Earth and the comet's orbit and noted that the shower's maximum would occur 12 days earlier than in the past, or on November 17.0, 1971 (essentially confirming the photographic maximum determined by Hawkins, Southworth and Stienon), at which time the radiant would be at α=26.2°, δ=+24.6° (20° south of the 19th century positions). Kresak noted that the closest approach of Earth to the comet's orbit was 0.05 AU, while Earth remained within 0.10 AU during November 6 to December 1.[23]

Bertil-Anders Lindblad conducted a computerized stream search during 1971 using 865 precise meteor orbits obtained during the Harvard Meteor Project. On this occasion, Lindblad used the D-criterion calculation and revealed the Andromedids to consist of two streams.[24] Two years later, however, a further investigation by Allan F. Cook, Lindblad, Brian G. Marsden, Richard E. McCrosky and Annette Posen revealed the 1971 study to have actually placed several members of the Andromedid stream in with the Piscids of September.

---

[21]Hawkins, Gerald S., Southworth, Richard B., and Stienon, Francis, *AJ*, **64** (June 1959), pp. 183-188.

[22]Marsden, Brian G., *IAU Circular*, No. 2347 (August 6, 1971).

[23]Marsden, Brian G., *IAU Circular*, No. 2362 (October 8, 1971).

[24]L1971A, p. 4.

The subsequent 24 photographic meteors were interpreted as indicating one very complex Andromedid stream, rather than two simple branches. The authors described the existence of "a systematic trend with the longitude of the sun (i.e., with that of the earth) such that the perihelion moves out from the sun, the inclination increases, and the node and argument of perihelion vary in such a way as to keep the longitude of perihelion unchanged."[25] The average orbit based on these 24 meteors is given in the "Orbit" section below. The Author's investigation into the photographic "Andromedids" of Cook et al, reveals the described orbital variations to be real, though it is obvious that the described transition is not smooth. A more direct explanation might be that the present Andromedid stream is composed of numerous filaments—each of which represents a ringlet of material left by Biela during previous evolutionary changes in its orbit.

As has already been pointed out, Biela underwent several close approaches to Jupiter and it seems likely that debris would have been left in each of the comet's previous orbits. It will be recalled that Weiss said these Jupiter encounters meant a steady decrease in the ascending node. This has continued, so that the date of maximum gradually moved from the second week of December back to mid-November. Comparing the observed orbit of Biela in 1772, with its hypothetical orbit of 1971 (see the "Orbit" section), it will be noted that perturbations over the last 200 years have also caused a steady decrease in the inclination and perihelion distance, as well as an increase in the argument of perihelion. For meteor observers this indicates that the current Andromedid activity of November comes from the newest orbits, while that of early December comes from the oldest. More importantly this also indicates that from November to December Earth encounters a series of filaments, the orbits of which gradually increase in perihelion distance, inclination, and ascending node, and decrease in the argument of perihelion—the same conditions noted in 1973 by Cook and his fellow researchers.

The 1970's saw a resurgence of interest in the Andromedid (or "Bielid") radiant as several amateur astronomers observed weak activity. During 2 hours on November 22, 1970, Martin Hale (Canisteo, New York) detected a rate of 1 Andromedid per hour,[26] while Mark Savill (Selsey, England) observed average rates of 4 per hour on November 21/22 and 23/24, 3.5 per hour on November 25/26, 1 per hour on November 26/27, and 2 per hour on December 4/5.[27] During November 12 and 14, 1971, A. Porter (Narragansett, Rhode Island) observed a total of 5 meteors from this radiant, all of which were described as

[25]C1973, p. 4.

[26]*MN*, No. 5 (March 1971), p. 7.

[27]*MN*, No. 6 (June 1971), p. 5.

red and of negative magnitudes. Porter commented, "This shower suffers from an inattention it does not deserve, because many amateurs hear somewhere that it's dead."[28]

Observers in England also observed the Andromedids in 1971. Robert Mackenzie, director of the British Meteor Society (BMS), said visual observers detected a maximum rate of 3-10 meteors per hour, while "BMS radio observations indicate a burst of faint meteors giving a ZHR of 35 meteors/hour." Mackenzie went on to describe annual Andromedid observations by the BMS extending from 1972 to 1975, in which maximum visual hourly rates attained 10, 2, 4, and 8, respectively.[29]

The most extensive recent observations of this stream comes from the Western Australia Meteor Section. Director Jeff Wood encouraged a survey of the Andromedid shower in 1979. Activity was noted during November 10-29 and a maximum ZHR of 3.73±1.86 came on November 27 from a radiant of $\alpha=28°$, $\delta=+38°$.[30] A total of 114 man hours were accumulated, but only 26 Andromedid meteors were observed. The average magnitude of these meteors was 3.42, while 3.8% left trains. The actual *hourly* rate of the activity peaked at 2 on November 26/27 and 1 on November 17/18.[31] Curiously, Wood's group compiled 76 man hours of observing during November 13-30, 1981 and revealed only 3 Andromedids.[32]

The Author concludes by stressing that the current Andromedid stream is a conglomeration of ringlets formed by the strong perturbations caused by repeated close approaches of the comet and its subsequent debris to Jupiter. Weiss and Newton provided the best evidence supporting the comet's orbital evolution, while observers of the last 100 years have revealed an increasing annual duration of the shower and the existence of several peaks of activity during the month of November. These peaks do not occur every year, and might indicate that the meteoric matter has still not had enough time to evenly distribute itself around the orbits of many of the ringlets.

Orbit

The first elliptical orbits to be computed for this stream came from Hawkins, Southworth and Stienon in 1959. They obtained two orbits using meteors photographed during 1950-1956. The first orbit only used photographic meteors

---

[28]*MN*, No. 25 (Mid-March 1975), p. 5.

[29]Mackenzie, Robert A., *Solar System Debris*. Dover: The British Meteor Society (1980), p. 27.

[30]Wood, Jeff, Personal Communication (October 15, 1986).

[31]*MN*, No. 49 (April 1980), p. 5.

[32]*MN*, No. 57 (April 1982), p. 10.

which possessed short trails, while the second was based on all 23 photographic meteors which had been isolated.

ω	Ω	i	q	e	a
242.7	225.5	7.5	0.777	0.732	2.90
245.4	228.1	6.3	0.783	0.728	2.88

The 24 photographic meteor orbits attributed as belonging to the Andromedid stream by Cook, Lindblad, Marsden, McCrosky and Posen in 1973 have been averaged by the Author to reveal the following orbit.

ω	Ω	i	q	e	a
261.9	200.3	6.1	0.616	0.794	2.990

During 1973, Allan F. Cook listed the following orbit which produced the Andromedids of 1885.[33]

ω	Ω	i	q	e	a
222	247	13	0.86	0.76	3.53

The following two orbits represent the observed orbital extremes shown by comet Biela from the year it was discovered until it was last seen.

	ω	Ω	i	q	e	a
1772	213.4	260.2	17.1	0.990	0.726	3.613
1852	223.2	247.3	12.6	0.861	0.756	3.525

According to a 1971 study by Marsden and Sekanina,[34] the orbit for comet Biela's (unobserved) 1971-1972 apparition would have been close to

	ω	Ω	i	q	e	a
1971	255.1	212.8	7.6	0.825	0.767	3.539

Leonids

Observer's Synopsis

The duration of this meteor shower covers the period of November 14-20. Maximum currently occurs on November 17 (λ=235°), from an average radiant of α=153°, δ=+22°. The maximum hourly rate typically reaches 10-15, but most notable are the periods of enhanced activity that occur every 33 years—events that are directly associated with the periodic return of comet Tempel-Tuttle. During these exceptional returns, the Leonids have produced rates of up to several thousand meteors per hour. The Leonids are swift meteors, which are best known for leaving a high percentage of persistent trains. The radiant's daily motion is +1.0° in α and −0.4° in δ.

---

[33]Cook, Allan F., *Evolutionary and Physical Properties of Meteoroids.* NASA SP-319, Washington, DC (1973), pp. 184-188.

[34]Marsden, B. G., and Sekanina, Z., *AJ*, **76** (December 1971), pp. 1139-1140.

History

The night of November 12-13, 1833, not only marks the discovery of the Leonid meteor shower, but sparked the actual birth of meteor astronomy. During the hours following sunset on November 12, some astronomers noted an unusual number of meteors in the sky, but it was the early morning hours of the 13th that left the greatest impression on the people of eastern North America. During the 4 hours which preceded dawn, the skies were lit up by meteors.

Reactions to the 1833 display are varied from the hysterics of the superstitious claiming Judgement Day was at hand, to the just plain excitement of the scientific, who estimated that a thousand meteors a minute emanated from the region of Leo. Newspapers of the time reveal that almost no one was left unaware of the spectacle, for if they were not awakened by the cries of excited neighbors, they were usually awakened by flashes of light cast into normally dark bedrooms by the fireballs.

At the time of the 1833 display, the true nature of meteors were not known for certain, but theories were abundant in the days and weeks which followed. The *Charleston Courier* published a story on how the sun caused gases to be released from plants recently killed by frost. These gases, the most abundant of which was believed to be hydrogen, "became ignited by electricity or phosphoric particles in the air."[35] The *United States Telegraph* of Washington, DC, stated, "The strong southern wind of yesterday may have brought a body of electrified air, which, by the coldness of the morning, was caused to discharge its contents towards the earth."[36] Despite these early, creative attempts to explain what had happened, it was Denison Olmsted who ended up explaining the event most accurately.

After spending the last weeks of 1833 trying to collect as much information on the event as possible, Olmsted presented his early findings in January 1834.[37] First of all, the shower was of short duration, as it was not seen in Europe, nor west of Ohio. His personal observations had shown the meteors to radiate from a point in the constellation of Leo, the coordinates of which were given as $\alpha=150°$, $\delta=+20°$. Finally, noting that an abnormal display of meteors had also been observed in Europe and the Middle East during November 1832, Olmsted theorized that the meteors had originated from a cloud of particles in space. Although the exact nature of this cloud was not explained properly, it did lead the way to a more serious study of meteor showers.

One of the more significant findings of the 1833 Leonid storm was the determination of the meteor shower's radiant. As mentioned above, Olmsted had

---

[35]*Charleston Courier*, Charleston, South Carolina (November 19, 1833).

[36]*United States Telegraph*, Washington, D.C. (November 13, 1833).

[37]Olmsted, Denison, *American Journal of Science Arts*, **25** (January 1834), p. 389.

obtained a position, but on the same morning, Professor A. C. Twining (West Point, New York) and W. E. Aiken (Emmittsburg, Maryland) obtained more precise estimates of α=148° 22', δ=+22° 20' and α=148° 10', δ=+23° 45', respectively.[38] This was the first time a shower radiant had ever been pinpointed more precisely than a simple direction in the sky or even a constellation.

New information continued to surface following the 1833 display which helped shed new light on the origin of the Leonids. First, a report was found concerning F. H. A. Humboldt's observation of thousands of bright meteors while in Cumana, South America during November 12, 1799. Further digging around this date in other publications revealed the spectacle was visible from the Equator to Greenland. Next, in November 1834, the Leonids reappeared and, although they were not as plentiful as in the previous year, they did demonstrate that some annual activity might be present from this region. In the years that followed, Leonid displays continued to weaken. In 1837, Heinrich Wilhelm Matthias Olbers combined all of the available data and concluded that the Leonids possessed a period of 33 or 34 years. He predicted a return in 1867.

The interest of the astronomical world began focusing on the predicted return of the Leonids as the decade of the 1860's began. Most important was Hubert A. Newton's examination of meteor showers reported during the past 2000 years. During 1863, he identified previous Leonid returns from the years 585, 902, 1582 and 1698.[39] During 1864, Newton further identified ancient Leonid displays as occurring during 931, 934, 1002, 1202, 1366 and 1602. He capped this study with the determination that the Leonid period was 33.25 years and predicted the next return would actually occur on November 13-14, 1866.[40]

The expected meteor storm occurred in 1866 as predicted, with observers reporting hourly rates ranging from 2000 to 5000 per hour. The 1867 display had the misfortune of occurring with the moon above the horizon, but observers still reported rates as high as 1000 per hour, meaning the shower may have actually been stronger than in the previous year. Another strong appearance of the Leonids in 1868 reached an intensity of 1000 per hour in dark skies.

The year 1867, marked an important development in the understanding of the evolution of the Leonids. On December 19, 1865, Ernst Wilhelm Liebrecht Tempel (Marseilles, France) had discovered a 6th-magnitude, circular object near Beta Ursae Majoris. After an independent discovery was made by Horace Tuttle (Harvard College Observatory, Massachusetts) on January 6, 1866, the comet took the name of Tempel-Tuttle. Perihelion came on January 12, 1866, afterwhich the comet began fading so rapidly, that it was not seen after February

---

[38]D1899, p. 255.

[39]Newton, H. A., *AJS (2nd series)*, **36** (1863), p. 147.

[40]Newton, H. A., *AJS (2nd series)*, **37** (1864), pp. 377-389.

9. Orbital calculations shortly thereafter revealed the comet to be of short period, and, as 1867 began, Theodor von Oppolzer had more precisely calculated the period to be 33.17 years.[41] Using observations from the 1866 Leonid display, Urbain Jean Joseph Le Verrier computed an accurate orbit for the Leonids,[42] and Dr. C. F. W. Peters, Giovanni Virginio Schiaparelli and von Oppolzer independently noted a striking resemblance between the comet and meteor stream.[43]

After a final notable display on November 14, 1869, when hourly rates reached 200 or more,[44] the following years were notable only due to a fairly consistent rate ranging from 10 to 15 Leonids per hour. Numerous confident predictions were put forth that the Leonids would next be at their best in 1899, and an early sign of returning enhanced activity was detected in 1898, when hourly rates reached 50-100 in the United States on November 14.[45]

What Charles P. Olivier called "the worst blow ever suffered by astronomy in the eyes of the public," was the failure of a spectacular meteor shower to appear in 1899.[46] Predictions had been made and newspapers in Europe and America made the public well aware that astronomers were predicting a major meteor storm. Although the "storm" failed to appear, the Leonids did possess maximum hourly rates of 40 on November 14—at least indicating some unusual activity. Later investigations revealed the stream to have experienced close encounters with both Jupiter (1898) and Saturn (1870), so that the stream's distance from Earth in 1899 was nearly double that of the 1866 return.

As it turned out, the actual peak of activity for the Leonids came on November 14-15, 1901. In the British Isles, Henry Corder (Bridgwater), E. C. Willis (Norwich) and others reported hourly rates as high as 25 before morning twilight interfered.[47] Several hours later, the Leonid radiant was well placed for observers in the United States, and it was apparent that the activity had increased. On the east coast, Olivier (Virginia) and Robert M. Dole (Massachusetts) independently obtained hourly rates of 60 and 37, respectively.[48] By the time the Leonids were visible over the western half of the United States, they had apparently reached their peak. At Carlton College (Minnesota) it was estimated that individuals could have counted about 400 per

---

[41]Von Oppolzer, T., *AN*, **68** (1867), p. 241.

[42]Le Verrier, U. J. J., *Comptes Rendus*, **64** (1867), pp. 94-99.

[43]Olivier, Charles P., *Meteors*. Baltimore: The Williams & Wilkins Company, 1925, p. 32.

[44]*Nature*, **1** (December 23, 1869), pp. 220-221.

[45]*Nature*, **59** (January 19, 1899), p. 279.

[46]Olivier, Charles P., *Meteors*. Baltimore: The Williams & Wilkins Company, 1925, p. 38.

[47]Denning, W. F., *Nature*, **65** (February 6, 1902), p. 332.

[48]Olivier, Charles P., *Meteors*. Baltimore: The Williams & Wilkins Company, 1925, p. 39.

hour. E. L. Larkin (Echo Mountain, California) estimated that rates reached a maximum of 5 per minute (300 per hour). By the time the British Isles had the radiant back in view, hourly rates had apparently declined to about 20. After analyzing the available data, William F. Denning concluded that the maximum of this shower came on November 15.48 Greenwich Mean Time (November 15.98 UT).[49]

The Leonids were barely detected in 1902, due to moonlight, but there was a reappearance in 1903. On November 16, Denning estimated a maximum hourly rate of 140, and said that for 15 minutes following 5:30 a.m. (local time) meteors were falling at 3 per minute. From plotted meteor paths, he found the radiant to have been 6° in diameter, centered at α=151°, δ=+22°.[50] John R. Henry (Dublin, Ireland) was also surprised by the intensity of the display, and he noted maximum rates near 200 per hour. Henry further noted that, at maximum, the Leonid meteors were pear-shaped and left rich trains. "Other members of the star shower dissolved in bright streaks, or made their appearance as vivid flashes of light...."[51] Finally, Alphonso King (Sheffield, England) did not begin observations until 5:57 a.m. He noted that 18 Leonids were seen in the first five and a half minutes, while only 16 were seen in the next half hour. King plotted 10 meteors which indicated a radiant of α=148°, δ=+22°.[52] From the above observations, it would seem the 1903 maximum came on November 16.2 UT.

The Leonids returned to normal in the years following 1903, with hourly rates ranging from 5 to 20 (average about 15). Despite having miscalculated the Leonid maximum in 1899, astronomers began to make predictions for the next return—the most likely date being 1932. Enhanced activity began early when, in 1928, maximum hourly rates reached 50 or more. During 1929, rates were lower, only 30 per hour, but moonlight was then a factor. During this latter year, members of the American Meteor Society (AMS) made fairly extensive observations, and Olivier's analysis revealed a radiant diameter of 5°-6° and a shower duration of 8-10 days.[53]

The Leonids began to show great strength in 1930. Professor C. C. Wylie (Iowa City, Iowa) estimated maximum hourly rates of 120 shortly before dawn on November 17. Olivier said the shower contained "many brilliant meteors with long enduring trains."[54] His analysis showed Leonids were first observed on November 13/14 and last seen on the 22nd. He confirmed that rates were

---

[49]Denning, W. F., *Nature*, **65** (February 6, 1902), pp. 332-333.

[50]Denning, W. F., *Nature*, **69** (November 19, 1903), p. 57.

[51]Henry, John R., *Nature*, **69** (November 26, 1903), p. 80.

[52]King, Alphonso, *Nature*, **69** (December 3, 1903), p. 105.

[53]Olivier, Charles P., *FOR*, No. 8 (1931), p. 1.

[54]Olivier, Charles P., *FOR*, No. 8 (1931), p. 35.

"considerably over 100 per hour, despite moonlight...."[55] The 1931 display showed a slight increase over 1930, but certainly not as great as expected considering the lack of moonlight. Olivier's analysis of AMS observations revealed rates between 130 and 190 per hour for observers in the United States during the pre-dawn hours of November 17.[56]

The predicted meteor storm of 1932 was looked for with great anticipation by astronomers, but it had been realized that moonlight would interfer with observations. Nevertheless, the first detection of the rapid rise to maximum came at Helwan Observatory (Egypt) during the pre-dawn hours of November 17. P. A. Curry was one of seven observers keeping a lookout for the expected storm, and the greatest hourly rates reached 51; however, it should be noted that the 5-minute counts showed a steady rise to 9 at 4 a.m.—amounting to 108 per hour—followed by a rapid decrease in numbers thereafter.[57] Members of the British Astronomical Association (BAA) were best placed for maximum, which came just a few hours after the Helwan observations. J. P. M. Prentice obtained the highest rates of 240 per hour.[58] Unfortunately, even after taking moonlight into account, it was obvious that a meteor storm comparable to those of 1833 and 1866 did not occur.

The Leonids seemed to decline slower than normal after 1932, as maximum rates remained between 30 and 40 meteors per hour from 1933 through 1939. This meant that greater than normal activity persisted from 1928 to 1939, or 12 years. The previous periods of enhanced activity occurred during 1898-1903, 1865-1869 and 1831-1836, which amounted to only 5 or 6 years.

Throughout the 1940's and 1950's hourly rates retained their "normal" character of 10-15 per hour. However, the period was highlighted by a new advance in astronomy—radar studies. Jodrell Bank Radio Observatory was the first station to detect the Leonids, with maximum observed rates being 24 in 1946, but only 3 to 11 during the period of 1947 to 1953.[59] Unfortunately, due to the weakness of the Leonids during the 1950's, the increasing sophistication of the equipment still could not obtain information such as radiant positions or radiant diameters.

Visual observers generally ignored the Leonids during the late 1950's, and this state of neglect caused many to completely miss the unexpected arrival of enhanced activity in 1961. Dennis Milon was one of five amateur astronomers observing outside Houston, Texas, when 51 Leonids appeared between 3:10 and

---

[55]Olivier, Charles P., *FOR*, No. 12 (1932), p. 1.

[56]Olivier, Charles P., *FOR*, No. 12 (1932), p. 39.

[57]Curry, P. A., *MNRAS*, **93** (January 1933), p. 191.

[58]Crommelin, A. C. D., *JBAA*, **43** (1933), p. 99.

[59]Lovell, A. C. B., *Meteor Astronomy*. London: Oxford University Press, 1954, p. 339.

4:10 a.m. on November 16 (about November 16.4 UT). The next morning the greatest one-hour interval produced a rate of 54 Leonids (about November 17.4 UT), bringing the Texas group to believe maximum had probably occurred late on the 16th. Similar rates were reported elsewhere. Norman D. Petersen (California) commented that the Leonids were blue-white, very rapid, and often left long-enduring trains 10° in length.[60]

The 1962 and 1963 displays were about normal with hourly rates of 15 or 20, while the 1964 display perked up with enhanced rates of 30 per hour. During 1965, observers in Hawaii and Australia were treated to one of the best displays since 1932. From the Smithsonian tracking station at Maui (Hawaii) hourly rates were near 20 on November 16.56 UT, but increased to about 120 by November 16.64 UT. Meanwhile, observers at the Smithsonian tracking station at Woomera (Australia) reported 38 Leonids of an average magnitude of –3 between November 16.65 and 16.77 UT.[61]

Although astronomers were still just one year away from the predicted Leonid maximum, optimism did not run high concerning the appearance of a meteor storm. Judging by the 1899 and 1932 returns, the stream orbit had obviously been perturbed so that a close encounter with Earth's orbit seemed no longer possible. About as far as astronomers were willing to gamble was to say that rates would probably be greater than 100 per hour. For much of the world, this is the best that was seen, but for the western portion of the United States, it was a night to be remembered.

On the night of November 17, 1966, expectations were high worldwide, but few observers got to see the Leonids as well as Dennis Milon and a dozen other amateur astronomers situated under the clear skies of Arizona. Observations began at 2:30 a.m. (November 17.35 UT) and 33 Leonids were detected during the next hour. After a short break, the next hour began at 3:50 a.m., with 192 Leonids being observed. The team had been keeping magnitude estimates during the early part of the shower, but this ended around 5:00 and, by 5:10, the observers were detecting 30 meteors every minute, but the display was far from over. Rates at 5:30 were estimated as several hundred a minute and the team estimated a peak rate of 40 per *second* was attained at 5:54 (November 17.50 UT)! The activity declined thereafter, and by 6:40 it was down to 30 per minute, despite the fact that astronomical twilight had begun 9 minutes earlier.[62] To sum up, it would seem the 1966 return of the Leonids was one of the greatest displays in history, with maximum rates being 2400 meteors per minute or 144,000 per hour.

---

[60]*Sky & Telescope*, **23** (February 1962), p. 65.

[61]Gingerich, Owen, *IAU Circular*, No. 1941 (November 19, 1965).

[62]*Sky & Telescope*, **33** (January 1967), pp. 4-5.

The major peak of the 1966 display was also enjoyed by observers in New Mexico and Texas. Observers in the eastern portion of the United States did report rates of several hundred per hour, but other countries reported rates generally less than 200 per hour, since maximum had occurred during daylight. An exception was observers at a USSR polar arctic station, who were able to monitor the shower at its peak. With the radiant only 8° above the horizon, the report from two observers said, "there was a continuous flight of meteors in a single direction, from north to south. Some appeared in the zenith and curved over the southern horizon, some appeared from the northern horizon and disappeared in the zenith, and some flew across the entire horizon, leaving behind a bright trail." R. L. Khotinok's analysis of the complete report revealed an observed maximum rate of 20,000 per hour, while a correction for the low altitude gave a rate of 130,000 per hour—agreeing quite well with the Arizona observations.[63]

In the years following the 1966 display, hourly rates for the Leonids remained high. From 1967 through 1969, observers continued to detect rates of 100-150 per hour. After a return to normality in 1970 (15 per hour), rates jumped to 170 per hour in 1971 and 40 in 1972. The Leonids have remained between 10 to 15 per hour at maximum ever since.

One of the first Leonid studies involving an analysis of observational data, was published in 1932 by Alphonso King. The study was basically a look at his observations made during 1899-1904 and 1920-1931. King noted the diameter of the radiant to generally be less than 4°, and he determined a radiant ephemeris which indicated a daily motion of +1.0° in α and −0.4° in δ.[64]

Some of the more interesting recent studies of the Leonids involved extensive observations by professional and amateur astronomers in the Soviet Union during 1971 and 1972. The first set of observations were made at Sudak and Simferopol during November 15-19, 1971. Although numerous observers participated, it was the more experienced observations of N. V. Smirnov and Yu. V. Lyzhin which were evaluated. Some of the various observed aspects of the meteors included 553 meteors with an average magnitude of 3.40 and 171 color estimates indicating 74% were green, 20% were white, 1% were blue and 1% were orange. One of the most striking discoveries was the detection of multiple radiants. Although six radiants were determined, the most active was the long-known radiant at α=151.7°, δ=+22.9° (based on 222 plotted meteors) and the authors noted that the total plots indicated activity primarily came from an area 2.5° x 8° centered on this radiant.[65]

---

[63]Khotinok, R. L., *SSR*, **1** (1967), p. 50.

[64]King, A., *MNRAS*, **93** (November 1932), pp. 109-111.

[65]Kremneva, N. M., Martynenko, V. V., and Smirnov, N. V., *SSR*, **10** (1976), pp. 91-94.

The 1972 visual survey was conducted during November 16-18, from the same locations given above. A magnitude breakdown was not given strictly for the Leonids, but for all meteors observed at Sudak. The average brightness ended up as 3.01 for 576 meteors, of which 335 were Leonids. On this occasion, six radiants were again determined from plots, with the main center being at α=151.9°, δ=+22.7° (based on 185 meteors).[66] The radiants were generally grouped into an area about 10° across; however, it should be noted that two radiants within this area were distinctly detected in both years—one near Mu Leonis (α=150°, δ=+28°) and the other between Gamma and Eta Leonis (α=151°, δ=+17°).

During 1967, one of the first mathematical surveys of the perturbations suffered by the Leonid meteor stream was conducted. Using the orbit determined for the 1866 Leonid shower, E. I. Kazimirchak-Polonskaya, N. A. Belyaev, I. S. Astapovich and A. K. Terent'eva examined 12 hypothetical meteor groups situated around the orbit. One of the major findings was that Jupiter and Saturn were primarily responsible for altering the encounter conditions between Earth and the meteor stream. Earth itself was even found to have a strong effect on meteor bodies passing within several thousand kilometers of its surface by shortening the revolution period by several years, strongly altering the eccentricity and even changing the inclination.[67]

The most ambitious study of the relationship between Tempel-Tuttle and the Leonids was published in 1981. Donald K. Yeomans (Jet Propulsion Laboratory, California) mapped out the dust distribution surrounding Tempel-Tuttle by "analyzing the associated Leonid meteor shower data over the 902-1969 interval." He noted that most of the ejected dust lagged behind the comet and was outside its orbit, which was directly opposite to the theory of outgassing and dust ejection developed to explain the comet's deviation from "pure gravitational motion." Yeomans suggested this indicated "that radiation pressure and planetary perturbations, rather than ejection processes, control the dynamic evolution of the Leonid particles."[68] Concerning the occurrence of Leonid showers, Yeomans said "significant Leonid meteor showers are possible roughly 2500 days before or after the parent comet reaches perihelion but only if the comet passes closer than 0.025 AU inside or 0.010 AU outside the Earth's orbit." He added that optimum conditions will be present in 1998-1999, but that the lack of uniformity in the dust particle distribution still makes a prediction of the intensity of the event uncertain.

---

[66]Kremneva, N. M., Martynenko, V. V., and Frolov, V. V., *SSR*, **11** (1977), pp. 92-95.

[67]Kazimirchak-Polonskaya, E. I., Belyaev, N. A., Astapovich, I. S., and Terent'eva, A. K., *SA*, **11** (November-December 1967), pp. 490-500.

[68]Yeomans, D. K., *Icarus*, **47** (1981), p. 492.

Orbit

The orbit of the Leonid stream, based on 12 photographic meteors obtained from W1954, H1959 and MP1961, as well as 3 radar meteors obtained from S1970, is as follows:

ω	Ω	i	q	e	a
173.1	235.4	161.0	0.983	0.901	9.956

The 1965 orbit of comet Tempel-Tuttle, according to Yeomans' 1981 study published in *Icarus* is

ω	Ω	i	q	e	a
172.6	234.4	162.7	0.982	0.904	10.272

November Monocerotids

Observer's Synopsis

The maximum of this meteor shower occurs on November 21 (λ=238.7°) from a radiant of α=109°, δ=–6°. Although weak annual activity may be present during the period of November 13 to December 2, strong activity of perhaps 100 meteors per hour may return every 10 years on the indicated date of maximum. The daily motion of the radiant is +0.8° in α and –0.4° in δ.

History

Our present knowledge of this meteor shower is primarily based on visual observations obtained in three separate years, but the implications are that this shower is of short duration and possesses an apparent period of almost exactly ten years.

The actual discovery should be credited to F. T. Bradley (Crozet, Virginia), who observed on the night of November 20/21, 1925. He began his observations at 11 a.m. (local time) and counted 37 meteors between 11:02 and 11:15. A 10-minute break was taken to obtain star charts for plotting, but when he resumed observations at 11:25 the outburst had ended. Without having had the opportunity to plot any of the meteors, no definite radiant could be determined, but judging by the meteors' tendency to move from east to west, Bradley estimated the radiant was below Orion. Fortunately, Charles P. Olivier had been working in the observatory at the University of Virginia that same night. He said he stepped outside for a few moments and "saw 3 bright meteors about 11:05 a.m. The paths of two of these were mentally noted quite accurately, the path of the third being too poorly seen, though it was parallel to that of the second." The deduced radiant was α=97.5°, δ=+8.5°, but Olivier admitted that the position was not very accurate. After uncovering several additional reports of enhanced activity, none of which acted to shed light on the position of the radiant, Olivier concluded that the meteors "were of various colors, bright, slow, and left trains."[69]

No additional activity was noted from this region until November 21.75, 1935, when Professor Mohd. A. R. Khan (Begumpet, India) witnessed "a fine shower of meteors whose radiant appeared to be near γ Monocerotis." Overall, over 100 meteors were noted in the first 20 minutes, while 11 were counted in the next 20 minutes.[70] A few months later, Khan wrote to Olivier with a more precise radiant position of $\alpha=110°$, $\delta=-5°$ (he revealed that the star γ Monocerotis was the same as α Monocerotis on other star charts). The parabolic orbit computed by Olivier is given in the "Orbit" section. The activity and radiant of the shower were apparently confirmed by the commanding officer of the USS *Canopus*, then anchored in Manila harbor, who noted that meteors appeared about once every 30 seconds during one 30-minute interval.[71]

Olivier was confident enough to say the strong returns of 1925 and 1935 were not only related, but pointed towards a probable return in 1945. Although he was the first to point out a possible ten-year period, Olivier did not rule out the possibility of this being an annual shower whose short duration and "apparently very narrow cross-section," would make it easy to miss completely. He cited an American Meteor Society observation of November 19.17, 1904 (UT), as possibly representing a previous appearance (AMS radiant number 165, with a position of $\alpha=95.4°$, $\delta=+10.9°$).[72]

If any attempts were made to observe activity from this stream in 1945 conditions would have been very poor due to the appearance of a full moon late on November 19. Conditions would have been a little better in 1955, with full moon coming on November 29, but no apparent searches or accidental observations were made.

In 1958 Lubor Kresak (Czechoslovakia) examined the 1925 and 1935 events, claiming the 1935 shower represented "a unique and highly interesting example of an extremely condensed meteor stream...." Kresak estimated the maximum ZHR reached about 2000 per hour, which he said made it "the most conspicuous meteoric event observed in the present century, with the exception of the two richest returns of the October Draconids." Kresak said the cometary character of the two showers was "beyond doubt," and that their very short durations indicated a very recent departure from their parent body. In trying to determine what comet was responsible, he compared the solar longitudes of the

---

[69]Olivier, Charles P., *PA*, 34 (March 1926), pp. 167-168.

[70]Olivier, Charles P., *PA*, 44 (February 1936), p. 89.

[71]Olivier, Charles P., *PA*, 44 (June-July 1936), pp. 327-328.

[72]Olivier, Charles P., *PA*, 44 (February 1936), p. 89.

1925 and 1935 displays (given as 238° 41' and 238° 44', respectively) and, after adding details on the theoretical orbit, he said only comet van Gent-Peltier-Daimaca (1944 I) came closest to representing the meteor stream orbit. Kresak admitted to some large discrepancies between the comet and stream orbits and concluded that if comet 1944 I was not responsible, the stream "must have been generated by a body too faint to be discovered by the present means."[73]

This region produced no notable activity during the five decades following 1935, but on the night of November 21, 1985, two independent discoveries were made by observers in California. The first was Keith Baker, night assistant at Lick Observatory. He stepped outside around 3:00 a.m. (local time) and observed 18 meteors in 7 minutes coming from a region near Canis Minor. The meteors were of magnitude 2 to 4, rapid, of short duration, and left no trains. From Capitola, Richard Ducoty observed 27 meteors during 3:41 and 3:45 a.m., 5 during 3:45 and 3:49 a.m., 2 during 3:49 and 3:53 a.m., and 2 during 3:53 and 3:57 a.m. His estimate of the radiant position was $\alpha=109°$, $\delta=-7°\pm5°$. He said, "The brightest meteors were 0 to –2. Their speed was quite fast, a little slower than the Leonids."[74]

The 1985 radiant estimate by Ducoty provided an excellent confirmation of Khan's 1935 radiant determination. The Author has utilized these observations, as well as the apparent ten-year period, to calculate an elliptical orbit. A search was then conducted among published observations of the last 100 years to see if other observations could be located. The earliest possible observation appears to be that of W. Doberck (Hong Kong Observatory), who plotted five meteors from a radiant of $\alpha=102.5°$, $\delta=-12°$ during November 19-25, 1895.[75] Although this radiant seems to support the ten-year period, Doberck gave no indication of strong activity and said all the meteors were between magnitude 4 and 5.

The next possible visual observations were made by R. M. Dole (East Lansing, Michigan) and Cuno Hoffmeister (observing during an expedition to South-West Africa). Dole plotted three meteors from $\alpha=111°$, $\delta=-11.2°$ during November 17.8, 1923,[76] while Hoffmeister plotted several meteors from $\alpha=112°$, $\delta=-10°$ during November 26, 1937.[77] Neither radiant satisfies the ten-year period, but neither radiant produced very many meteors. The Author suggests these observations might indicate the presence of weak activity for at least two years before and after the appearance of the main shower and that the stream is wide enough to produce a duration of at least November 17-26.

---

[73]Kresak, L., *BAC*, **9** (1958), pp. 88-96.

[74]*MN*, No. 73 (April 1986), p. 6.

[75]Doberck, W., *AN*, **140** (June 15, 1896), pp. 375-380.

[76]Olivier, Charles P., *FOR*, No. 4 (1929), p. 21.

[77]H1948, p. 244.

The Author used a D-criterion value of 0.20 and less to compare photographic and radio meteor orbits, with his computed elliptical orbit. Altogether, only four orbits were revealed: two photographic meteors from the Harvard Meteor Project (1952-1954) and two radio-echo meteors from the first session of the Radio Meteor Project (1961-1965). These meteors do not offer much evidence to support either the ten-year period or the short-duration suggested, although the photographic meteors do seem to offer additional proof supporting the persistence of activity for a period of two years before and after the expected dates of maxima. What is interesting is that the implied duration of activity is extended to November 13-December 2. The dates of the appearance of the meteors (as well as their radiant positions) are November 13, 1953 (α=101.7°, δ=–2.1°), November 26, 1954 (α=115.4°, δ=–7.4°), November 16, 1962 (α=109.1°, δ=–5.8°), and December 2, 1963 (α=118.0°, δ=–9.4°). Combining the meteor radiants with the visual observations of 1935, 1937 and 1985, the Author finds a daily motion of +0.8° in α and –0.4° in δ.

Orbit

Using a radiant given by Professor Khan for 1935, Olivier computed the following parabolic orbit.

ω	Ω	i	q	e	a
85.5	58.5	114.5	0.46	1.0	∞

The Author has computed the following elliptical orbit based on the stream's apparent period of ten years.

ω	Ω	i	q	e	a
101.1	58.7	108.8	0.428	0.908	4.64

Using two photographic (H1959 and JW1961) and two radio meteor orbits obtained from the 39,145 orbits computed by Sekanina during the two sessions of the Radio Meteor Project, the Author finds the following average orbit.

ω	Ω	i	q	e	a
95.7	59.4	112.3	0.464	0.941	7.831

The orbit of comet van Gent-Peltier-Daimaca (1944 I) is

ω	Ω	i	q	e	a
33.1	57.9	136.2	0.874	1.0	∞

Taurids

Observer's Synopsis

There are two branches of the Taurids active during its long duration in the Autumn months (or Spring months in the Southern Hemisphere). The Northern

Taurids are active from October 12 to December 2. Maximum is also of long duration and extends over November 4-7 (λ=221°-224°) from an average radiant of α=54°, δ=+21°. The radiant's daily motion is +0.78° in α and +0.19° in δ. The Southern Taurids are active during September 17 to November 27. They reach maximum during October 30 to November 7 (λ=216°-224°) from an average radiant of α=53°, δ=+12°. This radiant's daily motion is +0.99° in α and +0.28° in δ. Both showers possess maximum hourly rates near 7.

History

In 1940, Fred L. Whipple commented that the "multiplicity of radiants, the uniformity and the long endurance of the Taurid stream of meteors have disguised its character as one of the more important known showers."[78] Indeed, today's Taurids seem almost lost amidst a collection of strong showers producing activity during the months of October and November. But such was not always the case.

I. S. Astapovich and A. K. Terent'eva conducted a study of fireballs appearing between the 1st and 15th centuries and revealed the Taurids to have been "the most powerful shower of the year in the 11th century (with 42 fireballs belonging to them) and no shower, not even the great ones, could be compared with them as to activity." The authors said both branches of the stream were active: the Northern Taurids possessed a duration of October 20-November 18, with an average radiant of α=56°, δ=+24°, while the Southern Taurids had a duration of October 25-November 17 and an average radiant of α=54°, δ=+8°. The northern stream was the strongest of the two branches and possessed a radiant measuring 6°x1°. The southern shower was only half as active as the northern and possessed a radiant 3° in diameter.[79] The existence of the Taurid streams cannot be accounted for between the 11th and 19th centuries.

The modern-day discovery of both streams was made in 1869. The Northern Taurids were observed by Giuseppe Zezioli (Bergamo, Italy) during November 1-7, when he plotted 11 meteors from a radiant of α=56°, δ=+23°. The Southern Taurids were observed by T. W. Backhouse (Sunderland, England) on November 6, when 5 meteors were plotted from α=54°, δ=+14°, and possibly by G. L. Tupman (Mediterranean Ocean) on November 12, when 8 meteors were plotted from α=52°, δ=+12°.[80] Although the Southern Taurids

---

[78]Whipple, F. L., *Proceedings of the American Philosophical Society*, **83** (October 1940), p. 742.

[79]Astapovich, I. S., and Terent'eva, A. K., *Physics and Dynamics of Meteors*. eds. Lubor Kresak and Peter M. Millman. Dordrecht: D. Reidel Publishing Company, 1968, pp. 316-317.

[80]D1899, p. 238.

were rarely detected during the remainder of the 19th century, the Northern Taurids were frequently observed. Unfortunately, no one was recognizing that a shower was being observed almost annually from the Taurus region in early November.

It was not until 1918 that the loose ends were finally tied together when Alphonso King (Ashby, England) announced the existence of a new meteor shower. His interest was sparked by the observation of six meteors by T. F. Cranidge and himself within 24 minutes on November 12, 1918. King said the hourly rate for two observers would have been 15, but correcting for bright moonlight would have made this much higher. One additional meteor was seen later in the evening and King's analysis of the seven plots revealed a radiant of α=54°, δ=+24°. As King was a believer in the stationary radiant theory, his search for previous appearances of this radiant revealed a duration extending from August to December; however, half of the radiants he uncovered occurred during the first half of November. These radiants follow:[81]

Northern Taurid Radiants

Date (UT)	RA(°)	DEC(°)	Meteors	Observer
1870, Nov. 13	55	+25	—	Greenwich
1871-4, Nov. 6-12	56	+24	—	Greg & Herschel
1872, Nov. 1-3	56	+24	13	Tupman
1876, Nov. 7-10	55	+24	—	Corder
1890, Nov. 4-16	53	+26	8	Booth
1891, Nov. 13	53	+24	4	Corder
1896, Nov. 13	52	+23	7	Blakeley & Corder
1900, Nov. 9-27	56	+28	6	Herschel

Observers in the above table were Greenwich Observatory (Manchester, England), R. P. Greg (England), A. S. Herschel (Ashby, England), G. L. Tupman (Mediterranean Ocean), H. Corder (Writtle, England), D. Booth (Leeds, England) and E. R. Blakeley (Dewsbury, England).

The Southern Taurids were first discussed in depth in 1920. William F. Denning pointed out that the majority of the "considerable number of fireballs" which appeared in early November of 1920 came from a Taurid shower at α=59°, δ=+12°. He said he had noted the radiant "to have been very active on November 2-3, 1886, when 17 of its meteors were seen at Bristol, and they indicated a diffused radiant situated a few degrees west of the Hyades."[82]

In the years following 1920, observations of both radiants were fairly abundant, though both were rarely seen by the same observer in the same year.

---

[81]King, A., *MNRAS*, **79** (December 1918), pp. 158-159.

[82]Denning, W. F., *The Observatory*, **43** (December 1920), p. 432.

For example, J. P. M. Prentice plotted 10 Northern Taurids from a radiant of α=60°, δ=+22° during November 8-10, 1921 (the observation was even confirmed by A. Grace Cook, who plotted 11 meteors from α=63°, δ=+22° during November 6-10);[83] however, he detected strong Southern Taurid activity during November 13 & 15, 1922, from α=53°, δ=+14°,[84] and November 2-17, 1923, from α=54.8°, δ=+11.6°.[85] In a 1924 article briefly discussing the Taurid radiants, Denning pointed out that the Southern Taurids (referred to as the Lambda Taurids) exhibited "marked variation in strength in different years."[86]

In the midst of an ambitious study of 5406 visual radiants, Hoffmeister reexamined the Taurid streams in his 1948 book *Meteorströme*. He failed to recognize the Southern Taurids, but he uncovered 91 radiants representing the Northern Taurids. He indicated the stream's duration to extend from September 27 (λ=183.6) to December 10 (λ=256.9°), with the radiant steadily moving from α=27.2°, δ=+16.8° to α=78.8°, δ=+20.0°. Maximum activity was represented by 16 radiants and fell on November 4 (λ=221.9°), when the radiant was at α=51.1°, δ=+21.8°.[87] *[The Author would like to point out that on page 82 of Hoffmeister's book, there is a radiant matching the general description of the Southern Taurids. It was based on only 4 visual observations and occurred on November 3 from α=57°, δ=+8°.]*

Although observers and researchers tended to agree with the notion that the region of Taurus and Aries contained several active radiants during October and November, it was the photographic analysis of F. W. Wright and Whipple that made the first elaborate attempt to isolate these. Altogether they found four radiants: Northern Taurids, Southern Taurids, Northern Arietids, and Southern Arietids. The two Arietids streams were not well represented in the data and the authors contemplated that the southern branch might form a continuous stream with the Southern Taurids.

Wright and Whipple's analysis of the two Taurid streams was quite complete. They found 49 double-station and single-station photographic meteors which represented the Southern Taurids. These indicated a duration extending from October 26 to November 28, and a radiant moving from α=46° 56', δ=+13° 24' to α=67° 00', δ=+16° 16'. The mean date of photographic activity came on November 8.69, at which time the radiant was at α=55° 13', δ=+14° 29'. The 24 double-station and single-station meteors representing the Northern Taurids revealed a duration of October 17-December 1, with the radiant moving from

[83]Denning, W. F., *The Observatory*, 44 (December 1921), p. 376.
[84]Denning, W. F., *The Observatory*, 46 (January 1923), p. 25.
[85]Denning, W. F., *The Observatory*, 47 (January 1924), p. 31.
[86]Denning, W. F., *The Observatory*, 47 (August 1924), p. 255.
[87]H1948, pp. 140-141.

α=44° 35', δ=+19° 01' to α=67° 32', δ=24° 27'. The mean date of photographic activity came on November 10.72 from a radiant of α=56° 55', δ=+22° 25'. The authors said the hourly rate of the Southern Taurids seemed to rise abruptly to an early November maximum, slowly decline, and then rise again to a secondary maximum on November 11, while the Northern Taurids showed only a flat maximum around mid-November. They concluded that the differences and character of the activity of the two streams indicated the Southern Taurids were less diffuse than the northern branch and, therefore, may have developed more recently.[88]

Other photographic surveys were conducted during the 1950's and 1960's by astronomers in both the United States and the Soviet Union. The subsequent analyses of these photographic meteor orbits tended to reveal similar orbital details, but the utilization of fewer meteors did not allow a determination of the radiant, date of maximum and daily motion that could compare with that of Wright and Whipple. The Author has combined the double-station photographic orbits obtained from all of these surveys and has produced orbits for both Taurid streams (see "Orbit" section); however, the data is still inadequate to enable an accurate determination of even the date of maximum. The reason for this is that photographic Taurids are in greater abundance in October than in November. Whether this is due to mass distribution within the streams or just an inadequate use of cameras during November can not be determined at this time. The Author has, however, determined the daily motion of both Taurid radiants, with values of +0.78° in α and +0.19° in δ being determined for the northern stream, and +0.99° in α and +0.28° in δ being determined for the southern stream.

Radio-echo studies became a powerful addition to the arsenal of astronomers in the mid-1940's. Unfortunately, even the best equipment then available, which was located at Jodrell Bank, possessed a resolution so low it was impossible to separate the two Taurid streams. Thus, from 1946 to 1958 radio-echo details revealed only a general picture of the Taurid shower. For 1946 and 1947, not even a radiant could be determined, but radio-echo rates of 18 per hour were detected on November 9 of the former year, while rates reached 9 per hour on November 6 of the latter year. In 1950, 109 echoes were detected on November 9, which revealed a radiant of α=55°, δ=+25°. The maximum hourly rate reached 14.[89]

For the period of 1951-1953, the Jodrell Bank survey obtained four radiant determinations. In 1951, 57 echoes detected on November 7, revealed a 4°-diameter radiant of α=61°, δ=+25°, while the maximum hourly rate reached 25.

---

[88]Wright, F. W., and Whipple, F. L., *Technical Reprint of Harvard College Observatory*, No. 6 (1950).

[89]HA1952, p. 223.

Two radiants were detected in 1952. The first was a 3°-diameter radiant located at $\alpha=52°$, $\delta=+24°$ on November 5, while the second was a 6°-diameter radiant located at $\alpha=59°$, $\delta=+17°$ on November 10. The maximum hourly rates attained 7 and 14, respectively. In 1953, a 3°-diameter radiant was detected at $\alpha=58°$, $\delta=+25°$ on November 9. The maximum hourly rate reached 8.[90] The additional survey years of 1954-1958 closely reflected the results obtained during 1950-1953.

As the 1960's began, radio equipment had been set up in other areas of the world—equipment which was more sensitive than that at Jodrell Bank. For the first time, astronomers had the means to precisely detect meteors at magnitudes fainter than what photographic methods offered. During 1960, B. L. Kashcheyev and V. N. Lebedinets (Kharkov Polytechnical Institute, USSR) succeeded in splitting the Taurids into two distinct streams, despite the fact that the equipment did not operate beyond October 23. Southern Taurids were detected during September 20-October 22, during which time 73 meteors were detected from an average radiant of $\alpha=27.2°$, $\delta=+8.6°$. The Northern Taurids were detected during October 11-23, during which time 13 meteors were detected from an average radiant of $\alpha=33.5°$, $\delta=+18.2°$. The authors determined orbits for each stream, based on velocity measurements, and concluded that both streams were in good agreement with the orbits determined by photographic methods.

The next step in the evolution of radio equipment possessed the capability of detecting meteors far below naked-eye visibility. They uncovered a very interesting bit of information on the Taurid stream: the orbital planes of the northern and southern streams were so similar at this level, computer analysis was unable to distinguish a difference between the two streams. Zdenek Sekanina, director of the Radio Meteor Project of the 1960's, noted, "The gap between the two branches, so striking in the case of bright photographic meteors, is no longer seen in the radio sample. Also, the radio Taurids appearing on the same day as the bright photographic Taurids have their radiants, on an average, shifted eastward." Sekanina said the most notable difference in the orbital elements was in the longitude of perihelion, which varied from the photographic orbits by nearly 10°. He concluded that the separation between the photographic and radio data "may suggest a difference in the mean age between the two groups of meteors."[91]

With observations of both Taurid radiants becoming more numerous as the 20th century progressed, certain facts about the streams became known. Two of the most notable characteristics were the long durations and the slow daily

---

[90]B1954, pp. 80-81.

[91]S1970, p. 482.

motion of each streams' radiant. This led to the 1930s conclusions of O. Knopf[92] and Cuno Hoffmeister[93] that the Taurids were of interstellar origin rather than a product of the solar system. This conclusion was challenged in 1940, when Fred L. Whipple published a list of fourteen photographic meteors detected by the northern stations of Harvard Observatory during 1937-1938. Orbits were computed for six of the meteors simultaneously photographed by two cameras, and this allowed Whipple to discovery that the Taurids possessed unusually short periods. He concluded that the semimajor axis, eccentricity and longitude of perihelion all pointed to a possible association with periodic comet Encke, and that the observed 10°-15° difference in the planes of the meteor orbits and the comet could be explained as the result of 14,000 years worth of perturbations from Jupiter.[94]

The origin of the Taurids was reexamined by Whipple and S. Hamid during 1950. They calculated the effects of secular perturbations by Jupiter on the orbital inclination and longitude of perihelion of nine photographic meteor orbits and found the orbital planes of four of the meteors to coincide with that of comet Encke 4700 years ago. Three other orbits coincided with one another, but *not* with comet Encke 1500 years ago. The authors theorized "that the Taurid streams were formed chiefly by a violent ejection of material from Encke's Comet some 4700 years ago, but also by another ejection some 1500 years ago, from a body moving in an orbit of similar shape and longitude of perihelion but somewhat greater aphelion distance...." It was suggested that this unknown body had separated from Encke some time in the past.[95]

Whipple's 1940 paper discussed more than the Taurids and their link to comet Encke. He said the stream's apparent spread of 0.2 AU meant Mercury, Venus and Mars were also likely to encounter it. He also noted that the stream could produce a post-perihelion shower for Earth which would occur in late June and early July during daylight.[96] Of course it will be some time before the "Taurids" of Mercury, Venus and Mars are confirmed, but, in 1951, Mary Almond computed orbits for the daylight streams discovered at Jodrell Bank and found the Beta Taurids of June to be very similar to the Northern Taurids.[97]

---

[92]Knopf, O., *AN*, **242** (1931), p. 161.

[93]Hoffmeister, C., *Die Meteor*. Leipzig: Akademische Verlagsgesellschaft M. B. H. (1937), p. 46.

[94]Whipple, F. L., *Proceedings of the American Philosophical Society*, **83** (October 1940), pp. 711-745.

[95]Whipple, F. L., and Hamid, S., *AJ*, **55** (October 1950), pp. 185-186.

[96]Whipple, F. L., *Proceedings of the American Philosophical Society*, **83** (October 1940), p. 723.

[97]Almond, Mary, *MNRAS*, **111** (1951), p. 41.

Visual details of the Taurid meteors have not been lacking during the last two decades, though it is unfortunate that the two streams are rarely separated due to their closeness to one another. With astronomers attempting to estimate the ages of the northern and southern branches, it would be especially interesting to compare the characteristics of the respective meteor populations. The only recent attempt to visually separate the northern and southern streams came in 1983, when members of the AK Meteore (East Germany) observed 40 Southern Taurids and 69 Northern Taurids. The first stream was detected during September 26-December 3, with the average magnitude being estimated as 4.25, while the second stream was seen during September 26-December 4, with an average magnitude of 3.87.[98] Although the average magnitudes are unusually low (due to very favorable observing conditions), the difference between these values seems significant and, of course, implies that the Southern Taurids possess a fainter population of meteors than the northern branch. Support or denial of such a statement will, however, require additional observations. Meanwhile, researchers must be content with visual details on the Taurid stream as a whole, as given in the following table.

Taurid Magnitudes and Trains

Year(s)	Ave. Mag.	# Meteors	% Trains	Observer(s)	Source
1976	2.72	---	1.3	McLeod	*MN*, No. 36
1976	3.47	---	2.0	Martinez	*MN*, No. 36
1976	2.44	---	3.4	Matous	*MN*, No. 36
1978	2.33	284	10.9	WAMS	*MN*, No. 45
1983	2.92	206	4.4	WAMS	*WGN*, **12**, No. 3
1984	2.89	75	4.0	Miskotte	*WGN*, **13**, No. 1
1985	2.27	---	4.2	WAMS	*MET*, **16**, No. 4
1985	2.93	59	---	Roggemans	*WGN*, **13**, No. 6
1985	2.76	83	6.0	Finland	*WGN*, **14**, No. 2
1985	3.46	50	---	Hillestad	*WGN*, **14**, No. 2

Observers in the previous table were Norman W. McLeod, III (Florida), Felix Martinez (Florida), Bert Matous (Missouri), Western Australia Meteor Section, Koen Miskotte (The Netherlands), Paul Roggemans (Belgium), 15 observers in Finland, Trond Erik Hillestad (Norway).

Estimates of the hourly rates of the Taurids reveal fairly weak activity which hardly allows the shower to stand out among the other active showers and the usual numbers of sporadic meteors. In 1980, Norman W. McLeod, III (Florida) saw the highest hourly rates reach 5-8 for the Southern Taurids and 2-4

---

[98]Rendtel, Jürgen, *WGN*, **12** (April 1984), p. 44.

for the Northern Taurids during November 5-7.[99] That same year, members of the Western Australia Meteor Section found the Southern Taurids to reach a maximum ZHR of 14.21±1.89 on November 6, while the Northern Taurids reached a maximum of 6.57±1.15 on November 10.[100] Members of AK Meteore (East Germany) observed the Taurids during 1983. According to Jürgen Rendtel, they noted the Southern Taurids to have began October with ZHRs around 2. This rate persisted until around October 20, when the ZHR climbed to over 3. After a maximum ZHR of about 4 was attained during October 27-November 9, activity levels quickly dropped to around 2 following November 11. For the Northern Taurids, the 1983 ZHR levels were typically between 2-4 from October 1 to December 4. A maximum ZHR of 4-6 came during October 27-November 9.[101]

The Western Australia Meteor Section has provided excellent determinations of the color of the Taurid meteors in recent years. During 1983, 206 meteors were observed, with the predominant colors being determined as white (46.7%) and yellow (44.0%). Other observed colors included orange (5.3%), blue (2.7%) and green (1.3%).[102] In 1978, they reported yellow and blue-green percentages of 33.1% and 6.3%, respectively,[103] while observations in 1979 revealed yellow and orange percentages of 23.2% and 4.3%, respectively.[104]

Orbit

The following orbits were determined by the Author using photographic meteors collected from W1954, MP1961, B1963 and BK1967. The Northern Taurids are based on 20 meteors, while the Southern Taurids are based on 25.

	ω	Ω	i	q	e	a
N.	295.6	224.2	3.2	0.343	0.843	2.186
S.	114.3	27.3	5.0	0.370	0.806	1.910

During 1960, Kashcheyev and Lebedinets operated radio-echo equipment at the Kharkov Polytechnical Institute (USSR) and established the following orbits for the northern and southern branches of the Taurid stream. Since the equipment was not operated after October 23 (thus missing the maximum of each stream) the values given for the ascending nodes are probably considerably smaller than they should be.

---

[99]*MN*, No. 53 (April 1981), p. 5.

[100]Wood, Jeff, Personal Communication (October 15, 1986).

[101]Rendtel, Jürgen, *WGN*, **12** (April 1984), p. 44.

[102]Wood, Jeff, *WGN*, **12** (June 1984), p. 75.

[103]*MN*, No. 45 (April 1979), p. 12.

[104]*MN*, No. 49 (April 1980), p. 6.

	ω	Ω	i	q	e	a
N.	294.6	205.4	5.5	0.36	0.84	2.17
S.	118.2	14.6	2.2	0.33	0.84	2.08

As noted earlier, the northern and southern streams were not separated by the very sensitive radio-echo surveys of the Radio Meteor Project. Thus, the following orbits should be considered as averages for the Taurid stream as a whole.

	ω	Ω	i	q	e	a
S1970	114.1	49.7	1.4	0.385	0.770	1.679
S1976	293.6	217.2	0.0	0.398	0.750	1.596

As can be seen, it could be possible to argue that the southern stream had the greater influence on the first orbit, while the second orbit was influenced for the northern stream.

The orbits of comet Encke and the Beta Taurids are given here for comparison. The orbit of the Beta Taurids comes from S1973.

	ω	Ω	i	q	e	a
Encke	186.0	334.2	11.9	0.340	0.847	2.218
β Taurids	239.2	274.5	0.3	0.274	0.834	1.653

Additional November Showers

Alpha Pegasids

The Alpha Pegasids were discovered in 1959, while Richard E. McCrosky and Annette Posen were examining the orbits of meteors photographed during the Harvard Meteor Project of 1952-1954. Three meteors were found, all of which had been photographed on November 12, 1952 (the date had been incorrectly published as November 11). McCrosky and Posen referred to this stream as the "Mu Pegasids" and gave the average radiant as $\alpha=340°$, $\delta=+23°$.[105] The orbit was given as follows:

ω	Ω	i	q	e	a
199.4	229.7	8.3	0.965	0.778	4.35

McCrosky and Posen suggested an association with the lost periodic comet Blanpain (1819 IV), based on the similarity of the longitude of perihelion. They said the "comet was subject to repeated perturbations by Jupiter, so it is not unreasonable that the nodes of the orbit of the meteor stream should be the reverse of those of the original comet orbit..." The orbit of the comet was

ω	Ω	i	q	e	a
350.1	77.4	9.1	0.892	0.699	2.96

---

[105]MP1959, p. 26.

In 1971 Bertil-Anders Lindblad conducted a computerized stream search using 2401 photographic orbits detected during the Harvard Meteor Project. In addition to the three meteors originally detected by McCrosky and Posen, two additional meteors were added: one from November 12, 1952, and one from October 29, 1953. The five meteors revealed a duration of October 29-November 12 and an average radiant of α=344°, δ=+19°.[106] The following orbit was revealed.

ω	Ω	i	q	e	a
200.2	227.0	6.8	0.966	0.718	3.512

The strongest support for this stream's existence obviously comes from the photographic data and the suggestion of a link to comet Blanpain. On the other hand, since four of the five photographic meteors were detected on November 12, 1952, Lindblad suggested "caution should perhaps be exercised" when considering this stream as an annual shower producer. These words have taken on a stronger meaning to the Author as he attempted to locate visual observations of this shower. Although the very first observation of this radiant seems to have been made by R. de Kövesligethy (Hungary) on November 9, 1885, when five meteors were plotted from α=344°, δ=+19°,[107] no trace of the radiant is present among over 6000 American Meteor Society radiants, made during 1900-1984 and over 5400 German observations made during 1908-1938 (H1948).

In recent years a negative report came from Norman W. McLeod, III (Florida), who observed only one *possible* meteor from this stream during nearly 26 hours of observation between November 4 and 17, 1978.[108] On the other hand, a positive report came from members of the Western Australia Meteor Section, who observed a maximum ZHR of 2.20±0.12 from α=336°, δ=+19° during November 1-2, 1980.[109]

---

[106]L1971B, 16-17.

[107]D1899, p. 285.

[108]*MN*, No. 39 (January 1978), p. 6.

[109]Wood, Jeff, Personal Communication (October 15, 1986).

Chapter 12: December Meteor Showers

Coma Berenicids

Observer's Synopsis

This shower's duration extends from December 8 to January 23. The precise date of maximum is not known, but probably falls within the period of December 20-29, at which time the radiant is at $\alpha=165°$, $\delta=+30°$. The stream's activity is very weak, but numerous meteors have been photographed in the United States and the Soviet Union.

History

The discovery of this meteor shower came about as a result of stream searches among the photographic meteors detected during the Harvard Meteor Project of 1952-1954. First mention of this stream's existence was made by Richard E. McCrosky and Annette Posen in 1959. Their find was based on 6 photographic meteors which indicated a duration of January 13 to 23.[1] In 1973, Allan F. Cook, Bertil-Anders Lindblad, Brian G. Marsden, McCrosky, and Posen isolated 7 photographic meteors from the same collection of Harvard meteors that indicated another stream with a duration of December 12 to 17, which they called the December Leo Minorids. They commented that the "orbit bears a strong resemblance to that of the Coma Berenicids in January."[2]

Indeed the orbits of both streams are so similar that their mutual existences cannot be due to mere chance. Unfortunately, a gap in the photographic records that spans the period of December 17 and 29 eliminates data that could have provided a vital link between these two streams. In addition, there have been no coordinated efforts to study this shower more completely since the stream's announcement—the only available observations being accidental prediscovery sightings made while observing other showers of December and January. The Author believes these "accidental" sightings not only provide proof that the two

---

[1]MP1959, p. 26.
[2]C1973, pp. 1-5.

streams are one and the same, but probably point to a late December or early January maximum. A few of these observations are listed below.

*During December 20-21, 1878, T. W. Backhouse (Sunderland, England) observed seven meteors from a radiant located at α=157°, δ=+35°.[3]

*On December 29, 1940, Professor Mohd. A. R. Khan (Begumpet, India) plotted 6 meteors from a radiant of α=171°, δ=+31°. Designated AMS radiant no. 4060, Khan's estimate of the radiant's reliability was given as "good".[4]

*On December 29.35, 1948, Jeremy H. Knowles (Marblehead, Massachusetts) plotted 4 meteors which seemed to originate from α=170°, δ=+22.5°. Designated AMS radiant no. 3036, Knowles estimated this as a "good" observation.[5]

What makes this minor meteor stream especially interesting is the similarity of its orbit to the orbit of an "unconfirmed" comet that was reported in 1912-1913. The comet was officially designated 1913 I and was discovered in the early morning hours of December 30, 1912, by B. Lowe, an amateur astronomer in South Australia. Lowe managed to continue observations on January 2, 4, 5, and 9, before the comet entered morning twilight; however, an unfortunate positional error in Lowe's announcement telegram, which had been sent to Adelaide Observatory on January 7, caused searches made on January 8 and 11 to reveal nothing. Using four positions that could be derived from Lowe's rough observations, both A. C. D. Crommelin and M. Viljev were able to independently compute nearly identical orbits, except for a variation of 40° in the orbital inclination. *[Recent calculations by the Author reveal Viljev's orbit to more precisely represent the observed positions.]*

First mention of this curious relationship was made in 1954 by Fred L. Whipple. In an article showcasing 144 photographic meteor orbits obtained during 1936 to 1951, by Harvard College Observatory cameras, Whipple identified one meteor (designated 1918) with an orbit so similar to that of 1913 I, so as to make a comet-meteor association "fairly probable except for an uncertainty in the inclination of the comet's orbit."[6] Whipple added that "if future meteor observations confirm the association, we may learn more about the orbit of Comet Lowe by meteor studies than we can glean from the direct

---

[3]D1899, p. 256.

[4]Olivier, Charles P., *PA*, **49** (December 1941), p. 550.

[5]Olivier, Charles P., *PA*, **57** (December 1949), p. 504.

[6]W1954, p. 211.

observation." The suggestion that the Coma Berenicids were associated with comet Lowe was also suggested by McCrosky and Posen in 1959.

Beginning with the original 13 meteors isolated in the earlier computerized stream searches, the Author began examining other sources and soon compiled a database consisting of 23 photographic meteors. As a whole, these meteors indicate an orbital period of 72.4 years, but what is most striking is that this orbit seems to be composed of two fairly distinct filaments with periods of 27.3 years and 157.1 years. In addition, the orbital inclinations differ very little, while the perihelion distances vary by 0.1 AU and the argument of perihelion differs by 13°. A comparison is given in the "Orbit" section below.

Recent observations have been made by amateur astronomers in Western Australia, but the failure to detect the stream annually indicates either an irregular distribution of this stream's meteors or very weak annual activity which is occasionally detected only by very diligent surveys. During the period of 1969-1980, Michael Buhagiar (Perth, Western Australia) plotted 20,974 meteors for the purpose of compiling a list of southern hemisphere meteor radiants. On two occasions during those 12 years, he detected a radiant near Delta Leonis. Given a duration of December 20-23 and a maximum of December 21, the activity radiated from α=167°, δ=+27° with a maximum hourly rate of 2.[7] The Western Australia Meteor Section, led by Jeff Wood, managed to detect activity from this radiant during January 5-6, 1980. A maximum ZHR of 6.44±3.22 was detected on January 6 from α=171°, δ=+24°.[8]

Orbit

From 23 photographic meteor orbits obtained from W1954, HS1958, H1959, HS1961, JW1961, MP1961, B1963 and BK1967 the following orbit was derived by the Author for the Coma Berenicids.

ω	Ω	i	q	e	a
264.9	277.8	134.4	0.547	0.969	17.379

As mentioned in the text, these photographic meteors can also represent two fairly distinct orbits with the following elements:

ω	Ω	i	q	e	a
260.2	279.7	135.4	0.582	0.980	29.115
273.6	274.3	132.5	0.481	0.947	9.075

The orbit of comet Lowe (1913 I), as computed by M. Viljev in 1913,[9] is

	ω	Ω	i	q	e	a
1913 I	280.3	304.2	120.5	0.405	1.0	∞

---

[7]Buhagiar, Michael, *WAMS Bulletin*, No. 160 (1981).

[8]Wood, Jeff, Personal Communication (October 15, 1986).

[9]*The Observatory*, **36** (August 1913), p. 347.

Geminids

Observer's Synopsis

This meteor shower is active during the period December 6 to December 19. Upon reaching maximum activity during December 13 to 14 (λ=261.3°), hourly rates are typically near 80, while the radiant is at α=112.5°, δ=+32.6°. The daily motion is about +0.83° in α and –0.28° in δ. The meteors are described as rapid and yellowish, with about 4% displaying persistent trains. They possess an average magnitude near 2.4.

History

The appearance of this meteor shower seems to have been fairly sudden during the 1860s. It was first noted in 1862, when Robert P. Greg (Manchester, England) found a radiant at α=100°, δ=+33° for the period of December 10-12.[10] B. V. Marsh and Prof. Alex C. Twining (United States) independently discovered the activity in 1862,[11] while Alexander S. Herschel noted one very probable radiant at α=105°, δ=+30°, during December 12 to 13, 1863, as well as three fireballs from near the same radiant in 1863 and 1864.[12] During the 1870s, observations of the Geminids became more numerous as astronomers realized a new annual shower had been discovered.

The first estimate of the strength of the Geminids came in 1877, when the hourly rate was given as about 14. The same rate was also given by observers in England during 1892, but it was noted that almost twice as many bright meteors were present than had been seen in 1877. In addition, the 1892 observations also revealed three radiants to be active in Gemini, with the most active being located near Pollux (Beta Geminorum). In 1896, English observers gave hourly rates near 23 and pointed out that the greatest activity came from near Castor (Alpha Geminorum). They also observed "a number of bright pale green meteors from the radiant...."[13] During the beginning of the 20th century, the hourly rate of the Geminid stream was being reported as 15 to 30 per hour—with the average being over 20.

Hourly meteor rates during the 1930s ranged from 40 to 70, and, although these rates continued to rise during the next 50 years, the increase was not as dramatic as it had been between 1890 and 1930. For the 1940s and 1950s, hourly rates averaged about 60. During the 1960s, they were near 65 and the

---

[10]King, A., *MNRAS*, **86** (1926), pp. 638-641.

[11]King, A., *MNRAS*, **86** (1926), pp. 638-641.

[12]D1899, p. 249.

[13]Mackenzie, Robert A., *Solar System Debris*. Dover: The British Meteor Society, 1980. pp. 28-30.

1970s brought rates near 80. Rates between 1980 and 1985 have ranged from 60 to 110.

As with other major showers, the first person to begin sorting through the visual data was William F. Denning. As early as 1885, Denning had evidence that the radiant moved slightly westward as each day went by, and, in 1923, he published a radiant ephemeris.[14] His analysis revealed a daily motion of +1.25° in α and –0.10° in δ. In June 1926,[15] Alphonso King essentially confirmed Denning's findings as he published an ephemeris that revealed a daily motion of +1.23° in α and –0.10° in δ. Despite the fact that two researchers had arrived at similar conclusions, the matter of the motion of the radiant came under fire in 1931. Vladimir A. Maltzev criticized King's conclusions—claiming an inadequate treatment of the basic data.[16] His subsequent ephemeris revealed a daily motion of +1.05° in α and –0.06° in δ. The general correctness of Maltzev's daily motion has since been confirmed by Allan F. Cook and Robert A. Mackenzie. Cook published a paper in 1973 which revealed a motion of +1.02° in α and –0.07° in δ after an examination of photographic meteors,[17] while Mackenzie correlated visual observations of the British Meteor Society and found a motion of +0.97° in α and –0.08° in δ.[18]

Visual observations have shown this shower to possess a very sharp peak of activity, with hourly rates remaining above a value of half the maximum for about two days.[19] Although visual evidence of this shower indicates activity persists from December 6 to 19, definite photographic members of this shower have been detected as early as December 4,[20] while radar studies have shown activity as early as November 30 and as late as December 29.[21]

One of the most complete studies of the average magnitude of the Geminid shower was conducted in 1982, by George H. Spalding. Using meteor magnitude estimates made by members of the British Astronomical Association during the period 1969 to 1980, Spalding showed that for solar longitudes of 254° to 255° (December 7) the magnitude is about 2.14. It brightens slightly to 1.63 by the time the sun reaches longitudes of 256° to 257° (December 9), then

---

[14]Denning, W. F., *MNRAS*, **84** (1923), p. 46.

[15]King, A., *MNRAS*, **86** (1926), pp. 638-641.

[16]Lovell, A. C. B., *Meteor Astronomy*. London: Oxford University Press, 1954, pp. 310-311.

[17]Cook, A. F., *Evolutionary and Physical Properties of Meteoroids*. Washington, D.C.: NASA, 1973, p. 186.

[18]Mackenzie, Robert A., *BMS Radiant Catalogue*. Dover: The British Meteor Society, 1981, p.38.

[19]Spalding, George H., *JBAA*, **92** (1982), pp. 227-233.

[20]MP1961, pp. 15-84.

[21]S1970, pp. 475-493.

proceeds to steadily fade to a magnitude of 2.41 at longitudes of 260° to 261° (December 13). Maximum occurs shortly thereafter, and the magnitude brightens during the next several days, so that by the time of solar longitude 265° to 266° (December 18), the average magnitude is near 1.60. Spalding said "in the two days before maximum there is a moderate concentration of small particles, but ... the Earth then moves into a region of larger particles."[22]

In 1984, P. B. Babadzhanov and Yu. V. Obrubov also stressed the correlation between the solar longitude and the magnitude of the meteors. According to their calculations, which they say agree with observations, they found that meteors of magnitude 6 reach maximum at a solar longitude 0.9° earlier than the maximum of meteors of magnitude 1. Meteors of magnitude –4 tend to reach maximum 1.3° later than meteors of magnitude 1.[23] This survey tends to confirm the British study, except for the fact that Spalding said the Geminids produced brighter meteors on December 9 than on the 7th or 13th.

Generalized estimates of the average magnitude have been published over the years, as well as the estimated percentages of meteors showing trains. Some examples are given in the following table.

Geminid Magnitudes and Trains

Year(s)	Ave. Mag.	# Meteors	% Trains	Observer(s)	Source
1971-1984	2.83	4325	----	McLeod	Personal Comm.
1950	2.62	50	----	Regina	*JRASC*, **46**, p. 37
1954	2.38	24	----	Montreal	*JRASC*, **49**, p. 171
1955	2.49	2951	----	Czechs	*BAC*, **9**, p. 13
1974	2.11	151	1.9	Simmons	*MN*, No. 25
1976	2.43	----	1.2	Martinez	*MN*, No. 36
1976	2.66	----	1.2	Matous	*MN*, No. 36
1980	2.29	449	----	Lunsford	Personal Comm.
1982	2.56	893	2.4	Lunsford	Personal Comm.
1982	2.10	1101	3.0	NMS	*WGN*, **12**, No. 2
1983	2.39	604	9.1	Lunsford	Personal Comm.
1983	2.78	3036	7.0	WAMS	*WGN*, **12**, No. 3
1985	2.87	4960	4.0	International	*WGN*, **14**, No. 2

The observers cited are Norman W. McLeod, III (Florida); Regina Astronomical Society (Canada); Montreal Centre of Royal Astronomical Society of Canada; 7 observers in Czechoslovakia; Karl Simmons (Florida); Felix Martinez (Florida); Bert Matous (Missouri); Robert Lunsford (California); Nippon Meteor Society and the Western Australia Meteor Section.

---

[22]Spalding, George H., *JBAA*, **92** (1982), pp. 230-231.

[23]Babadzhanov, P. B., and Yu. V. Obrubov, *SA*, **28** (Sept.-Oct. 1984), pp. 589-590.

A major advance in the understanding of the already mentioned intricacies of this meteor stream was made in 1947. Fred L. Whipple had been involved in the Harvard Meteor Project, a photographic survey aimed at better understanding meteors and their origins by obtaining data that could be used to calculate orbital elements. While analyzing meteors associated with the Geminids he found an orbital period of only 1.65 years, as well as a high eccentricity and a low inclination.[24] Such an orbit attracted the attention of Miroslav Plavec (Prague), who began investigating the effects of perturbations on the orbit.

Plavec found that only two planets effect the orbit of the Geminids—Earth and Jupiter, though the former was considered negligible compared to the effects of the giant planet. "From the observer's point of view," he wrote," the most important phenomenon is the rapid backward shift of the node." The degree of this shift was calculated to cause the date of maximum to occur one day earlier every 60 years. Another interesting conclusion involved the point of intersection between the stream's orbit and the ecliptic. For the year 1700, it was found that the intersection point was placed 0.1337 AU inside Earth's orbit. For 1900, the intersection point was located 0.0178 AU inside Earth's orbit and in 2100, the point would be 0.1066 AU *outside* of Earth's orbit.[25] Thus, Plavec not only showed why the activity of the Geminids was steadily increasing, but he also demonstrated that the activity would eventually decline and that sometime in the future Earth would no longer contact the stream's orbit.

Despite Plavec's calculations, the fate of the Geminid stream was still considered a matter that was up for grabs. In 1967, during the International Astronomical Union's Symposium No. 33, I. S. Astapovich and A. K. Terent'eva submitted a paper entitled "Fireball Radiants of the 1st-15th Centuries."[26] They discussed their determination of the radiants of 153 meteor showers. According to their findings, a total of 14 fireballs were detected between 1038 and 1099 AD from a radiant similar to the Geminids', while additional fireballs were noted in 381 and 1163. They remarked that the "fireballs of the 11th century gave a definite radiant α=103°, δ=+26° (December 6-18)." They said the 11th century radiant was situated south and east of the present radiant and concluded that the radiant indicated that "apparently there has occurred a secular increase of the orbital inclination and a change in the line of apsides." They added, "the node of the orbit remained practically unchanged in the course of nine centuries."

---

[24] Whipple, F. L., *Proceedings of the American Philosophical Society*, **91** (1947), p. 189.

[25] Plavec, Miroslav, *Nature*, **165** (March 4, 1950), pp. 362-363.

[26] Astapovich, I. S., and A. K. Terent'eva, *Physics and Dynamics of Meteors*. eds. Lubor Kresak and Peter M. Millman. Dordrecht: D. Reidel Publishing Company, 1968, pp. 308-319.

Controversy over the Geminids' past continued throughout the 1970s, though astronomers generally seemed to favor the work of Plavec. In 1982, Ken Fox, Iwan P. Williams and David W. Hughes published a paper entitled, "The evolution of the orbit of the Geminid meteor stream."[27] They essentially confirmed Plavec's findings of a nodal retrogression rate of about 1.6°/century, as well as his recognition of the relative newness of the shower in historical records—thus, eliminating the link to the fireballs of the 11th century (see the December Monocerotids). However, the confirmation of the nodal retrogression rate led to a problem: the observations did not confirm the predicted change in the date of maximum that amounted to one day in about 60 years. The authors theorized that the predicted change was actually being altered and proceeded to analyze several possibilities.

Since the orbit of the Geminid stream passed through the asteroid belt, the British researchers looked for an asteroid that may periodically pass near the stream's orbit. They found that asteroid 132 Aethra actually passed only 0.0003 AU from the Geminid orbit; however, they quickly discovered that for the asteroid to account for the variations noted would require it to possess a mass only slightly less than that of Jupiter! Another possibility was that of general relativity—an affect noted in several planetary orbits—but the result of the calculations was a slight *increase* in the nodal retrogression rate, rather than the expected slowing down.

The final possibility considered was "the shape of the cross-section of the intersection of the meteor stream with the ecliptic plane." A computer simulation predicted the meteor rate profile was skew. Fox, Williams and Hughes further elaborated on this distribution in a paper published in 1983. "At the present time the Geminid shower slowly builds up to maximum rate and then drops away from maximum relatively sharply. About 50 yr ago the skewness should have been exactly the opposite with a sharp build up to maximum rate and a much slower falling away." The proposed model indicated Earth's orbit would intersect the Geminid stream only between 1800 and 2100. It also explained the currently observed mass segregation within the stream.[28]

A major question concerning the Geminid stream involves its origin. It was long known that no parent comet for this stream was present in current catalogs, but, since the exact size and shape of the stream were not known until 1947, few conjectures were made. In 1950, Plavec theorized about the Geminid stream's parent body and pointed out that the "existence of a parent comet in such a short-period orbit, even in the past, seems to be not very probable. Planetary perturbations could scarcely have reduced the semimajor axis so much. More

---

[27]Fox, Ken, Iwan P. Williams and David W. Hughes, *MNRAS*, **199** (1982), pp. 313-324.
[28]Fox, Ken, Iwan P. Williams and David W. Hughes, *MNRAS*, **205** (1983), pp. 1155-1169.

probably, the Geminids were separated from a parabolic comet by the close approach of the comet to the sun."[29] Concerning a possible candidate for the parabolic comet mentioned, Plavec considered the great comet of 1680 (after a suggestion made in 1931 by Maltzev) and concluded that the close approach of the two orbits at a point slightly beyond the Geminid perihelion point, made a possible connection impossible to exclude.

Lubor Kresak strengthened the comet link to this meteor stream's formation, but instead of offering a theory as exotic as Plavec's, he favored a more direct formation of the Geminids. In 1972, he wrote that the parent comet "must have previously occupied the present orbit."[30] He stressed that the compact nature of the stream would eliminate the possibility of it having formed in a different orbit and then been perturbed into the present orbit. Eleven years later, Kresak's theory would gain considerable strength.

On October 11, 1983, during a search for moving objects amidst the data gathered by the Infrared Astronomical Satellite (IRAS), Simon Green and John K. Davies found a rapidly moving asteroid in Draco. The next evening, Charles Kowal (Palomar Observatory, California) confirmed the body by photographing it with the 48-inch Schmidt telescope. The asteroid received the preliminary designation 1983 TB.

As early orbital calculations were being made, the International Astronomical Union *Circular* for October 25, 1983, relayed the opinion of Fred L. Whipple that this asteroid possessed an orbit almost identical to that of the Geminid meteor stream.[31] Additional observations confirmed the link and the asteroid eventually received the permanent designation of 3200 Phaethon. The excitement of having found the parent body of the Geminid stream was almost dwarfed by another realization, this was the first time an asteroid had been definitely linked to a meteor shower and it subsequently serves as an important link between comets and meteor streams.

Orbit

Based on 37 photographic meteors obtained from W1954, B1963, TH1976 and D1980, the following orbit was revealed:

ω	Ω	i	q	e	a
324.1	261.3	23.9	0.143	0.896	1.612

The orbit of Apollo asteroid 3200 Phaethon is as follows:

ω	Ω	i	q	e	a
321.7	265.0	22.0	0.139	0.890	1.434

---

[29]Plavec, pp. 362-363.

[30]Kresak, L., *SSR*, **6** (1972), p. 65.

[31]Marsden, Brian G., *IAU Circular*, No. 3881 (October 25, 1983).

Sigma Hydrids

Observer's Synopsis

This shower is visible during the period of December 4 to 15. It reaches maximum on December 11 from an average radiant of α=127°, δ=+2°. Its typical ZHR at maximum is usually around 3 to 5.

History

The discovery of this stream should be attributed to Richard E. McCrosky and Annette Posen. During 1961 they published a list of 2529 photographic meteor orbits that had been computed from double-station photographs obtained during the Harvard Meteor Project of 1952-1954. They identified 7 of these orbits as indicating a stream which produced a radiant near Sigma Hydri.[32] Later that year, Luigi G. Jacchia and Fred L. Whipple published an analysis of 413 precise photographic orbits also obtained during 1952-1954 and identified 3 meteors as belonging to this stream.[33] The average radiant indicated by the precise orbits was α=126.7°, δ=+1.8°.

Confirmation of a December shower from this radiant was made by observers at Waltair, India, during a survey for minor meteor showers conducted by M. Srirama Rao, P. V. S. Rama Rao and P. Ramesh during 1961-1967.[34] During December 12-15, 1963, 5 meteors were detected from α=127°, δ=+4°. It was calculated that the radiant had an hourly rate of 6.6. During December 11-14, 1964, the radiant was again detected. On this occasion 31 meteors were detected from α=123°, δ=+7°. The hourly rate was given as 6.0. Together, these two radiants formed what was designated as Minor Shower No. 5.

A search through previously published accounts listing visual radiants reveals occasional indications of this shower's presence in the latter half of the 19th century and well into the 20th century. But the most striking series of observations extracted from these older records, are the 1937 observations made by Cuno Hoffmeister in Southwest Africa (these are radiants 4582, 4612, 4621, 4642 and 4668 in H1948). Between December 6 and 13, five radiants were noted on various days which possess D-criterions that indicate certain membership with the Sigma Hydrid stream. The average position of these radiants is α=129°, δ=+3°.

Recent observations of this meteor shower have revealed some interesting data. Members of the National Association of Planetary Observers Meteor Section in Australia have revealed activity to occur during the period December 3

---

[32]MP1961, pp. 78-83.

[33]JW1961, pp. 112-113.

[34]RR1969, p. 768 & 772.

to 19, with a maximum ZHR of 3-5 meteors radiating from α=128°, δ=+4° on December 11.[35] During 1978, the same group monitored this shower during December 2 to 10. They found the hourly rate to be highest on the latter date at 5. Based on the 49 meteors observed, the average magnitude was determined as 3.06, while 6.1% of the meteors left persistent trains. The meteors were said to have a "Perseid type velocity" and 21.4% were yellow.[36]

On the whole, northern hemisphere observers have rarely observed rates greater than 1 meteor per hour during 1978 to 1984. However, during 1983, Norman W. McLeod, III noted rates rose to 6 per hour on December 8/9, and was still at 4 per hour on December 10/11.[37] During December 10/11-13/14, 1982, Robert Lunsford (California) determined the average magnitude of the Sigma Hydrids as 3.6.[38]

Orbit

Based on 8 photographic meteors obtained from W1954, HS1958, HS1961, JW1961, MP1961 and M1976, the following orbit was obtained:

ω	Ω	i	q	e	a
120.6	78.9	127.5	0.248	0.991	26.053

Using an orbit similar to the one above, Ken Fox (Queen Mary College, England) projected the orbit of this stream backward and forward for 1000 years.[39] The following two orbits were obtained:

	ω	Ω	i	q	e	a
950	120.5	78.1	125.6	0.25	0.99	29.28
2950	121.5	81.0	124.0	0.25	0.99	30.13

The orbit of the Sigma Hydrids thus seems fairly stable. In 950, the date of maximum was unchanged from that of the present, while the radiant position was α=125.8°, δ=+1.1°. In 2950, maximum will occur about one day later than at present and the radiant will be at α=127.0°, δ=+1.6°.

December Monocerotids

Observer's Synopsis

This shower has a duration which persists from November 9 to December 18. Maximum occurs around December 11, with the radiant then being at α=101°,

---

[35]Wood, Jeff, Personal communication (October 24, 1985).

[36]Wood, Jeff, *MN*, No. 45 (April 1979), pp. 11-12 & *MN*, No. 48 (January 1980), p. 7.

[37]*MN*, No. 65 (April 1984), p. 10.

[38]*MN*, No. 61 (April 1983), p. 7.

[39]F1986, pp. 523-525.

δ=+10°. Though some photographic data has been obtained, the overall visual rate of this shower is only 1 to 2 per hour; however, radar studies indicate the shower makes annual appearances, so that optical aid may be necessary to better observe the activity produced by this stream.

History

This stream was discovered by Fred L. Whipple in 1954, during a search through 144 meteor orbits detected during the photographic surveys of Harvard College Observatory during 1936 to 1951. Two meteors (designated 2313 and 2405) had been photographed on December 13 and 15, 1950, and possessed very similar orbits that indicated a radiant at α=103°, δ=+8°.[40] What made the find especially significant was the fact that the indicated orbit was very similar to that of comet Mellish (1917 I)—a comet with an orbital period of about 145 years (uncertainty of 0.6 years).[41]

In 1961, Richard E. McCrosky and Annette Posen, both of Harvard College Observatory, published a list of 2529 meteor orbits obtained during the Harvard Meteor Project during February 1952 to July 1954.[42] They found 3 meteors, photographed between December 10 and 17, that belonged to the December Monocerotid stream, thus offering the first confirmation that a stream with the orbital elements suggested by Whipple was present.

During the 1960's, the true nature of this stream began to be realized as four major radar surveys revealed further evidence supporting the stream's existence. The first was conducted by C. S. Nilsson at Adelaide, Australia, during the period of December 1960, to December 1961.[43] Although the computer technique designed to process the 2101 radar meteor orbits did not reveal a stream with the orbit determined by the photographic data, Nilsson studied the radiant data and located 2 meteors from December 1960, and 6 meteors from December 1961. The latter meteor orbits allowed a good determination of the stream's orbit. It also indicated a duration of December 5 to 12, with an average radiant of α=101.6°, δ=+9.6°.

Interestingly, Nilsson's computer technique did reveal a stream during both 1960 and 1961, which possessed a similar orbit to the photographic determinations, but with an inclination some 10° to 15° lower. This resulted in an average radiant of α=95°, δ=+15°—meaning it was shifted nearly 9° to the northwest. What made this stream particularly significant was the fact that it

---

[40]W1954, p. 210.

[41]Marsden, Brian G., *Catalog of Cometary Orbits*. New Jersey: Enslow Publishers, 1983, pp. 18 & 50.

[42]MP1961, pp. 78-84.

[43]N1964, pp. 226-229 & 251.

seemed to possess a twin during September 23-29, at α=162°, δ=+14°. Nilsson noted that "care must be taken not to confuse the separate and newly determined radiant at 95°, +15° with the Monocerotid radiant at 102°, +10°. The values of geocentric velocity are similar, and the orbit of the former is also near parabolic; however, the inclination of the orbit is definitely smaller than that of the Monocerotid stream." He added the the similarity of the orbits might suggest a connection.

In 1973, Zdenek Sekanina published the results of the Harvard Radio Meteor Project conducted at Havana, Illinois, during 1961-1965.[44] This project revealed a very distinct stream with an average radiant at α=91.4°, δ=+15.0°—very similar to the strong, lower inclination stream noted by Nilsson. Sekanina commented that "the photographic stream associated with P/Mellish has both its mean optical radiant and its theoretical radiant in Monoceros, but the mean radiant of the rich radio stream detected in our sample, which is beyond any doubt also related to the comet, lies in Orion...." A further surprise revealed by this radar study was that the indicated duration of this stream was from November 9 to December 18. During 1968-1969, Sekanina revived the Radio Meteor Project,[45] this time with a greater sensitivity than that earlier attained, and among the 275 streams detected was the same stream detected during 1961-1965. The indicated duration was November 16 to December 14 and the average radiant was α=93.8°, δ=+14.4°. Due to the similarity of these streams to the meteor orbits found in the photographic studies, Sekanina referred to the streams as the Monocerotids.

Up to this point the radio meteor surveys seemed to indicate that the most prominent shower in this part of the sky during the first half of December was 9° away from the radiant indicated by photographic meteors; however, in 1975, G. Gartrell and W. G. Elford published their results of a radar survey conducted during 1968-1969.[46] They wrote that "although no Monocerotids were found to be associated by the systematic stream searches, three meteors were observed with radiants and velocities corresponding to this stream." Their radiant was α=106°, δ=+6°.

The first study of this meteor stream was conducted by M. Kresáková during 1974, as she studied comet Mellish and its possible relationship to the Geminids.[47] While studying comet Mellish and the lists of photographic meteors, it was noted that two apparent minor streams were present. The first, designated "component A", possessed nearly identical orbital elements to the

---

[44]S1973, pp. 254, 257, & 260.

[45]S1976, pp. 290 & 302.

[46]GE1975, pp. 596 & 603.

[47]Kresáková, M., *BAC*, 25 (1974), pp. 20-33.

Monocerotids first suggested by Whipple. It was based on 16 meteors acquired from both American and Russian sources. The second stream was composed of 9 photographic meteors and was identified with the 11 Canis Minorids first announced by Keith B. Hindley in 1969 (see the 11 Canis Minorids later in this chapter). The Monocerotids of Kresáková lay very close in radiant position to that predicted for P/Mellish.

A preliminary examination of previously known data by the author had seemed to indicate two distinct streams being present with a separation of about 10° in inclination; however, the photographic meteors listed by Kresáková and the original radar orbits acquired during the two Radio Meteor Projects directed by Sekanina, seem to show no sign of distinct streams, but, instead, one very diffuse stream. Nevertheless, one cannot ignore the fact that photographic and radar studies tend to indicate distinctly different orbital inclination averages. Thus, it would seem that the fainter and smaller radar meteors tend to possess more lower inclination orbits than the brighter, larger photographic meteors. Meteor showers visible to the naked eye will probably originate from the photographic radiant, while telescopic meteors will tend to lean toward the radar radiant.

Based on the number of meteors detected both by photographic and radio-echo techniques, it would seem the Monocerotid shower may be stronger telescopically than visually. Coordinated attempts to observe this stream visually have never been conducted, but *Meteor News* published details of many individual attempts to observe this shower between 1977 and 1985 (see issues 40, 49, 53, 57, 61, 65, 68 and 73). Typical observations indicate hourly rates of only one to two per hour. During December 11/12-13/14, 1982, Robert Lunsford (San Diego, California) observed ten Monocerotids and gave the average magnitude as 3.2.[48]

Surprisingly, the December Monocerotids may be responsible for the solving of a major puzzle in meteor astronomy. During the International Astronomical Union's Symposium No. 33 in 1967, I. S. Astapovich and A. K. Terent'eva submitted a paper entitled "Fireball Radiants of the 1st-15th Centuries."[49] They discussed their determination of 153 meteor radiants and pointed out a collection of 14 fireballs that emanated from a radiant similar to that of the Geminids during the period 1038 to 1099 AD. They remarked that the "fireballs of the 11th century gave a definite radiant α=103°, δ=+26° (December 6-18), the early fireball of 381, December 13 passing 5° to the South of it."

---

[48]*MN*, No. 61 (April 1983), p. 7.

[49]Astapovich, I. S., and A. K. Terent'eva, *Physics and Dynamics of Meteors*. eds. Lubor Kresak and Peter M. Millman. Dordrecht: D. Reidel Publishing Company, 1968, pp. 308-319.

Their linking of this fireball radiant to the Geminid shower has caused much controversy (see the Geminids earlier in this chapter), but in 1985, Ken Fox and Iwan P. Williams offered a reasonable solution.

Beginning with an orbit possessing the angular elements determined by Kresáková and borrowing the semimajor axis of P/Mellish, Fox and Williams examined the orbital evolution of a particle between 783 AD and 3183 AD.[50] Perturbations from Jupiter, Saturn, Uranus and Neptune were taken into account. Their results were that the ascending node varied by less than 5° during the 2400 years examined, while the nodal distance from Earth's orbit "remained fairly constant." The final conclusions of the study indicated that "a Monocerotid shower would still be seen during December in the eleventh century," as well as 1000 years from the present time. Also, Fox and Williams showed that, of the currently known meteor showers, only the Monocerotids were capable of producing a shower close to the radiant of the eleventh century fireballs, though "there is always the possibility that the ancient fireballs came from a stream which is not observable at present."

Orbit

Based on 12 of the 16 photographic meteors listed by Kresáková, the following orbit was determined:

	ω	Ω	i	q	e	a
	133.7	74.4	29.4	0.154	1.000	∞

The most precise radar study conducted to date has been the 1968-1969 Radio Meteor Project conducted by Sekanina. It was based on 30 radio meteors.

	ω	Ω	i	q	e	a
S1976	135.8	71.8	22.3	0.153	0.975	6.199

For comparison purposes, the orbit of comet Mellish (1917 I) is given below. It was calculated by Asklöf in 1932, and is based on 170 observations. With a period of 145 years, it has an uncertainty of 0.6 year.[51]

	ω	Ω	i	q	e	a
1917 I	121.3	88.0	32.7	0.190	0.993	27.647

Using the orbit of comet Mellish, Ken Fox (F1986) computed the orbit of the Monocerotid stream for periods 1000 years in the past and future with the following results.

	ω	Ω	i	q	e	a
950	121.8	86.9	32.8	0.19	0.99	27.29
2950	119.9	90.0	32.4	0.19	0.99	27.30

---

[50]Fox, Ken, and Iwan P. Williams, *MNRAS*, **217** (1985), pp. 407-411.

[51]Marsden, Brian G., *Catalog of Cometary Orbits*. New Jersey: Enslow Publishers, 1983, pp. 18 & 50.

In 950, the shower's maximum would have occurred only one-half day earlier than at present from a radiant of α=102.0°, δ=+6.0°. In 2950, maximum will occur 1.1 days later than at present, with the radiant being α=103.3°, δ=+5.2°. As can be seen, the orbit of comet Mellish is very stable, so any associated stream must also be stable.

Chi Orionids

Observer's Synopsis

The Northern Chi Orionids are active during November 16-December 16, with a maximum ZHR of two coming on December 10 (λ=258°) from α=82°, δ=+23°. The Southern Chi Orionids are active during December 2-18, with a maximum ZHR of three coming on December 10 (λ=258°) from α=88°, δ=+20°. The meteors of both streams tend to be bright, with about 14% leaving trains.

History

The Chi Orionids have been around for at least 100 years, though its weak activity at a time when the Geminids are nearing prominence should probably be blamed for its frequently being overlooked by observers. As early as the 1890's, William F. Denning included this shower in his annual list of active radiants published in *The Observatory*. The most important discovery came during the photographic surveys of the 1950's, when the Chi Orionids became one of several annual meteor showers recognized as split into northern and southern branches.

The earliest observations of activity from this stream extend back to the last quarter of the 19th century, when Denning (Bristol, England) observed activity during 1876 and 1877. In the former year he plotted 13 meteors during December 6-7 emanating from a radiant of α=80°, δ=+23°. In the latter year, 6 meteors were plotted during December 6-8 from α=80°, δ=+25°. The meteors were described as slow on both occasions. The first person to confirm Denning's radiant was E. F. Sawyer (Cambridge, Massachusetts), who plotted 5 meteors during December 7-12, 1879 from α=82°, δ=+23°.[52]

Similar observations continued to be made during the 1880's and 1890's, but this was clearly referring to the Northern Chi Orionids. The Southern Chi Orionids were not in the same degree of prominence with only two apparent 19th century observations: Denning plotted 11 meteors during November 30-December 10, 1885 from a radiant of α=88°, δ=+19°, while Denning combined the plots of E. Weiss (Hungary) and Giuseppe Zezioli (Bergamo, Italy), and

---

[52]D1899, p. 245.

found 17 meteors revealing a radiant of α=86°, δ=+19° for the period of December 9-13.[53]

The Southern Chi Orionids were not well observed during the first half of the 20th century, but numerous observations were made of the northern branch. Most notable was the inclusion of the Northern Chi Orionid radiant in a table of "third grade radiants" in Cuno Hoffmeister's 1948 book *Meteorströme*. Ten radiants were isolated which were active around December 11 (λ=259°) from an average radiant of α=83°, δ=+22°.[54] The Author's examination of these radiants revealed an average diameter of 3°.

One of the more important events in the acquisition of knowledge about the Chi Orionids was the Harvard Meteor Project of 1952-1954. A few early analyses of this photographic data (JW1961 and MP1961) revealed members of the stream, but it was the computerized analysis of Bertil-Anders Lindblad (Lund Observatory, Sweden) which provided the most complete data.

The Southern Chi Orionids were best represented among the data collected by Lindblad. Eight meteors indicated a duration of December 7-14 and an average radiant of α=85°, δ=+16°. The Northern Chi Orionid data consisted of four meteors indicating a duration of December 4-13 and an average radiant of α=83°, δ=+26°.

The Author has further examined the photographic meteor orbits obtained during the 1940's, 1950's, and 1960's, in the United States and Soviet Union, and has arrived at more refined orbits for each branch of the Chi Orionid stream using stricter values of the D-criterion than used by Lindblad. The southern branch was again best represented with 12 meteors. The stream's duration was December 4-14, while the average radiant was α=83.9°, δ=+16.8°. Curiously, the stricter D-criterion shortened the duration of the northern branch to December 11-17. The average radiant was α=84.4°, δ=+26.1°.

Radio-echo surveys conducted during the 1960's were successful in detecting the Chi Orionids, but not in isolating the northern and southern branches. C. S. Nilsson (Adelaide Observatory, South Australia) detected four radio meteors during December 7-12, 1960, and gave the average radiant as α=80.0°, δ=+16.8°.[55] Zdenek Sekanina detected the Chi Orionids during both sessions of the Radio Meteor Project. During the 1961-1965 session, the duration was given as December 2-18. The date of the nodal passage was given as December 9.9 (λ=257.5°), at which time the radiant was at α=87.1°, δ=+20.6°.[56] The 1968-1969 session indicated a duration of November 16-

[53]D1899, p. 246.

[54]H1948, p. 82.

[55]N1964, pp. 226-229 & 233.

[56]S1973, pp. 257 & 260.

December 16. The date of the nodal passage was given as December 9.9 (λ=257.5°), at which time the radiant was at α=81.5°, δ=+23.4°.[57]

The Western Australia Meteor Section (WAMS) has recently provided very complete visual observations of both the northern and southern branches of the Chi Orionid stream.[58]

> *1977—The Northern Chi Orionids were detected during December 2-13. A maximum ZHR of 2.84 came on the 11th from α=84°, δ=+25°. The Southern Chi Orionids were detected during December 9-14. A maximum ZHR of 3.16 came on the 9th from α=85°, δ=+14°.
>
> *1979—The Northern Chi Orionids were detected during December 7-16. A maximum ZHR of 2.30 came on the 12th from α=88°, δ=+22°. The Southern Chi Orionids were detected during December 7-13. A maximum ZHR of 1.91 came on the 10th from α=85°, δ=+13°.
>
> *1980—The Northern Chi Orionids were detected during December 4-14. A maximum ZHR of 2.17 came on the 11th from α=83°, δ=+25°. The Southern Chi Orionids were detected during December 5-15. A maximum ZHR of 4.05 came on the 10th from α=88°, δ=+12°.

The WAMS also detected the Chi Orionid streams during 1978. By combining the 48 meteors observed, Jeff Wood (section director) was able to provide additional details about the Chi Orionid meteors in general. The stream seems to produce fairly bright meteors with an average magnitude of 1.60, while 14.3% of the meteors produced trains. Concerning colors, it was determined that 31.4% of the meteors were yellow, while 4.5% were blue or green.[59]

Orbit

The Author has collected photographic meteor orbits from W1954, MP1961 and BK1967. The Northern Chi Orionid orbit was based on 5 meteors, while the Southern Chi Orionid orbit was based on 12 meteors.

	ω	Ω	i	q	e	a
N.	275.5	261.0	2.4	0.528	0.739	2.02
S.	101.5	77.9	6.0	0.460	0.798	2.28

---

[57]S1976, pp. 290 & 302.

[58]Wood, Jeff, Personal Communication (October 15, 1986).

[59]Wood, Jeff, *MN*, No. 45 (April 1979), p. 12.

The stream was also detected during three radio-echo surveys of the 1960's, though the northern and southern branches could not be separated. The first two orbits were obviously influenced by the Southern Chi Orionids, while the last orbit was influenced by the Northern Chi Orionids.

	ω	Ω	i	q	e	a
N1964	93.7	78.6	4.6	0.556	0.70	1.852
S1973	109.0	77.5	2.6	0.420	0.765	1.790
S1976	278.7	257.5	0.2	0.515	0.711	1.783

Phoenicids

Observer's Synopsis

The duration of this shower extends from November 29 to December 9. Estimates of the ZHR seem to remain above one during December 2-7, while a sharp rise to five occurs on December 5 (λ=253°). The radiant is located at α=15°, δ=–52° at the time of maximum. The meteors possess an average magnitude of 3.27, while only 2% leave trains. The shower is most notable for producing a rate of 100 meteors per hour in 1956—the year marking its discovery. The stream is probably produced by the lost period comet Blanpain (1819 IV).

History

On December 5, 1956, observers in New Zealand, Australia, the Indian Ocean, and South Africa detected strong activity from the southern constellation Phoenix. Investigations by H. B. Ridley revealed maximum visual hourly rates of 100 and a radiant at α=15°, δ=–45°,[60] while C. A. Shain estimated visual hourly rates greater than 60 and a radiant of α=15°, δ=–58°.[61]

A complete investigation of the visual observations was made by Ridley during 1962. He said the activity was first detected by R. Lynch (Auckland, New Zealand) at 10:10 UT, and that it continued to become visible "as sunset progressed westwards," until last detected by S. C. Venter (Pretoria, South Africa) at 22:45 UT. Ridley's analysis could not clearly pinpoint the time of maximum, but he estimated "it must have lain somewhere between 17^h and 21^h UT." For orbital calculations, he adopted 19^h UT (λ=253° 33') as the time of maximum. One of the most impressive features of the shower was the apparent presence of exploding fireballs, as observers were frequently comparing the meteors "to the Moon, Venus, Jupiter, Sirius, etc." The meteors were also

[60]Ridley, H. B., *BAA Circular*, No. 382 (1957).

[61]Shain, C. A., *The Observatory*, 77 (1957), p. 27.

frequently described as reddish and yellow.[62] From a table listing the magnitude distribution of 61 meteors, the Author finds an average magnitude of 2.39.

Radio-echo observations were also made from Adelaide Observatory (South Australia) on December 5, 1956, with A. A. Weiss determining a maximum rate of 30 per hour and a radiant of $\alpha=15°\pm2°$, $\delta=-55°\pm3°$. The hourly rate was considered a puzzle by Weiss. For major showers, such as the Delta Aquarids and Geminids, the radio-echo rate was always greater than the visual rate, while the radio-echo Phoenicids were only one-third the strength shown by visual observations. One possible theory put forth was that "Earth was still on the fringe of the stream" when radio observations were made. Another suggested explanation was the possibility that slow meteors possessed a lower ionizing efficiency that faster meteors—a condition never satisfactorily tested prior to the appearance of the Phoencids.[63]

A search through the major catalogs of Southern Hemisphere meteor stream radiants (M1935, H1948) reveals no prior observation of the Phoenicids. For the 30 years following the shower's sudden appearance in 1956, no return of the prominent activity was noted; however, visual observations were been made on numerous occasions since the early 1970's. According to Robert A. Mackenzie (director of the British Meteor Society), activity was detected by observers in the Southern Hemisphere during 1972, 1973, 1976, and 1977, with maximum ZHRs of 4, 5, 2, and 5, respectively.[64]

The Western Australia Meteor Section (WAMS) has made excellent observations of this stream in recent years. In 1977, Phoenicids were noted during December 2-5, with a maximum ZHR of 4.68±1.91 coming on December 5 from a radiant of $\alpha=15°$, $\delta=-57°$. In 1979, activity was observed during December 5-6, with a maximum ZHR of 5.67±4.01 coming on December 5 from $\alpha=15°$, $\delta=-51°$. In 1980, activity was detected during November 29-December 9, with a maximum ZHR of 2.58±0.37 coming on December 4 from $\alpha=17°$, $\delta=-52°$.[65] During 1983, 17 amateur astronomers compiled 62 man-hours and detected Phoenicids during December 1-10. The ZHR was above 1 during December 2-7, with a maximum of 5.6±1.3 coming on the night of December 4/5. The meteors had an average magnitude of 3.27, while only 2% left trains.[66] Another extensive observation program was conducted in 1985, as 25 observers compiled 122 man-hours over 9 nights. The maximum ZHR reached 8, while the

---

[62]Ridley, H. B., *JBAA*, **72** (1962), pp. 267-268.

[63]Weiss, A. A., *AJP*, **11** (1958), pp. 114-117.

[64]Mackenzie, Robert A., *Solar System Debris.* Dover: The British Meteor Society (1980), p. 28.

[65]Wood, Jeff, Personal Communication (October 15, 1986).

[66]Wood, Jeff, *WGN*, **12** (June 1984), pp. 75-76.

average magnitude was 2.38 and 4.8% of the meteors possessed trains.[67] Finally, during 1986, 16 members of the WAMS observed for 40 man-hours during December 2-7, with a maximum ZHR of 4.6±1.6 coming on December 5/6. The average magnitude was 2.88, while 5.3% of the meteors left trains.[68]

One of the most intriguing aspects of the Phoenicid stream is its apparent link to the lost periodic comet Blanpain (1819 IV). The link was first recognized by Ridley, during 1957, after he had computed a parabolic orbit. Weiss essentially confirmed the results later that year, but the small difference between the orbits was caused by the respective radiants Ridley and Weiss chose as representing the shower's activity—the adopted radiants differing by 10° in declination. Ridley indicated that the small change in the longitude of perihelion between the stream and comet orbits, as well as the smaller ascending node of the stream, logically fit the expected evolution of a low-inclination comet orbit.[69] Unfortunately, the comet was only observed for 59 days during 1819-1820, so that an error of several months exists in the revolution period. This uncertainty, combined with probable perturbations by Jupiter, makes it quite impossible to establish how different the current comet's orbit is from the Phoenicid orbit.

Orbit

Ridley computed the first orbits for this stream based on a radiant of α=15°, δ= –45°. The first orbit is parabolic, but after noting a similarity to the orbit of the lost periodic comet Blanpain (1819 IV), Ridley computed the second orbit with an assumed period of 5.1 years.

ω	Ω	i	q	e	a
358.9	73.6	16.5	0.985	1.0	∞
358.7	73.6	13.4	0.987	0.667	2.963

Weiss computed the following two orbits based on the radio-echo determination of the radiant position as α=15°, δ=–55°. The first orbit parabolic, while the second orbit was computed with an assumed period of 5.1 years.

ω	Ω	i	q	e	a
0.1	73.4	19.3	0.985	1.0	∞
0.3	73.4	15.9	0.985	0.667	2.958

Visual observations of weak activity from this radiant in recent years reveals Weiss' radiant and calculations to best represent the stream's orbit.

The 1819 orbit of comet Blanpain was

ω	Ω	i	q	e	a
350.2	79.2	9.1	0.892	0.699	2.962

---

[67]*Meteoros*, **16** (August 1986), p. 69.
[68]*MN*, No. 79 (October 1987), p. 6.
[69]Ridley, H. B., *JBAA*, **72** (1962), p. 270.

Ursids

Observer's Synopsis

Occurring primarily between December 17 and 24, this meteor shower reaches maximum on December 22 (λ=270°) with a radiant at α=217°, δ=+76°. The maximum hourly rate is usually between 10 and 15, except in 1945, when rates exceeded 100. Meteors belonging to this stream are typically faint.

History

This meteor shower seems to have been discovered by William F. Denning, who, during several years around the turn of the century, observed a radiant at α=218°, δ=+76°, which endured during December 18-22.[69] Further radiants have been listed by Cuno Hoffmeister (H1948) for 1914, 1931 and 1933. Coordinated studies did not commence until Dr. A. Becvár accidentally observed a strong display in 1945.

Becvár was observing at the Skalnate Pleso Observatory (Czechoslovakia) on December 22, 1945, when he, and other observers, noted meteors falling at a rate of 169 per hour.[70] Many meteors were photographed and M. Dzubák computed a preliminary radiant of α=233°, δ=+82.6°. A reinvestigation of the data by Zdenek Ceplecha revealed a ZHR of 108 and a photographic radiant of α=217.1°, δ=+75.9°.[71]

Coordinated studies of this meteor shower were finally begun in 1946. Becvár confirmed his detection of an active radiant, but found the maximum hourly rates to reach only 11, on the night of December 22. Bochnícek and Vanysek confirmed Becvár's observations and both observers also established radiants for December 22.9. Bochnícek estimated it as α=213°, δ=+75°, while Vanysek found it to be α=217.8°, δ=+76.7°.

In 1947, visual studies of this stream were carried out by J. P. M. Prentice, of the British Astronomical Association. During 1 hour 43 minutes on December 22, only 1 meteor from the Ursid radiant was seen, but in 25 minutes on December 23, 8 meteors were detected—making the hourly rate about 20.[72] Four of the eight meteors detected on the latter date were plotted and revealed a radiant of α=207°, δ=+74°. The radiant diameter was less than one degree.

In addition to visual observations of this shower, 1947 was also notable for the first detection of the radiant by radio-echo observations.[73] Aerials at

---

[69]Denning, W. F., *MNRAS*, **84** (November 1923), p. 52.

[70]Strömgren, Elis, *IAU Circular*, No. 1026 (January 24, 1946).

[71]Ceplecha, Z., *BAC*, **4** (1951), p. 156.

[72]Prentice, J. P. M., *JBAA*, **58** (May 1948), p. 140.

[73]Clegg, J. A., Hughes, V. A. and Lovell, A. C. B., *JBAA*, **58** (May 1948), pp. 134-139.

Jodrell Bank first got a bearing on this shower at 3 hours UT on December 22. Beginning at 9 hours UT, hourly rates could be determined and, until 11 hours UT on December 23, these rates averaged 15 (equivalent visual rate of about 10). Thereafter, activity dropped sharply. The pointing of the aerial to different directions allowed the radiant to be determined as α=195±8°, δ=+78±5°.

Radio-echo observations became the primary means of studying this shower during the period 1948 to 1953, with observers at Jodrell Bank determining the following details of what appears to be a very consistent shower—except for the observations in 1945:

Ursid Radio-Echo Radiants

Year	Date of Max.	Hourly Rate	Solar Longitude	R.A.(°)	Decl.(°)
1948	Dec. 21.3	15	269.4	210±10	+82±8
1949	Dec. 22.3	13	270.2	207.1±8	+77.6±3
1950	Dec. 22	20	269.8	199±8	+77±3
1951	Dec. 23	13	270.5	200	+77
1952	Dec. 22	9	270.4	?	?
1953	Dec. 23	11	271	?	?

The Ursids went through another period of neglect after 1953, but this finally ended in 1970, when the British Meteor Society (BMS) began a period of annual observations. From observations spanning the period 1970 to 1976, the BMS found an average radiant of α=217°, δ=+76°. Maximum occurred at a solar longitude of 270.66° (about December 22), with the duration being established as December 17-24. Hourly rates were 10 in 1970, 22 in 1971, 16 in 1972, 18 in 1974, 9 in 1975 and 4 in 1976.[74] The moon interfered in 1973, but during daylight hours on December 22, BMS radio observers detected a short, 1-hour burst of activity that produced a corresponding visual hourly rate of 30.

Observations have been numerous in the United States, but they indicate the shower is far from what observers saw during the late 1940s and early 1950s. The following table lists analyses that have appeared in *Meteor News*.

Year	ZHR	Observers	Hours	Duration	Issue
1970	2.5	7	91	20-28	05
1971	2.9	4	29	20-25	10
1974	1.2	4	21	22-23	25
1982	2.0	2	8.6	19-25	61

[74]Mackenzie, Robert A., *Solar System Debris*. Dover: The British Meteor Society, 1980. pp. 28-30.

Observers in Japan tend to confirm the United States' finding of weak activity. They detected ZHRs of 5.7 in 1970 and 2.4 in 1971.[75] There does, however, seem to be occasional strong displays of this shower. Observers in Sogne, Norway, noted a strong display during 2 hours of observations on December 22, 1979, with estimated ZHRs of 25 and 27.[76] Veteran meteor observer, Norman W. McLeod, III (Florida) has commented that the Ursids "must be a compact stream like the Quadrantids. You have to be within 12 hours of maximum to see much."[77]

The first half of the 1980's showed no particularly impressive activity from the Ursids. Thus, it was particularly surprising that when an unexpected outburst of activity occurred on December 22, 1986, there were several observers monitoring the shower. Luc Gobin (Koszalin, Poland) reported unexpected "very high rates" while operating radio equipment at 66.17 MHz.[78] Gobin's equipment had detected average hourly echo rates of 60 to 68 during December 19, 20, 21, and 23, but found rates of 171 on the 23rd.[79] This enhanced activity was also noted visually, and seems to have peaked during the nighttime hours over Europe. George Spalding (director of the British Astronomical Association Meteor Section) reevaluated his observations of December 22 and arrived at a ZHR of 87±29.[80]

Trond Erik Hillestad (Norwegian Meteor Section) reported the observations of Kai Gaarder and Lars Trygve Heen. The former observer detected 94 Ursids in 4 hours, including 37 in the hour following December 22.83 (ZHR=64±11), and reported an average magnitude of 1.90. The latter observer saw 75 Ursids in 2 hours, including 54 in the hour following December 22.88 (ZHR=122±17), and reported an average magnitude of 2.61. These two observers saw only 4 and 2 Ursids, respectively, during a one-hour interval on the night of December 23. Of the 175 Ursids seen, 17.1% possessed a persistent train. Of the 66 Ursids of magnitude 2 and brighter, 51.5% were white, 33.3% were yellow, 7.6% were red, 2.3% were green, and 5.3% were blue.[81]

As noted above, the Norwegian observers reported Ursid rates of only 2-4 per hour on December 23.8, 1986—thus, indicating a very rapid decline in activity. But it should also be pointed out that the rise to maximum was also very great. R. Koschack, R. Arlt, and Jürgen Rendtel (Arbeitskreis Meteore, East

---

[75]*MN*, No. 10 (March 1972), p. 5.

[76]*MN*, No. 51 (October 1980), p. 5.

[77]*MN*, No. 35 (March 1977), p. 8.

[78]Steyaert, Chris, Personal Communication (January 19, 1987).

[79]Van Wassenhove, Jeroen, *WGN*, **15** (February 1987), pp. 12-13.

[80]Roggemans, Paul, *WGN*, **15** (April 1987), p. 50.

[81]Hillestad, Trond Erik, *WGN*, **15** (April 1987), pp. 59-60.

Germany) detected a ZHR of less than 5 on December 21.8,[82] while R. Taibi (Temple Hills, Maryland) observed 2-3 per hour around December 22.4.[83]

From observations obtained early in the 20th century, Denning had suggested that this shower was associated with "Mechain-Tuttle's Comet"—now known simply as comet Tuttle.[84] Upon Becvár's announcement of his discovery of the shower in 1945, he mentioned that a connection with periodic comet Tuttle "is highly presumable."[85]

The correlation between the observed activity rates of the Ursids and the perihelion passages of periodic comet Tuttle is particularly strange. In fact, the strong shower of 1945 actually occurred six years after the comet's perihelion passage—thus, placing the comet near aphelion! No rate estimates are available prior to 1945, and, unfortunately, observations virtually ceased during the latter half of the 1950's and early 1960's, so that the delay cannot be confirmed following the 1953 perihelion passage. Comet Tuttle next passed perihelion on March 31, 1967. No apparent observations were made in that year, but from 1968 onward, the rates generally remained between 10-15 per hour until 1973, when, as noted earlier, radio equipment operating in England detected activity corresponding to visual rates of 30 per hour during a one-hour period in daylight. This increase came six years after the comet's perihelion passage. The comet next passed perihelion on December 14, 1980. Hourly rates appeared normal in the following years, until December 22, 1986, when European observers detected very high rates while using both visual and radio-echo methods. Thus, three of the last four perihelion passages of comet Tuttle have been followed six years later by strong meteor activity.

Orbit

The 1961-1965 session of the Radio Meteor Project did not operate during the period of December 21-27; however, during December 29-January 3, eight meteors were seen emanating from a radiant probably associated with the Ursids. The established orbit was given as follows:

	ω	Ω	i	q	e	a
S1970	194.7	280.9	63.0	0.968	0.761	4.046

Zdenek Sekanina admitted that the maximum of the shower was missed, but noted that "several meteors with orbits somewhat resembling that of P/Tuttle were detected...."[86]

---

[82]*MN*, No. 78 (July 1987), p. 6.

[83]*MN*, No. 77 (April 1987), p. 7.

[84]Denning, W. F., *The Observatory*, 44 (January 1921), p. 31.

[85]Strömgren, Elis, *IAU Circular*, No. 1026 (January 24, 1946).

[86]S1970, p 488.

The Author has found two photographic meteors in W1954 and M1976, which indicate the following orbit:

ω	Ω	i	q	e	a
208.4	267.3	52.6	0.929	0.848	6.112

Orbits computed by other researchers from visual observations indicate that the photographic orbit is actually closer to the stream's true orbit, though it is possible that the fainter radio meteors are in a slightly different ellipse.

Two samples of comet Tuttle's fairly stable orbit follows:

	ω	Ω	i	q	e	a
1790	207.0	270.8	54.2	1.045	0.819	5.775
1980	206.9	269.9	54.5	1.015	0.823	5.720

Using an orbit similar to the photographic one above, Ken Fox (F1986) computed the orbit of the Ursid stream for periods 1000 years in the past and future. Although Earth seems to have been in contact with the Ursid stream 1000 years ago, no contact will apparently be possible 1000 years in the future.

	ω	Ω	i	q	e	a
950	205.4	276.1	53.4	0.99	0.83	5.86

In 950, the shower's maximum would have occurred six days later than at present from a radiant of α=214.5°, δ=+73.2°.

Additional December Showers

Delta Arietids

First recognition of a possible December radiant near Delta Arietis was made by Richard E. McCrosky and Annette Posen in 1959, while analyzing the photographic meteor orbits obtained by the Harvard Meteor Project during 1952-1954. Seven meteors indicated a shower was active during December 8-13, with a probable maximum on the 8th. The average radiant was given as α=51°, δ=+21°.[87] In 1971, a computerized stream search of the same 1952-1954 photographic meteors by Bertil-Anders Lindblad, revealed the Delta Arietids to be split into northern and southern branches; however, among the 14 meteors, which suggested a duration of December 8-January 2, only 2 were from the southern branch. The average radiant of the northern branch was α=54°, δ=+25°.[88]

No trace of this stream seems to be present in records covering the 19th century, but the first appearance of shower members may have occurred early in the 20th century, when several fireballs appear in various sources. The first

---

[87]McCrosky, Richard E., and Annette Posen, *AJ*, **64** (February 1959), pp. 25-27.
[88]L1971B, pp. 16-18.

really good visual radiants from this stream appear in Cuno Hoffmeister's Meteorströme during 1912 and 1932. The most detailed observation obtained thus far was by observers in Waltair, India, on December 8, 1964. M. Srirama Rao, P. V. S. Rama Rao and P. Ramesh plotted 10 meteors from $\alpha=57°$, $\delta=+22°$ and concluded that the radiant produced a rate of 7.5 meteors per hour.[89] However, numerous observers' attempts to observe the shower during the 1970's and 1980's have never shown hourly rates greater than one at the time of the predicted maximum.

The orbits of the northern and southern branches as determined from 12 meteors and 3 meteors, respectively, are as follows:

	ω	Ω	i	q	e	a
N.	226.6	262.8	3.3	0.865	0.635	2.370
S.	53.2	80.2	5.2	0.828	0.666	2.479

11 Canis Minorids

The discovery of this radiant came about on the morning of December 11, 1964, while Keith B. Hindley was conducting a telescopic meteor watch so as to "record positional details of Geminid meteors." The watch lasted 3 hours and 45 minutes, during which 26 telescopic meteors were detected. Six of these meteors proved to be true Geminids, but five others clearly emanated from an area 35 arc minutes across centered at $\alpha=117.0°$, $\delta=+12.8°$. The meteors ranged in brightness from 6.0 to 11.0, and were described as white and swift. On the night of December 13/14, Hindley observed for 4.5 hours, but no trace of activity was noted from the 11 Canis Minorid radiant, which brought Hindley to conclude that the shower possessed a very short duration.[90] Hindley computed the following parabolic orbit.

ω	Ω	i	q	e	a
157.4	78.0	107.9	0.038	1.0	∞

Hindley noted a similarity between this orbit and that of comet Nicollet-Pons (1821 I), although "the difference of 28° in the longitude of the ascending node made it unlikely that these two orbits are in fact related."[91] In 1970, Hindley and M. A. Houlden suggested that comet Mellish (1917 I) might have produced the stream after their adopted radiant of $\alpha=115°$, $\delta=+12°$ produced the following parabolic orbit.[92]

ω	Ω	i	q	e	a
158	78	89	0.04	1.0	∞

---

[89]RR1969, p. 768.

[90]Hindley, Keith B., *JBAA*, **79** (February 1969), pp. 138-139.

[91]Hindley, Keith B., *JBAA*, **79** (February 1969), p. 141.

[92]Hindley, K. B., and Houlden, M. A., *Nature*, **225** (1970), p. 1232.

New light was shed on the stream during 1974, while M. Kresáková was investigating photographic meteor orbits in an attempt to suggest a link between comet Mellish (1917 I) and the Geminids. Along the way nine meteors were found which were tentatively referred to as "short-period component B," but later identified with the 11 Canis Minorids. The suggested duration was December 4-15. The average radiant was given as α=109.3°, δ=+12.4°, while the daily motion was given as +0.53° in α and −0.37° in δ.[93] The established orbit was

ω	Ω	i	q	e	a
150.9	78.8	29.1	0.092	0.942	1.64

The orbit of comet Mellish was given as

ω	Ω	i	q	e	a
121.3	88.0	32.7	0.190	0.993	27.64

Kresáková concluded that the 11 Canis Minorids might be part of a chain association, whereas comet Mellish produced the December Monocerotids, which produced the 11 Canis Minorids, which subsequently produced the Geminids. Kresáková theorized that the chain could have began following a disruption of comet Mellish.

Visual observations of this radiant were made by the Western Australia Meteor Section during 1979. The overall duration was given as December 7-9, while a maximum ZHR of 1.27±0.40 occurred on the 9th from a radiant of α=116°, δ=+14°.[94]

Alpha Puppids

The discovery of this stream should be attributed to Ronald A. McIntosh (Auckland, New Zealand), who listed this stream in his 1935 paper, "An Index to Southern Meteor Showers." Based on two visual radiants detected in the period 1927-1934, the position was determined as α=117.5°, δ=−40.5°, while the duration was given as December 3-4.[95]

Cuno Hoffmeister spent most of 1937 in South-West Africa, where he was part of an expedition to isolate meteor radiants in southern skies. Four possible Alpha Puppid radiants seem to have been detected: the first on December 6 (λ=253.5°) from α=125°, δ=−36°, another on the 8th (λ=255.6°) from α=124°, δ=−38°, a third on the 9th (λ=257.0°) from α=118°, δ=−45° and the final radiant on the 10th (λ=257.6°) from α=115°, δ=−40°.[96] These observations seem to have offered the first hint that this stream produces a fairly diffuse radiant.

---

[93]Kresáková, M., *BAC*, 25 (1974), pp. 26-27.

[94]Wood, Jeff, Personal Communication (October 15, 1986).

[95]M1935, p. 712.

[96]H1948, pp. 244-245.

During the operation of radar equipment at Christchurch (New Zealand) during 1956, C. D. Ellyett and K. W. Roth detected activity from this radiant during 10 nights between November 17 and December 8. The resolution capability of the equipment was not especially high so that the right ascension appeared to irregularly vary from 112° to 126°, while the declination fluctuated between –41° and –49°. The authors gave average radiant estimates of α=122°, δ=–45° for November 17-21, and α=120°, δ=–43° for November 27-December 8.[97]

The Western Australia Meteor Section (WAMS) has recently obtained several excellent observations of this stream, which they refer to as the "N Puppids." In 1977, stream members were detected during December 2-6, with a maximum ZHR of 6.89±1.15 coming on December 2 from α=120°, δ=–43°.[98] During 1978, activity was detected during November 25-December 6 and a maximum of 10 meteors per hour came on December 4/5 from an average radiant of α=118°, δ=–43°. The meteors were described as fast, with an average magnitude of 3.29 and a total of 2.2% left trains. The meteor colors tended to be blue-white or white.[99]

The WAMS continued to observe this radiant during 1979 and 1980. In the former year, the duration of activity extended over the period of November 24-December 9. A maximum ZHR of 9.39±0.58 came on December 1 from α=116°, δ=–42°. In the latter year, observations revealed activity during November 29-December 15. The maximum ZHR reached 7.39±1.24 on December 5 from a radiant of α=128°, δ=–45°.[100]

---

[97]Ellyett, C. D., and Roth, K. W., *AJP*, **17** (1964), p. 505.
[98]Wood, Jeff, Personal Communication (October 15, 1986).
[99]Wood, Jeff, *MN*, No. 45 (April 1979), p. 11.
[100]Wood, Jeff, Personal Communication (October 15, 1986).

Appendix 1: Definitions

Aberration: An effect that is caused by the rotation of Earth on its axis. This becomes a very important factor when calculating the true radiants and orbits of meteors and meteor showers.

Apparent Radiant: The point from which meteors are observed to radiate before being corrected for zenithal attraction and aberration.

Bolide: A bright meteor which is observed to fragment or explode. If close enough the observer can expect to hear sounds of the explosion.

Daylight Stream: A meteor stream which is above the horizon during the same time as the sun. It is detectable only by radio and radar.

Double Radiant: A meteor stream that is affected by planetary perturbations can become split into two or more branches. Several major meteor streams visible near the ecliptic—and therefore more susceptible to perturbations—possess double radiants, usually referred to as northern and southern components.

Fireball: A meteor which is brighter than any planet or star, i.e. brighter than magnitude -4.

Hourly Rate: The number of meteors seen in an hour. Abbreviated "HR".

Meteor: The streak of light produced when a meteoroid encounters Earth's atmosphere and burns up due to the friction of air resistance. Meteors are popularly known as "shooting stars" and "falling stars."

Meteorite: A meteor that is large enough to survive its passage through the atmosphere and reach the ground.

Meteor Shower: A shower of meteors occurs when Earth's orbit intersects the orbit of a meteor stream. The meteors usually fall at a rate less than one per minute as a result of their being separated by hundreds of kilometers.

Meteor Storm: This is a rare event that occurs when Earth encounters closely grouped meteors in the orbit of a meteor stream. During the time of this encounter, meteors can fall at rates greater than 1000 per minute.

Meteor Stream: This represents the orbit of meteoroids as they travel about the sun. Meteors are the by-product of comets, so it is possible for the parent comet to be traveling in the same orbit—if it still exists.

Meteor Swarm: A cluster of meteoroids traveling within a meteor stream that is separated by much smaller distances than normal. If encountered by a planet, a meteor storm will occur.

Minor Showers: Meteor showers that produce less than 10 meteors per hour at the time of maximum activity.

Path: The trajectory of a meteor projected against the sky from the observing site.

Persistent Train: Train luminosity that lasts more than a second.

Radiant: The point from which a meteor appears to emanate. For meteor showers the radiant is actually a by-product of a perspective effect, similar to what is seen when looking down railroad tracks—the meteors appear to be moving in all directions, but they are actually parallel to one another.

Radiant Drift: The movement of a meteor shower's radiant against the star background. This characteristic is common to all meteor showers and is caused by Earth's passage through a meteor stream.

Solar Longitude: The longitude of the sun as given in geocentric coordinates. The evaluation of meteor data strongly relies on this figure rather than a conventional date.

Sporadic Meteor: A meteor which does not belong to an active shower radiant. The number visible per hour on any given night with the naked eye averages 6 after sunset and 18 just before dawn.

Stationary Radiant: A radiant which remains in one position for days, weeks, or months. Such a radiant is a mathematical impossibility.

Telescopic Meteors: A meteor below the limit of naked-eye visibility that is only visible in telescopes or binoculars.

Terminal Burst: The flare at the end of a meteor's path.

Toroidal Stream: A stream with an orbit of low eccentricity and high inclination which surrounds the Earth's orbit.

Train: A trail of ionized dust and gas that remains along the path of a meteor.

Wake: Train luminosity lasting only a fraction of a second.

Zenithal Attraction: The displacement of a radiant's location due to the gravitational attraction of Earth. The displacement becomes more pronounced as a radiant approaches the horizon, so that corrections must be applied.

Zenithal Hourly Rate (ZHR): This is the rate a meteor shower would possess if seen by an observer with a clear, dark sky and with the radiant in the zenith.

Appendix 2: Shower Associations

As indicated in the Introduction, it has long been known that meteor showers were the debris left by comets. In addition, recent years have brought about the realization that the Apollo group of asteroids, or asteroids which cross Earth's orbit, are probably the inactive nuclei of "dead" comets. The table below is a list of currently accepted meteor shower associations—all but one of which is discussed in this book. The Author has listed the possible parent bodies (either comets or asteroids) in bold-faced text, and the possible resulting meteor shower, or showers, in plain text.

	ω	Ω	i	q	e	a
Alcock (1959 IV)[1]	124.7	159.2	48.3	1.15	1.00	∞
Alpha Ursa Majorids (radar)	125.6	145.0	47.7	0.85	0.64	2.33
Biela (1852 III)[2]	223.2	247.3	12.6	0.86	0.76	3.53
Andromedids (1885—visual)	222	247	13	0.86	0.76	3.53
Andromedids (photo)	245.4	228.1	6.3	0.78	0.73	2.88
Blanpain (1819 IV)[3]	350.2	79.2	9.1	0.89	0.70	2.96
December Phoenicids (visual)	358.9	73.6	16.5	0.99	1.0	∞
Encke (1971 II)[4]	186.0	334.2	12.0	0.34	0.85	2.22
Daytime Beta Taurids	239.2	274.5	2.2	0.33	0.83	1.85
S. Taurids (radar)	114.1	49.7	1.4	0.39	0.77	1.68
S. Taurids (photo)	111.6	29.5	5.4	0.39	0.80	1.97
N. Taurids (radar)	293.6	217.2	0.0	0.40	0.75	1.60
N. Taurids (photo)	296.8	222.5	3.0	0.34	0.85	2.24

---

[1]See pages 170-171.
[2]See pages 211-212 & 216-220.
[3]See pages 241-242 & 261-263.
[4]See pages 117-118 & 238-241.

Giacobini-Zinner (1946 V)[5]	171.8	196.3	30.7	1.00	0.72	3.51
October Draconids (radar)	184.0	192.2	51.5	1.00	0.62	2.65
Grigg-Skjellerup (1972 II)[6]	359.3	212.7	21.1	1.00	0.66	2.97
Pi Puppids (visual)	359	212	25	1.01	1.0	∞
Halley (1835III)[7]	110.7	56.8	162.3	0.59	0.97	17.98
Eta Aquarids (radar)	79.5	44.9	161.2	0.47	0.83	2.82
Eta Aquarids (visual)	100.3	44.3	162.7	0.59	1.0	∞
Orionids (radar)	87.4	27.3	164.5	0.56	0.85	3.63
Orionids (photo)	80.0	28.0	163.7	0.59	1.01	
Ikeya (1964 VIII)[8]	290.8	269.3	171.9	0.82	0.99	53.51
Epsilon Geminids (radar)	223.0	203.0	175.0	0.88	0.75	3.58
Epsilon Geminids (photo)	234.3	205.7	172.8	0.79	0.98	35.96
Kiess (1911 II)[9]	110.4	158.0	148.4	0.68	0.99	184.62
Aurigids (visual)	126.7	158.7	149.0	0.81	1.0	∞
Mellish (1917 I)[10]	121.3	88.0	32.7	0.19	0.99	27.65
Monocerotids (radar)	135.8	71.8	22.3	0.15	0.98	6.20
Monocerotids (photo)	133.0	81.0	34.7	0.16	0.99	23.29
Metcalf (1919 V)[11]	185.8	121.4	46.4	1.12	1.00	∞
Omicron Draconids (radar)	192.2	114.7	46.2	1.01	0.77	4.33
Pajdusakova (1951 II)[12]	68.6	310.5	87.9	0.72	1.0	∞
Pajdusakovids (visual)	71.2	310.9	92.5	0.67	1.0	∞
Pons (1804)[13]	332.0	178.8	56.5	1.07	1.0	∞
Delta Mensids (radar)	346	178	58.3	0.98	0.87	10.00

---

[5]See pages 57-59.
[6]See pages 57-59.
[7]See pages 69-73 & 200-205.
[8]See page 197.
[9]See pages 175-176.
[10]See pages 254-258 & 268-269.
[11]See page 142.
[12]McIntosh, R.A., *Journal of B.A.A.*, **62** (March 1952), pp. 144-145. (Not listed in book, since only detected on one occasion.)
[13]See page 38.

Pons-Winnecke (1915 III)[14]	172.4	99.8	18.3	0.97	0.70	3.26
June Bootids (visual)	183	96	27	1.02	1.0	∞
S-W 3 (1930 VI)[15]	192.3	77.1	17.4	1.01	0.67	3.09
Tau Herculids (photo)	195.7	78.5	18.1	1.00	0.62	2.61
Swift-Tuttle (1862 III)[16]	152.8	138.7	113.6	0.96	0.96	24.33
Perseids (radar)	152.5	139.7	110.2	0.96	0.88	8.04
Perseids (photo)	150.5	138.3	112.8	0.94	0.90	9.84
Tempel-Tuttle (1965 IV)[17]	172.6	234.4	162.7	0.98	0.90	10.27
Leonids (photo)	173.1	235.4	161.0	0.98	0.90	9.96
Thatcher (1861 I)[18]	213.5	31.2	79.8	0.92	0.98	55.68
Lyrids (radar)	215.3	32.0	76.9	0.92	0.80	4.51
Lyrids (photo)	217.2	31.6	78.6	0.88	0.96	25.81
Tuttle (1939 X)[19]	207.0	269.8	54.7	1.02	0.82	5.70
Ursids (radar)	194.7	280.9	63.0	0.97	0.76	4.05
1566 (Icarus)[20]	31.0	87.6	22.9	0.19	0.83	1.08
Arietids (radar)	25.9	76.9	25.0	0.09	0.94	1.38
3200 (Phaethon)[21]	321.7	265.0	22.0	0.14	0.89	1.43
Geminids (photo)	324.1	261.3	23.9	0.14	0.90	1.61
1937 UB (Hermes)[22]	90.7	35.3	6.2	0.62	0.62	1.64
October Cetids (photo)	95.4	15.2	7.3	0.53	0.75	2.08

---

[14]See pages 89-94 & 98-99.
[15]See pages 95-99.
[16]See pages 163-164 & 169.
[17]See pages 222 & 228-229.
[18]See pages 51-53.
[19]See pages 266-267.
[20]See page 87.
[21]See page 251.
[22]See page 209.

Appendix 3: The "D-Criterion"

At the dawn of the study of meteor showers the methods involved in comparing one radiant to another were basically limited to a visual comparison of the radiant positions. Even though the mathematics were available to at least compute parabolic orbits for these radiants, they were ignored by most astronomers in the field. Subsequently, numerous misconceptions arose, of which the greatest involved the idea of stationary radiants—radiants whose positions remained unchanged for periods of up to 9 or 10 months. That was the legacy of the 19th Century and the beginning of meteoritics.

The 20th Century has been the age of a mathematics boom and the study of meteor showers has benefitted greatly. In 1922, Charles P. Olivier dealt the biggest blow against the concept of stationary radiants by calculating orbits for various days the Epsilon Arietid shower was supposed to be active. He showed that this shower—one of William F. Denning's most prominent stationary radiants—was actually composed of several individual showers.[1]

Orbital calculations reached a greater prominence during the 1940s and 1950s as meteor data was acquired by radio and photographic methods. Suddenly individual meteor orbits were in great number, bringing Richard B. Southworth and Gerald S. Hawkins to note that it was possible for two meteors to originate from two separate comets and yet have orbits that were quite similar. They said that if these origins could not be recognized the meteors would be incorrectly classed as belonging to the same stream. In 1963 Southworth and Hawkins developed the "D-criterion" to enable researchers to analyze the orbits of both meteors and meteor showers. This utilized the following equations.[2]

$$(1)\ (D(A,B))^2=(e_B-e_A)^2+(q_B-q_A)^2+(2*\sin(I_{AB}/2))^2+(((e_A+e_B)/2)*2*\sin(\Pi_{AB}/2))^2$$

$$(2)\ (2*\sin(I_{AB}/2))^2=(2*\sin((i_B-i_A)/2))^2+\sin(i_A)*\sin(i_B)*(2*\sin((\Omega_B-\Omega_A)/2))^2$$

$$(3)\ \Pi_{AB}=\omega_B-\omega_A+2*\arcsin(\cos((i_A+i_B)/2)*\sin((\Omega_B-\Omega_A)/2)*\sec(I_{AB}/2))$$

---

[1]Olivier, C. P., *MNRAS*, **83** (December 1922), pp. 87-92.

[2]SH1963, pp. 261-285.

These equations have become a standard for the evaluation of the membership of meteors and meteor radiants to particular meteor streams. There have been modifications suggested by other astronomers since 1963—notably by Zdenek Sekanina (1970) and Jack D. Drummond (1979)—but these generally produce no dramatic alterations to existing stream memberships.

Of major significance to the proper use of the D-criterion has been the determination of boundries to sort stream members from the sporadic background. Perfect stream membership is marked by a quantitative value of 0.00, but Southworth and Hawkins demonstrated that it was safe to accept a value as high as 0.20 to denote stream membership. They also demonstrated that the boundry was probably somewhat dependent on stream characteristics. For instance, the Taurid-Arietid stream of September and October is fairly diffuse in nature. Its D-criterion *average* is 0.19, and it is not unusual for some of their members to possess values close to 0.30.

In 1968, V. Porubcan published a study which involved the "problem of separation of diffuse meteor streams from the sporadic background...." Examining the Taurid and Sigma Leonid streams, he established a series of levels based on probability. The levels were as follows:[3]

Interval of D	Probability	Stream Membership
$0.00 \leq D \leq 0.18$	1.0	sure
$0.19 \leq D \leq 0.30$	0.8	very probable
$0.31 \leq D \leq 0.35$	0.6	probable
$0.36 \leq D \leq 0.38$	0.4	improbable
$0.39 \leq D \leq 0.40$	0.2	very improbable
$D \geq 0.40$	0.0	impossible

This probability approach was very beneficial in demonstrating the diffuse character of the Taurids. In fact, it doubled the existing membership of that stream! On the other hand, the Sigma Leonids suffered a devastating blow. Of the 27 members established by Southworth and Hawkins, approximately *90%* were shown to be sporadic. It should be noted that Southworth and Hawkins had suspected that the Sigma Leonids were composite and that an improved test would "separate this group into independent parts."

From investigations by the Author, it would seem safe to accept D-criterion values as high as 0.30 for *non-ecliptic* meteor streams only. Since most meteor activity clusters about the ecliptic, there are numerous instances of an overlapping effect. This reaches a peak during the summer months when activity emanates from the regions of Iota Aquarii, Delta Aquarii and Alpha Capricorni. It

---

[3]Porubcan, V., *Bulletin of the Astronomical Institute of Czechoslovakia*, **19** (1968), p. 335.

has been known for years that these are independent streams, however, upon examining photographic meteors from these streams it becomes apparent that their radiant boundries are somewhat diffuse. For this reason researchers should not use a D-criterion greater than 0.18.

For meteor streams that are far from the ecliptic it seems fairly safe to use D-criterions with values up to 0.30. In fact, it has been this acceptance that has allowed the author to confirm the true complexity of meteor streams. Reported multiple radiants in the Quadrantid and Perseid regions of January and August, respectively, have been noted in the past. These have been confirmed. In addition, complexities have been noted around the Geminids and numerous weak radiants have been found around well-known minor radiants. This fits with the mechanics of the solar system very well. Planetary perturbations are constantly at work on all bodies of the solar system; however, in the case of older, well-dispersed meteor streams, certain portions of the orbit may be pulled into slighter larger or smaller orbits at any one time. Thus, it can be said that the more stable the annual rates of a meteor shower are, the more diffuse the radiant must be.

The D-criterion is an important tool to the present study of meteor streams. If specific guidelines are followed carefully it will continue to enable astronomers in the field of meteoritics to understand and better isolate the various areas of activity that are present in our night sky.

Appendix 4: Source Abbreviations

ABH1952...Almond, Mary, Bullough, K., and Hawkins, G. S., *Jodrell Bank Annals*, **1** (December 1952), pp. 13-21.

ACL1949 ...Aspinall, A., Clegg, J. A., and Lovell, A. C. B., *Monthly Notices of the Royal Astronomical Society*, **109** (1949), pp. 352-358.

B1954.......Bullough, K., *Jodrell Bank Annals*, **1** (June 1954), pp. 68-97.

B1963.......Babadzhanov, P., *Smithsonian Contributions to Astrophysics*, **7** (1963), pp. 287-291.

BK1967.....Babadzhanov, P. B., and Kramer, E. N., *Smithsonian Contributions to Astrophysics*, **11** (1967), pp. 67-79.

C1958.......Ceplecha, Z., *Bulletin of the Astronomical Institute of Czechoslovakia*, **9** (1958), pp. 225-234.

C1964.......Ceplecha, Z., Jezková, M., Novák, M., Rajchl, J., and Sehnal, L., *Bulletin of the Astronomical Institute of Czechoslovakia*, **15** (1964), pp. 144-155.

C1973.......Cook, A. F., Lindblad, B.-A., Marsden, B. G., McCrosky, R. E., and Posen, A., *Smithsonian Contributions to Astrophysics*, **15** (1973), pp. 1-5.

C1977.......Ceplecha, Z., *Bulletin of the Astronomical Institute of Czechoslovakia*, **28** (1977), pp. 328-340.

CBJ1979....Ceplecha, Z., Bocek, J., and Jezková, M., *Bulletin of the Astronomical Institute of Czechoslovakia*, **30** (1979), pp. 220-225.

CF1973.....Cook, A. F., Forti, G., McCrosky, R. E., Posen, A., Southworth, R. B., and Williams, J. T., *Evolutionary and Physical Properties of Meteoroids*. eds. Hemenway, Curtis L., Millman, Peter M., and Cook, Allan F., Washington, D. C.: NASA (1973), pp. 23-44.

CHL1947...Clegg, J. A., Hughes, V. A., and Lovell, A. C. B., *Monthly Notices of the Royal Astronomical Society*, **107** (1947), pp. 369-378.

D1899.......Denning, W. F., *Memoirs of the Royal Astronomical Society*, **53** (1899), pp. 203-292.

D1912A.....Denning, W. F., *Monthly Notices of the Royal Astronomical Society*, **72** (1912), pp. 423-451.

D1912B.....Denning, W. F., *Monthly Notices of the Royal Astronomical Society*, **72** (1912), pp. 631-639.

D1916.......Denning, W. F., *Monthly Notices of the Royal Astronomical Society*, **76** (1916), pp. 219-239.

D1923A.....Denning, W. F., *Monthly Notices of the Royal Astronomical Society*, **84** (1923), pp. 43-56.

D1956.......Davidson, T. W., *Jodrell Bank Annals*, **1** (December 1956), pp. 116-125.

D1980.......Drummond, Jack D., Hill, Robert K., and Beebe, Herbert A., *Astronomical Journal*, **85** (1980), pp. 495-498.

E1949 ....Ellyett, C. D., *Monthly Notices of the Royal Astronomical Society*, **109** (1949), pp. 359-364.

E1960 ....Evans, G. C., *Jodrell Bank Annals*, **1** (November 1960), pp. 280-337.

F1986 ....Fox, K., *Asteroids, Comets, Meteors II*. eds. Rickman, H., and Lagerkvist, C.-I., Uppsala: University of Uppsala (1986), pp. 521-525.

GE1975.....Gartrell, G., and Elford, W. G., *Australian Journal of Physics*, **28** (1975), pp. 591-620.

H1948.......Hoffmeister, C., *Meteorstrome*. Leipzig: Verlag Werden und Werken Weimar (1948), pp. 1-286.

H1959.......Hughes, Robert F., *Smithsonian Contributions to Astrophysics*, **3** (1959), pp. 79-94.

HA1952.....Hawkins, G. S., and Almond, Mary, *Monthly Notices of the Royal Astronomical Society*, **112** (1952), pp. 219-233.

HS1958.....Hawkins, Gerald S., and Southworth, Richard B., *Smithsonian Contributions to Astrophysics*, **2** (1958), pp. 349-364.

HS1961.....Hawkins, Gerald S., and Southworth, Richard B., *Smithsonian Contributions to Astrophysics*, **4** (1961), pp. 85-95.

HT1977.....Harvey, Gale A., and Tedesco, Edward F., *Astronomical Journal*, **82** (1977), pp. 445-447.

JW1961.....Jacchia, Luigi G., and Whipple, Fred L., *Smithsonian Contributions to Astrophysics*, **4** (1961), pp. 97-129.

K1916.......King, A., *Monthly Notices of the Royal Astronomical Society*, **76** (1916), pp. 542-549.

K1955.......van Diggelen, J., and de Jager, C., *Meteors (A Symposium on Meteor Physics)*. Kaiser, T. R., editor. United Kingdom: Pergamon Press Ltd. (1955), p. 164.

KL1967.....Kashcheyev, B. L., and Lebedinets, V. N., *Smithsonian Contributions to Astrophysics*, **11** (1967), pp. 183-199.

KM1971 ....Korovkina, T. L., Martynenko, V. V., and Frolov, V. V., *Solar System Research*, **5** (1971), pp. 99-103.

L1971A .....Lindblad, Bertil-Anders, *Smithsonian Contributions to Astrophysics*, **12** (1971), pp. 1-13.

L1971B .....Lindblad, Bertil-Anders, *Smithsonian Contributions to Astrophysics*, **12** (1971), pp. 14-24.

M1935.......McIntosh, R. A., *Monthly Notices of the Royal Astronomical Society*, **95** (1935), pp. 709-718.

M1970.......Martynenko, V. V., *Solar System Research*, **4** (1970), pp. 39-43.

M1973.......Martynenko, V. V., *Solar System Research*, **6** (1973), pp. 222-224.

M1976.......McCrosky, R. E., Shao, C.-Y., and Posen, A., *Center for Astrophysics Preprint Series*, No. 665 (1976), pp. 1-13.

MP1959.....McCrosky, Richard E., and Posen, Annette, *Astronomical Journal*, **64** (February 1959), pp. 25-27.

MP1961.....McCrosky, Richard E., and Posen, Annette, *Smithsonian Contributions to Astrophysics*, **4** (1961), pp. 15-84.

N1964.......Nilsson, C. S., *Australian Journal of Physics*, **17** (1964), pp. 205-256.

Ö1934.......Öpik, Ernst, *Harvard College Observatory Circular*, No. 388 (1934), pp. 1-38.

PP1960......Plavcová, Z., and Plavec, M., *Bulletin of the Astronomical Institute of Czechoslovakia*, **11** (1960), pp. 226-227.

RM1974.....Rao, M. Srirama, and Gopalakrishna, A., *Australian Journal of Physics*, **27** (1974), pp. 679-686.

RR1969.....Rao, M. Srirama, Rao, P.V.S. Rama, and Ramesh, P., *Australian Journal of Physics*, **22** (1969), pp. 767-774.

S1970 ....Sekanina, Zdenek, *Icarus*, **13** (1970), pp. 475-493.

S1973 ....Sekanina, Zdenek, *Icarus*, **18** (1973), pp. 253-284.

S1976 ....Sekanina, Zdenek, *Icarus*, **27** (1976), pp. 265-321.

SH1963.....Southworth, R. B., and Hawkins, G. S., *Smithsonian Contributions to Astrophysics*, **7** (1963), pp. 261-285.

SK1975.....Smirnov, N. V., and Korovkina, T. L., *Solar System Research*, **8** (1975), pp. 97-100.

TH1976.....Tedesco, Edward F., and Harvey, Gale A., *Astronomical Journal*, **81** (November 1976), pp. 1010-1013.

W1954 ......Whipple, Fred L., *Astronomical Journal*, **59** (July 1954), pp. 201-217.

Name Index

NOTES

NOTES

NOTES

NOTES

NOTES